Wetlands and International Environmental Law

NEW HORIZONS IN ENVIRONMENTAL AND ENERGY LAW

Series Editors: Kurt Deketelaere, *University of Leuven, Belgium* and Zen Makuch, *Imperial College London, UK*

Environmental law – including the pressing considerations of energy law and climate change – is an increasingly important area of legal research and practice. Given the growing interdependence of global society and the significant steps being made towards environmental protection and energy efficiency, there are few people untouched by environmental and energy lawmaking processes.

At the same time, environmental and energy law is at a crossroads. The command-and-control methodology that evolved in the 1960s and 1970s for air, land and water protection may have reached the limit of its environmental protection achievements. New life needs to be injected into our environmental protection regimes – perhaps through the concept of sustainability in its environmental, economic and social forms. The same goes for energy policy and law, where liberalisation, environmental protection and security of supply are at the centre of attention. This important series seeks to press forward the boundaries of environmental and energy law through innovative research into environmental and energy law, doctrine and case law. Adopting a wide interpretation of environmental and energy law, it includes contributions from both leading and emerging international scholars.

For a full list of Edward Elgar published titles, including the titles in this series, visit our website at www.e-elgar.com.

Wetlands and International Environmental Law
The Evolution and Impact of the Ramsar Convention

Edited by

Royal C. Gardner

Hugh F Culverhouse Professor of Law and Director, Institute for Biodiversity Law and Policy, Stetson University College of Law, USA

Richard Caddell

Reader in Marine and Environmental Law, School of Law and Politics, Cardiff University, UK

Erin Okuno

Assistant Professor of Law, Stetson University College of Law, USA

NEW HORIZONS IN ENVIRONMENTAL AND ENERGY LAW

Edward Elgar PUBLISHING

Cheltenham, UK • Northampton, MA, USA

© The Editors and Contributors severally 2025

All rights reserved. No part of this publication may be reproduced, stored in a retrieval system or transmitted in any form or by any means, electronic, mechanical or photocopying, recording, or otherwise without the prior permission of the publisher.

Published by
Edward Elgar Publishing Limited
The Lypiatts
15 Lansdown Road
Cheltenham
Glos GL50 2JA
UK

Edward Elgar Publishing, Inc.
William Pratt House
9 Dewey Court
Northampton
Massachusetts 01060
USA

Authorised representative in the EU for GPSR queries only: Easy Access System Europe– Mustamäe tee 50, 10621 Tallinn, Estonia, gpsr.requests@easproject.com

A catalogue record for this book
is available from the British Library

Library of Congress Control Number: 2024951183

This book is available electronically in the Elgaronline
Law subject collection
https://doi.org/10.4337/9781802203028

ISBN 978 1 80220 301 1 (cased)
ISBN 978 1 80220 302 8 (eBook)

Printed and bound in Great Britain by
TJ Books Limited, Padstow, Cornwall

Contents

List of figures vii
List of contributors viii
Preface ix
List of abbreviations xi

Introduction to *Wetlands and International Environmental Law* 1
Royal C. Gardner, Richard Caddell, and Erin Okuno

PART I   THE RAMSAR CONVENTION –
               DEVELOPMENT AND INTERPRETATION

1   The Ramsar Convention at 50: Cultural narratives,
     enlightenment, and the Wetland Conservation Project   19
     Michael Bowman

2   The Ramsar Convention and the concept of consensus   52
     Royal C. Gardner

3   Ramsar Sites and their ecological character: A cornerstone
     of the Ramsar Convention on Wetlands   79
     Royal C. Gardner, Marcela Bonells, and Katherine Pratt

4   The 'wise use' of wetlands   105
     Edward J. Goodwin

5   Compensatory mechanisms and the loss of wetlands   131
     Jerneja Penca

6   Indigenous peoples and local communities: Wetlands
     management and the Ramsar Convention   154
     Simon Marsden

PART II WETLANDS AND THE INTERNATIONAL
         LEGAL ECOSYSTEM

7 Birds of a feather? The inter-relationship between the
   Ramsar Convention and the migratory waterbird regimes   180
   Melissa Lewis

8 The Ramsar Convention and the Convention on Biological
   Diversity: Proposals for enhanced cooperation, synergies,
   and interoperability for the Kunming-Montreal Global
   Biodiversity Framework   203
   Teresa Fajardo

9 The Ramsar Convention and general international water
   law: Complementary and mutually supportive regimes   231
   Owen McIntyre

10 The cultural values and services of wetlands: Evolution
   and obligations under the Ramsar Convention   265
   Evan Hamman

11 Wetlands, climate change, and international law   292
   An Cliquet

12 'Within and without': EU nature conservation law and the
   protection of wetlands   319
   Richard Caddell

13 Rights of wetlands and the Ramsar Convention   361
   Erin Okuno

Index   385

Figures

3.1	Sites number and area by year	86
3.2	Sites number and area by region	87
10.1	The relationship of social and cultural values (characteristics) and cultural (ecosystem) services under Ramsar to the concept of ecological character	277
10.2	Derivation of 'cultural services' of wetlands under Ramsar	280
10.3	The different classifications of ecosystem services provided by wetlands under Ramsar	282

Contributors

Marcela Bonells, former Science and Policy Officer, Ramsar Convention on Wetlands, Switzerland.

Michael Bowman, former Associate Professor and Director, University of Nottingham Treaty Centre, School of Law, University of Nottingham, United Kingdom.

Richard Caddell, Reader in Marine and Environmental Law, School of Law and Politics, Cardiff University, Wales, United Kingdom.

An Cliquet, Professor of International Environmental and Biodiversity Law, Department of European, Public and International Law, Ghent University, Belgium.

Teresa Fajardo, Associate Professor of Law, University of Granada, Spain.

Royal C. Gardner, Hugh F Culverhouse Professor of Law and Director, Institute for Biodiversity Law and Policy, Stetson University College of Law, United States.

Edward J. Goodwin, Associate Professor in Law and Co-director, History of Law and Governance Centre, University of Nottingham, United Kingdom.

Evan Hamman, Post-doctoral Fellow, Centre for Environmental Governance, University of Canberra, Australia.

Melissa Lewis, former Policy and Advocacy Programme Manager, BirdLife South Africa.

Simon Marsden, former Professor of Law, Curtin Law School, Australia.

Owen McIntyre, Professor of Law, School of Law, University College Cork, Ireland.

Erin Okuno, Assistant Professor of Law, Stetson University College of Law, United States.

Jerneja Penca, Senior Research Associate and Head, Mediterranean Institute for Environmental Studies, Science and Research Centre, Koper, Slovenia.

Katherine Pratt, Foreman Biodiversity Fellow, Stetson University College of Law, United States.

Preface

Wetlands are simultaneously among the most significant and imperilled natural ecosystems worldwide. Often deprecated as wastelands, wetlands are frequently misunderstood and underappreciated, yet provide vital habitats for a vast array of species and perform essential ecosystem services. Thankfully, the importance of wetlands has not been entirely overlooked by humans. In his classic poem *Inversnaid* (1881), Gerard Manly Hopkins advocated eloquently for the vitality of wetlands, beseeching:

> What would the world be, once bereft
> Of wet and of wildness? Let them be left,
> O let them be left, wildness and wet;
> Long live the weeds and the wilderness yet.

Ninety years later, the wistful pleas of an English poet would be tangibly advanced through a pioneering international treaty, with clear objectives and obligations for the enduring existence of our collective wetlands. In 1971, the Convention on Wetlands of International Importance especially as Waterfowl Habitat was concluded in Ramsar, Iran, in the light of longstanding concerns over the plight of wetland ecosystems across the planet. It would be the first major ecological treaty to be concluded in a vital and breathless decade for international wildlife law, and in many respects, its negotiation arguably heralded the dawn of modern international environmental law. Nevertheless, global wetland coverage has continued to recede precipitously and – much like wetlands themselves – the Ramsar Convention remains perhaps the least celebrated and understood of all the key biodiversity conventions. This Volume seeks to rectify this position and reflects on the development, operation, importance, and context of the Ramsar Convention as it passes its milestone 50-year anniversary.

The editors wish to express their thanks and appreciation to the extensive team of experts who contributed to this volume, who have collectively brought an array of different perspectives, approaches, and specialist knowledge to this project. The result is a rich panorama of analysis that sheds extensive new light on both the inner workings of the Ramsar Convention and its wider contribution to the further development of international environmental law. In keeping with the cooperative ethos that has long characterised the implementation of the Ramsar Convention, the authors also had the opportunity to present their initial thoughts at a writers' workshop and to share their knowledge on these contributions. We are grateful to Kees Bastmeijer, Kim Diana Connolly, and Marcus Carson, who also contributed their expertise and insights at this event.

This project was conceived and initiated during the height of the COVID-19 pandemic, and much of the work was conducted in the shadow of the accompanying impacts on family life and working and teaching conditions. We are deeply grateful to the contributors for their belief in and commitment to this project during a time of unprecedented personal and professional upheaval and challenges to conducting research, and for their exceptional work in developing their individual contributions.

We are also deeply appreciative of our own research networks, from which we have drawn encouragement and inspiration and benefitted from a nurturing environment within which to undertake scholarship of this nature (and on nature itself) – notably the Institute for Biodiversity Law and Policy at Stetson University College of Law and the Centre for Environmental Law and Policy at Cardiff University, as well as the extraordinary interdisciplinary alchemy of the Wales Governance Centre and its commitment to comradeship and academic rigour, which makes it such a special place in which to think and work. We are also grateful to the library team at Stetson University College of Law for their assistance.

We are especially indebted to Ben Booth, Amber Watts, and the production team at Edward Elgar Publishing, particularly Carolyn Boyle and Elizabeth Ruck, for their unstinting support and encouragement throughout the development, writing, and production of this Volume. Their commitment to and support for the editorial and writing team have been exceptional throughout the whole process – from our first tentative suggestion that the law of wetlands was a worthy theme to explore to final submission of the completed book – and demonstrates amply why Edward Elgar has become a vital natural habitat for the work of many environmental lawyers.

Ultimately, on a personal level, we are deeply grateful for the support of our families – especially Mary, Sasha, and Dan, who in the process have each learned far more about the minutiae of the Ramsar Convention and the plight of wetland ecosystems than they might ever have initially imagined or wanted.

The United Nations now recognises 2 February as World Wetlands Day to commemorate the occasion of the conclusion of the Ramsar Convention in 1971. World Wetlands Day offers an apt opportunity to reflect upon the value of these often maligned and deprecated ecosystems, and to highlight their critical importance to a vast array of biodiversity. Our hope is that this Volume does the same.

Royal C. Gardner, Richard Caddell, Erin Okuno
Gulfport, Florida, and Cardiff, Wales

Abbreviations

AEWA	Agreement on the Conservation of African-Eurasian Migratory Waterbirds
BBOP	Business and Biodiversity Offset Program
BLG	Liaison Group of the Biodiversity-related Conventions
CAF	Central Asian Flyway
CAFF	Conservation of Arctic Flora and Fauna
CBD	Convention on Biological Diversity
CDSN	European Committee for the Conservation of Nature and Natural Resources
CELDF	Community Environmental Legal Defense Fund
CEPA	Communications, Education and Public Awareness
CIC	International Council for Game and Wildlife Conservation
CITES	Convention on International Trade in Endangered Species of Wild Fauna and Flora
CJEU	Court of Justice of the European Union
CMS	Convention on the Conservation of Migratory Species of Wild Animals
COP	Conference of the Parties
CSR	Conservation Status Review
DaRT	Digital Data Reporting Tool
EAAFP	East Asian-Australasian Flyway Partnership
EC	European Community
ECHA	European Chemicals Agency
ECJ	European Court of Justice
ECOSOC	United Nations Economic and Social Council
EEC	European Economic Community
EIA	Environmental Impact Assessment
EU	European Union
ExCOP	Extraordinary Conference of the Parties
FCPF	Forest Carbon Partnership Facility
GARN	Global Alliance for the Rights of Nature

GBFF	Global Biodiversity Framework Fund
GWO	Global Wetland Outlook
FPIC	Free, Prior and Informed Consent
FPS	Fauna Preservation Society
HPAI	Highly Pathogenic Avian Influenza
IAS	Invasive Alien Species
IBA	Important Bird Area
ICBP	International Council for Bird Protection
ICJ	International Court of Justice
ICRW	International Convention for the Regulation of Whaling
ILC	International Law Commission
ILO	International Labour Organization
IPBES	Intergovernmental Science-Policy Platform on Biodiversity and Ecosystem Services
IPPC	International Plant Protection Convention
IRP	Implementation Review Process
ISWG	International Species Working Group
ITLOS	International Tribunal for the Law of the Sea
ITPGRFA	International Treaty on Plant Genetic Resources for Food and Agriculture
IUCN	International Union for Conservation of Nature
IWC	International Whaling Commission
IWMI	International Water Management Institute
IWRB	International Waterfowl and Wetlands Research Bureau
JWP	Joint Work Plan
LIFE	L'Instrument Financier pour l'Environnement
MA	Millennium Ecosystem Assessment
MAB	UNESCO Man and Biosphere Programme
MEA	Multilateral Environmental Agreement
MedWet	Mediterranean Wetlands Initiative
MOC	Memorandum of Cooperation
MOP	Meeting of the Parties
MOU	Memorandum of Understanding
MPA	Marine Protected Area
NBSAP	National Biodiversity Strategies and Action Plans
NDC	Nationally Determined Contribution

NGO	Non-governmental Organisation
OECM	Other Effective Area-Based Conservation Measure
OUV	Outstanding Universal Value
PES	Payments for Ecosystem Services
POW	Programme of Work
RAM	Ramsar Advisory Mission
RBO	River Basin Organisation
RCN	Ramsar Culture Network
REACH	Registration, Evaluation, Authorisation and Restriction of Chemicals
REDD	Reducing Emissions from Deforestation in Developing Countries
REIO	Regional Economic Integration Organization
RIS	Ramsar Information Sheet
R-METT	Ramsar Site Management Effectiveness Tracking Tool
RRI	Ramsar Regional Initiative
SAC	Special Area of Conservation
SDG	Sustainable Development Goals
SEA	Strategic Environmental Assessment
SEEA	System of Environmental-Economic Accounting
SPA	Special Protection Area
SSSI	Site of Special Scientific Interest
STRP	Scientific and Technical Review Panel
SWS	Society of Wetland Scientists
TEEB	The Economics of Ecosystems and Biodiversity
UDHR	Universal Declaration on Human Rights
UN	United Nations
UNCCD	UN Convention to Combat Desertification in those Countries Experiencing Serious Drought and/or Desertification, Particularly in Africa
UNDRIP	United Nations Declaration on the Rights of Indigenous Peoples
UNECE	United Nations Economic Commission for Europe
UNEP	United Nations Environment Programme
UNESCO	United Nations Educational, Scientific and Cultural Organization
UNFCCC	UN Framework Convention on Climate Change
UNGA	UN General Assembly
UNPFII	United Nations Permanent Forum on Indigenous Issues
VCLT	Vienna Convention on the Law of Treaties

WET	Wetland Extent Trends Index
WHC	World Heritage Convention
WHSRN	Western Hemisphere Shorebird Reserve Network
WOW	Wings Over Wetlands
WPE	Waterbird Population Estimates
WRI	World Resources Institute
WTO	World Trade Organization

Introduction to *Wetlands and International Environmental Law*

Royal C. Gardner, Richard Caddell, and Erin Okuno

1   WHY WETLANDS MATTER

Writing in 1964 as a contributor to a landmark publication on the protection of wetland ecosystems, celebrated Belgian naturalist and ornithological advocate Count Léon Lippens declared that: 'It is as stupid to drain the last of our great marshes, with their wealth of wildlife, as it would be to demolish the Cathedral of Chartres – to plant potatoes.'[1] While 60 years later, the famed French cathedral remains a masterpiece of gothic ecclesiastical architecture – and one free from encroachment by root vegetables – regrettably, the folly of global wetland eradication and degradation continues at a devastating rate.[2]

Wetlands are critical ecosystems that are found in every country and continent, however arid, frozen, or mountainous.[3] They encompass a variety of freshwater habitats (such as lakes, rivers, and marshes) and coastal and marine habitats (such as lagoons, estuaries, mangroves, and reefs).[4] While wetlands cover only a small portion of the planet's surface (at least 1.5–1.6 billion hectares globally),[5] they are of fundamental importance to the effective

---

[1]   Preface to GL Atkinson-Willes, *Liquid Assets* (Wildfowl Trust, IUCN, and IWRB, Slimbridge, 1964).

[2]   E Fluet-Chouinard et al, 'Extensive Global Wetland Loss Over the Past Three Centuries' (2023) 614 *Nature* 281.

[3]   Even Antarctica, whose wetlands have been the subject of scientific studies. Eg, MV Quiroga et al, 'The ecological assembly of bacterial communities in Antarctic wetlands varies across levels of phylogenetic resolution' (2022) 24 *Environmental Microbiology* 3486.

[4]   RC Gardner and CM Finlayson, *Global Wetland Outlook: State of the world's wetlands and their services to people 2018* (Ramsar Convention Secretariat, Gland, 2018) 11.

[5]   M Courouble et al, *Global Wetland Outlook: Special Edition 2021* (Ramsar Convention Secretariat, Gland, 2021) 15.

functioning of ecosystems worldwide, with broadly 40% of global biodiversity – especially birds – dependent upon these areas.[6]

Beyond their extraordinary ecological value, and despite often being erroneously and ignorantly derided as wastelands, '[w]etlands are vital for human survival'.[7] Wetlands provide a vast array of essential ecosystem services for humans, including crucial flood defences; coastal protection; continued supplies of freshwater; pollution control; goods and services related to food security; carbon sequestration and storage; disaster risk reduction; and recreational, cultural, and spiritual opportunities.[8] Humans also use wetlands for transportation, hydropower, genetic and medical resources, and raw materials, such as timber.[9] Yet wetlands are increasingly being destroyed across the planet, as the effects of climate change and poorly planned and ecologically unsympathetic development have dramatically decreased global wetland coverage. The rate of loss is alarming: according to one recent estimate, since 1700 the world has lost approximately 3.4 million square kilometres of inland wetlands, representing a 21% net loss of global wetland area.[10] The quality of remaining wetlands has likewise continued to deteriorate due to drivers such as pollution, drainage, unsustainable use, and climate change.[11] Wetland-dependent plant and animal species are also declining – approximately 25% are threatened with extinction.[12] It is against this somewhat troubling backdrop that this book considers the international conservation and management of wetlands and their dependent species, which remain significant subjects of international environmental law but have been chronically underexplored in the legal literature to date.

2   FROM PROJECT MAR TO THE RAMSAR CONVENTION

The origins of the Ramsar Convention on Wetlands of International Importance can be traced to the efforts of international organisations, individual scientists, and – perhaps somewhat surprisingly – waterfowl hunters. In 1960, Swiss ornithologist and naturalist Luc Hoffmann proposed that the International Union for Conservation of Nature (IUCN) establish an international programme with

---

[6] Gardner and Finlayson, note 4, 61.
[7] Ibid, 11.
[8] Courouble et al, note 5, 24, 35; Gardner and Finlayson, note 4, 5, 11, 12, 38.
[9] Gardner and Finlayson, note 4, 11, 12.
[10] Fluet-Chouinard et al, note 2, 283.
[11] Courouble et al, note 5, 6, 15.
[12] Gardner and Finlayson, note 4, 25.

the aim of conserving and managing wetlands.[13] IUCN recommended that other international organisations be invited to participate – specifically the International Waterfowl Research Bureau (IWRB) (now known as Wetlands International) and the International Council for Bird Protection (now known as Birdlife International).[14] In a nod to the programme's transboundary scope, it was christened 'Project MAR', in light of the fact that 'mar-' is the start of the names of different wetland types in English ('marsh'), French ('*marécage*') and Spanish ('*marisma*').[15] Hoffmann, who was appointed as Project MAR coordinator, organised a MAR Conference held in Saintes-Maries-de-la-Mer in the French region of Camargue in 1962.[16]

The MAR Conference brought together waterbird and wetland experts from 16 countries, although primarily from Europe.[17] As GVT Mathews recounts in his history on the Ramsar Convention, more than 50 papers were presented on a wide range of wetland-related topics, including the need for a coordinated international effort to properly manage wetlands and wetland-dependent species.[18] Baron Le Roy, representing the *Association Nationale des Chasseurs de Gibier d'Eau* (National Association of Waterfowl Hunters), is credited with suggesting that an international convention on wetlands be established.[19] Consequently, the conference adopted a recommendation:

> that IUCN compile a list, in accordance with an internationally agreed classification, of European and North African wetlands of international importance, together with detailed information on these areas; [and] … that the list be placed at the disposal of conservationists and those responsible for development schemes; and further … that this list may be considered as a foundation for an international convention on wetlands.[20]

---

[13] Hoffmann was also instrumental in establishing the World Wildlife Fund (WWF), serving as its first vice president. WWF, 'An environmental visionary and a father to WWF', 22 July 2016, *WWF* at: https://wwf.panda.org/es/?274010/WWF%2Dstatement%2Don%2Dthe%2Dextraordinary%2Dlife%2Dof%2DDr%2DLuc%2DHoffmann.

[14] GVT Matthews, *The Ramsar Convention on Wetlands: Its History and Development* (Ramsar Convention Secretariat, Gland, reissued 2013) 9.

[15] Ibid.

[16] Ibid.

[17] Ibid.

[18] Ibid.

[19] DA Stroud et al, 'Development of the text of the Ramsar Convention: 1965–1971' (2022) 73 *Marine and Freshwater Research* 1107, 1109.

[20] PJS Olney et al, *Project Mar: The conservation and management of temperate marshes, bogs and other wetlands* (IUCN, 1965) 7, at: https://portals.iucn.org/library/sites/library/files/documents/NS-005.pdf.

IUCN subsequently compiled and published a list of 236 wetland sites from 29 countries, which were mostly European, alongside sites in Algeria, Morocco, Tunisia, and Turkey.[21] IUCN noted that the countries on the list 'share an adequate part not only of the profit they take from the resource, but also of the responsibilities for its conservation'.[22]

With this nudge from the international organisations that had been involved in the MAR Conference, governments began to formally participate in the planning of follow-up meetings, beginning with the First European Meeting on Wildfowl Conservation convened at St Andrews, Scotland, in 1963.[23] At the second such meeting in 1966, held in Noordwijk aan Zee, the Netherlands, the governments considered the first draft of a proposed wetland convention, developed by IWRB.[24] The meeting participants requested that the Dutch government prepare a new draft convention based on input from the gathering.[25] The Dutch draft was slated for discussion at the third meeting in Leningrad in 1968, but due to the Soviet invasion of Czechoslovakia, most Western European countries boycotted the event.[26] Following the Leningrad meeting, the Soviet Ministry of Agriculture produced a separate draft convention.[27] IWRB reconciled the competing texts, and the 1969 meeting in Moscow of the International Union of Game Biologists provided an opportunity to thaw East-West relations, as well as consider the IWRB compromise draft.[28] Post-Leningrad, the Dutch government adopted the IWRB compromise draft as the official draft, with only minor revisions.[29] After additional input by governments, IWRB prepared a final draft for a conference scheduled to be held in Babolsar, Iran in 1971.[30]

Iran decided to shift the conference to the city of Ramsar on the coast of the Caspian Sea. There, delegates from 18 countries conducted further negotiations and adopted the text of the treaty on 2 February 1971.[31] Its full name – the Convention on Wetlands of International Importance especially as Waterfowl

---

[21] Although Albania was included, no sites were identified due to a lack of information. Ibid, 21. Germany was divided into an 'Eastern Part' and a 'Western Part', and Great Britain and Northern Ireland were listed separately. Ibid, 5.
[22] Ibid, 7.
[23] Matthews, note 14, 13.
[24] Stroud et al, note 19, 1109.
[25] Ibid.
[26] Ibid, 1110.
[27] Matthews, note 14, 20.
[28] Stroud et al, note 19, 1110.
[29] Ibid.
[30] Ibid.
[31] Matthews, note 14, 27.

Habitat – included a nod to the initial impetus for protecting wetlands as bird habitats. It was signed the following day.[32]

The Ramsar Convention broadly defines 'wetlands' as:

> areas of marsh, fen, peatland or water, whether natural or artificial, permanent or temporary, with water that is static and flowing, fresh, brackish or salt, including marine water the depth of which at low tide does not exceed six metres.[33]

The Convention sets forth three main obligations for contracting parties, known as the 'three pillars' of Ramsar. The first pillar is that, when joining the Convention, each contracting party must designate at least one wetland to be included on the List of Wetlands of International Importance.[34] The Convention specifies that the designation criteria take into account a wetland's 'international significance in terms of ecology, botany, zoology, limnology or hydrology', as well as its importance to wetland-dependent birds.[35] As of June 2024, Ramsar's 172 contracting parties had designated more than 2500 Wetlands of International Importance, referred to as 'Ramsar Sites'.[36] Ramsar Sites range in size from one hectare to massive designations such as Queen Maud Gulf in Canada (6.278 million hectares) and Ngiri-Tumba-Maindombe in the Democratic Republic of Congo (6.569 million hectares).[37] The United Kingdom has designated the most Sites (175),[38] while Brazil has designated the largest area (approximately 26.8 million hectares).[39] A contracting party has an obligation to promote the conservation of the Ramsar Sites that it desig-

---

[32] Ibid, 28.

[33] Convention on Wetlands of International Importance especially as Waterfowl Habitat, adopted 2 February 1971, entered into force 21 December 1975, 996 UNTS 247 (amended 1982 & 1987), Art 1.1.

[34] Ibid, Art 2.4.

[35] Ibid, Art 2.2. The contracting parties have approved nine criteria for the listing of Ramsar Sites, in two broad categories: sites that contain representative, rare, or unique wetland types; and sites that are internationally important for conserving biological diversity. For the latter category, there are criteria related to species' ecological communities, including specific criteria for waterbirds, fish, and other taxa. Ramsar Convention, *The Ramsar Sites Criteria*, at: https://www.ramsar.org/sites/default/files/documents/library/ramsarsites_criteria_eng.pdf.

[36] 'Ramsar Sites Information Service', at: https://rsis.ramsar.org/.

[37] Ibid.

[38] Mexico, second with 144 designations, nearly claimed the top spot as a result of the Scottish Independence Referendum in 2014, which failed to pass. Had Scotland become independent, the United Kingdom would no longer be able to count the 51 Ramsar Sites in Scotland.

[39] Ibid.

nates.[40] Additionally, the contracting party is expected to report and respond to any human-induced negative changes to a Ramsar Site's ecological character.[41]

The second pillar is an overarching obligation of 'wise use' that applies to all wetlands within the territory of a contracting party.[42] 'Wise use' of wetlands refers to 'the maintenance of their ecological character, achieved through the implementation of ecosystem approaches, within the context of sustainable development'.[43] Wise use contemplates the adoption of national laws and policies, the implementation of management plans, and public awareness and education activities.[44] As Goodwin points out in this book, the duty to maintain ecological character in the wise use context may be seen as an obligation of conduct, while the duty to maintain ecological character with respect to designated Ramsar Sites is an obligation of result.[45]

The Convention's third pillar relates to international cooperation. The text of the Convention calls for consultation regarding transboundary wetlands and shared water systems, such as river basins, as well as coordination of policies and regulations regarding shared wetland-dependent species.[46] Another aspect of international cooperation, recognised by the contracting parties as the universe of multilateral environmental agreements (MEAs) expanded, is participation in other biodiversity-related conventions,[47] such as the Convention on Biological Diversity,[48] the Agreement on the Conservation of African-Eurasian

---

[40] Convention on Wetlands, note 33, Art 3.1.
[41] Ibid, Art 3.2.
[42] Ibid, Art 3.1.
[43] Ramsar Convention, *Resolution IX.1 Annex A: A Conceptual Framework for the wise use of wetlands and the maintenance of their ecological character* (2005) 6.
[44] See generally Ramsar Convention Secretariat, *Handbook 1: Wise use of wetlands* (Ramsar Handbooks, 4th edition, Ramsar Convention Secretariat, Gland, 2010).
[45] Chapter 4 in this volume.
[46] Convention on Wetlands, note 33, Art 5.
[47] Ramsar Convention Secretariat, *Handbook 20: International cooperation* (Ramsar Handbooks, 4th edition, Ramsar Convention Secretariat, Gland, 2010) 21–23.
[48] Convention on Biological Diversity, adopted 5 June 1992, entered into force 29 December 1993, 1760 UNTS 79.

Migratory Waterbirds (AEWA),[49] the World Heritage Convention,[50] and the United Nations Framework Convention on Climate Change.[51]

The organisational structure of the Ramsar Convention follows a now-familiar format. The Conference of the Parties (COP), which typically meets every three years, is the ultimate decision-making body.[52] Fourteen such meetings have been held since the inaugural COP in Cagliari, Italy in 1980, along with three extraordinary meetings – one of which occurred during the COVID-19 pandemic. The 15th COP is slated to be held in Zimbabwe in 2025. During the inter-sessional period between the meetings of the COP, the Standing Committee – composed of representatives determined on a regional basis – acts on behalf of the contracting parties.[53] The Secretariat (originally called the Bureau), which is headed by a Secretary General, manages the day-to-day affairs of the Convention and is located in Gland, Switzerland, in the IUCN headquarters building.[54] A scientific advisory body – the Scientific and Technical Review Panel (STRP), created by a resolution adopted at the fifth COP – provides independent advice on a range of wetland-related matters, including emerging issues.[55]

When the Ramsar Convention was adopted more than five decades ago, it was a veritable pioneer: it was the first of the modern MEAs that attempted to be global in scope while focusing on a particular ecosystem. It also was the first to institute the concept of internationally recognised protected areas. Scientists and international organisations were instrumental in establishing the Ramsar Convention, and scientific endeavours and the Convention's 'International Organization Partners'[56] (IOPs) continue to play a critical role

---

[49] Agreement on the Conservation of African-Eurasian Migratory Waterbirds (AEWA), adopted 16 June 1995, entered into force 1 November 1999, 2365 UNTS 203.

[50] Convention Concerning the Protection of the World Cultural and Natural Heritage (World Heritage Convention), adopted 16 November 1972, entered into force 17 December 1975, 1037 UNTS 151.

[51] United Nations Framework Convention on Climate Change, adopted 9 May 1992, entered into force 21 March 1994, 1771 UNTS 107.

[52] RC Gardner, E Okuno, and D Pritchard, 'Ramsar Convention governance and processes at the international level' in P Gell, N Davidson, and CM Finlayson (eds), *Ramsar Wetlands: Values, Assessment, Management* (Elsevier, Amsterdam, 2023) 38.

[53] Ibid, 46–47.

[54] Ibid, 47–49.

[55] Ibid, 49–52.

[56] The IOPs currently recognised by the Convention are Wetlands International, Birdlife International, IUCN, WWF, International Water Management Institute (IWMI), and Wildfowl & Wetlands Trust. Ibid, 53–57.

in the Convention's work. For example, the STRP has produced the *Global Wetland Outlook*, which reports on the state of the world's wetlands,[57] and an IOP representative addresses the opening and closing sessions of each meeting of the COP.[58] But whether and to what extent the contracting parties are paying heed to the urgent calls for action from these groups are open questions – and ones that we examine further in the subsequent chapters of this book.

3   THE IMPLEMENTATION AND CONTEXT OF THE RAMSAR CONVENTION

The alarming decline in wetland ecosystems sets the scene for this book, which seeks to examine the pioneering work of the Ramsar Convention and to evaluate its wider context within the firmament of international environmental law. One of the great achievements of the ongoing development of international rules and bodies for the environment, especially over the previous half century, has been the emergence of specific and specialised regimes concerned with nature conservation.[59] Indeed, the Convention itself presaged the extraordinary legal developments of a crucial decade for the multilateral protection of the natural environment, having been concluded shortly before the watershed Stockholm Conference in 1972, which would frame global objectives for a new era in international law-making and institutions. Nevertheless, the Ramsar Convention has arguably proved to be something of an outsider within this framework, with its institutional architecture remaining steadfastly external to the wider United Nations (UN) institutional structures.[60]

Subsequent biodiversity treaties have drawn heavily upon the experience and structure of the Ramsar Convention. Despite this longstanding experience, however, the impact and importance of the work conducted under the auspices of the Ramsar Convention have not been systematically evaluated, and the lessons learned from five decades of concerted multilateral activity towards the conservation of wetlands and their dependent species have remained largely unexamined from a legal perspective. This is a surprising and disappointing oversight. The Ramsar Convention is thus perhaps the least under-

---

[57]   Gardner and Finlayson, note 4; Courouble et al, note 5.

[58]   Ramsar Convention, COP14, *Conference Report* (2022) [4] (referencing IUCN opening statement) and [431] (referencing IWMI closing statement), at: https://www.ramsar.org/sites/default/files/documents/library/cop14_report_e.pdf.

[59]   See further M Bowman, P Davies, and C Redgwell, *Lyster's International Wildlife Law* (2nd ed, Cambridge University Press, Cambridge, 2010).

[60]   See further Chapter 3 in this volume; BH Desai, *Multilateral Environmental Agreements: Legal Status of the Secretariats* (Cambridge University Press, Cambridge, 2010) 152–53.

stood of the great international wildlife treaties,[61] and its contribution to the wider development of international law has not always been fully appreciated. The milestone 50th anniversary of the Ramsar Convention therefore provides a timely and valuable opportunity to reflect upon the progress of the wetlands regime and its wider impacts and interactions within the landscape of international environmental law.

This book is divided into two substantive parts, reflecting two distinct but fundamentally intertwined lines of enquiry. The Ramsar Convention continues to provide an essential framework for the designation and protection of Wetlands of International Importance and for the promotion of wider protections for these ecosystems more generally. The chapters in Part I accordingly examine the core commitments and values of the Ramsar Convention and its constituent bodies, shedding new light on the emergence, maturation, and evolution of its operative principles and overarching conservation objectives. The chapters in Part II build on the contributions in Part I to ground the Ramsar Convention within its wider context.

As Bowman outlines,[62] the Ramsar Convention was in many respects the first example of a new style of international law-making, in which the often rigid and monolithic architecture of intergovernmental regulation was unable to provide the agile, flexible, and data-driven policy solutions required of environmental stewardship. An early example of this lies in the initial structure of the whaling treaties of the 1930s, which required hard-fought and periodic renegotiations to operate effectively.[63] An initial trailblazer in this regard was the 1946 International Convention for the Regulation of Whaling (ICRW)[64] – a concise treaty whose primary departure was the provision of a pioneering institutional structure comprising a management body, scientific advisory capacity, and an annexed schedule that was considered an 'integral part'[65] of the ICRW that could be rapidly updated and adjusted to respond to emerging environmental demands. The Ramsar Convention both followed and refined

---

[61] Trouwborst et al characterise the Ramsar Convention as one of the 'Big Five' global wildlife law instruments, along with the World Heritage Convention, the Convention on International Trade in Endangered Species of Wild Flora and Fauna, CMS, and CBD. A Trouwborst et al, 'International Wildlife Law: Understanding and Enhancing Its Role in Conservation' (2017) 67 *BioScience* 784, 785.

[62] Chapter 1 in this volume.

[63] See further M Fitzmaurice, *Whaling and International Law* (Cambridge University Press, Cambridge, 2015) 6–28.

[64] International Convention for the Regulation of Whaling, adopted 2 December 1946, entered into force 10 November 1948, 161 UNTS 72.

[65] Ibid, Art I(1).

this approach and, as the first major wildlife treaty of the latter half of the twentieth century, entrenched the broad institutional model deployed by MEAs that remains the hallmark of modern international environmental law-making.[66]

While many such bodies share a broad structural resemblance, they are nevertheless unique products of their individual negotiating contexts, with their own priorities, working arrangements, and regulatory approaches. In this regard, the hallmark of the Ramsar Convention has long been its tailored institutional values and the key tools it has developed with which to implement them. These central values of the Ramsar Convention are explored further in this book. Reflecting the negotiating history of the Convention itself, a lodestar of the Ramsar regime has been its ethos of consensus. The operative mandates of MEAs frequently venture into politically contentious territory, and the work of a number of organisations has been extensively waylaid by obstructive practices arising from rancorous disagreement[67] and derailed by the vagaries of domestic policies,[68] and has sometimes become the collateral damage of geopolitical conflict.[69] As Gardner outlines,[70] while the Ramsar Convention has not been immune to turbulence, the traditional ethos of consensus and collaboration remains central to its activities – even if this may at times prove to dilute aspects of the language of individual resolutions and to reward larger and more vociferous delegations with a greater degree of practical influence.

---

[66] See further RR Churchill and G Ulfstein, 'Autonomous Institutional Arrangements in Multilateral Environmental Agreements: A Little-Noticed Phenomenon in International Law' (2000) 94 *American Journal of International Law* 623; MJ Bowman, 'Beyond the "Keystone" COPs: The Ecology of Institutional Governance in Conservation Treaty Regimes' in M Fitzmaurice and D French (eds), *International Environmental Law and Governance* (Brill Nijhoff, Leiden, 2015) 6.

[67] For example, the longstanding impasse within the International Whaling Commission over commercial whaling, which resulted in significant institutional inertia. See further Fitzmaurice, note 63, 57–122.

[68] For instance, the obstructionist approaches of the Trump administration towards the global climate change regime, resulting in the withdrawal of the United States from the Paris Agreement. F Jotzo, J Depledge, and H Winkler, 'US and International Climate Policy under President Trump' (2018) 18 *Climate Policy* 813; J Pickering et al, 'The Impact of the US Retreat from the Paris Agreement: Kyoto Revisited?' (2018) 18 *Climate Policy* 818.

[69] Notably the ongoing suspension of the Arctic Council in the light of the Russian invasion of Ukraine. T Koivurova and A Shibata, 'After Russia's Invasion of Ukraine in 2022: Can We Still Cooperate with Russia in the Arctic?' (2023) 59 *Polar Record* e12.

[70] Chapter 2 in this volume.

Another core value of the Ramsar Convention lies in its capacity and zeal for outreach, education, and science. As Bowman observes, the Ramsar Convention has a key role to play in maintaining public concern over the conservation needs of wetlands. In an era of climate anxiety and incessant environmental doomscrolling, the plight of wetlands is often drowned out in the popular appreciation of ecological imperatives. The Ramsar Convention thus maintains a constant and consistent mandate to advance environmental education and to emphasise the values and contributions of wetlands. This remains an unheralded but crucial part of the work of any MEA, and the Ramsar Convention has proved to be not merely a global instrument to conserve wetlands but a powerful advocate for their values, functions, and management – not least through the *Global Wetland Outlook*, its flagship assessment of the state of the world's wetlands and their services to people.[71]

Moreover, this curation of wetland information also includes an understanding of the values of wetlands that extends beyond the laudable, yet partial, objective of attaining scientific knowledge. This creeping recognition of non-environmental values positions the Ramsar Convention as an instrument that has long been ahead of its time. Wetlands have values and significance beyond their role as habitats and ecosystems, holding importance for human spirituality in ways that have not always been fully appreciated. As Hamman elaborates,[72] cultural values formed part of the initial preambular acknowledgements of the Ramsar Convention but were largely marginalised in the early work of the Convention. This is perhaps unsurprising given that the World Heritage Convention,[73] a treaty of similar vintage, provided the most obvious vehicle for the elaboration of practices towards addressing the cultural value of natural features. While cultural appreciation remains a contested issue, Hamman illustrates that this has steadily gained recognition within the remit of the Ramsar Convention, with social and cultural values having been factored into designation practices in a growing number of instances. While there are clear thematic and political boundaries to the cultural remit of the Ramsar Convention – Hamman's extensive review of five decades of practice reveals that the wetland treaty is very clearly not a cultural heritage regime – there is growing recognition that cultural and traditional practices are inherent in the work of all multilateral environmental instruments to at least some degree. As Hamman's contribution indicates, the Ramsar Convention has been able to sympathetically incorporate these values, where appropriate, with the

---

[71] Reproduced at: https://www.global-wetland-outlook.ramsar.org/outlook.
[72] Chapter 10 in this volume.
[73] World Heritage Convention, note 50.

ecological case for the protection of individual sites – even if this remains a work in progress.

In a similar vein, the role of local and Indigenous communities has also found a degree of traction within the practice of the Ramsar Convention, as Marsden examines.[74] As the role of Traditional and Indigenous Knowledge continues to coexist amorphously within many multilateral structures, such values have also been advocated as an important element of international law generally and individual legal structures specifically. As with the cultural dimension, the attention given to local and Indigenous community interests is a rather less prominent element of the practice of the Ramsar Convention. Both Hamman's and Marsden's contributions reveal that these wider norms have been acknowledged under the aegis of the Ramsar Convention and identified as a prospectively significant element of wetland management and conservation in particular locations, but the more ephemeral and intangible role of wetlands to human cultures and spirituality has proved far less prominent than their ecological values. For better or worse, the Ramsar Convention can be seen as having remained resolutely true to its core ideals, tools, and approaches in promoting wetland conservation. It has also demonstrated an appreciation of the importance of wetlands to local and Indigenous communities, as it has become increasingly infused within its conservation mandate.

Indeed, for all its scientific, educational, and cultural work, the central contribution of the Ramsar Convention (like any other comparable MEA) should be weighed first and foremost against its ability to mobilise the national implementation of global commitments to facilitate discernible ecological improvements. To this end, at the heart of the Ramsar Convention lies an overarching commitment for contracting parties to advance the conservation and 'wise use' of wetlands. As Goodwin's historical excavation of these terms reveals,[75] the Ramsar Convention broke new ground – even in the fertile negotiating climate of the 1970s – in enshrining a commitment towards what would subsequently be articulated and categorised as 'sustainable development',[76] marrying conservation aspirations with a clear-sighted appreciation that many wetlands are mixed spaces in which habitat considerations and anthropogenic needs commingle and coexist.

---

[74] Chapter 6 in this volume.

[75] Chapter 4 in this volume.

[76] For instance, the European Union has directly equated the wise use of wetlands with the pursuit of sustainable development. European Commission, *Communication from the Commission to the Council and the European Parliament: Wise Use and Conservation of Wetlands* COM(95) 189 (final) 42.

This approach is reflected in the tools deployed by the Ramsar Convention, which initially provided an intriguing template for the modern law of protected areas. As Gardner, Bonells, and Pratt explore,[77] the Ramsar Convention's primary tool for wetland conservation is the designation by contracting parties of individual sites as Wetlands of International Importance ('Ramsar Sites'). This designation incorporates another key value of the Ramsar Convention: the need to maintain the ecological character of a Ramsar Site. In keeping with the ebb and flow of global environmental conditions, wetland ecologies are subject to natural variations, and regrettably, many such changes are anthropogenically driven contractions in coverage and quality. This in turn has spawned further tools, in the form of the Montreux Record to document changes and to prioritise corrective measures, and Ramsar Advisory Missions, which – in keeping with the collaborative ethos of the Ramsar regime – provide on-site assessments and guidance to facilitate positive conservation interventions.

A further tool, explored by Penca,[78] is the pioneering system of compensatory mechanisms, which has become increasingly prevalent in national zoning and planning practices. Compensation remains a contentious element of environmental practice, presupposing that vital ecosystems can be readily replicated, restored, and retrofitted elsewhere, and generate a seamless transition of vital habitats for an array of species. Increasingly, however, compensation has been viewed as a means by which important socio-economic projects can coexist, and – as with the broader notion of 'wise use' and its correlation with sustainable development – the Ramsar Convention represents a first multilateral articulation of this approach.

Despite these tools and values, however, the boldness of these innovations has not always been replicated in their implementation. As Gardner, Bonells, and Pratt lament, while there has been no shortage of protective designations, outreach initiatives, and collaborative assessments, Ramsar Sites often experience the shortcomings of many protected areas more generally, in that they are frequently unaccompanied by a comprehensive, deliverable, and targeted management plan. Wetlands are regrettably well represented within the dubious pantheon of so-called 'paper parks'. Similarly, as Penca ruefully observes, while the Ramsar Convention has demonstrated extraordinary versatility, creativity, and resilience, it has also presided over a distinctly mixed record of achievement. This raises important existential questions for the Ramsar Convention and its values and approaches. Thus, as Okuno explores,[79] the future of wetland protection may revolve less around paternalistic, incon-

---

[77] Chapter 3 in this volume.
[78] Chapter 5 in this volume.
[79] Chapter 13 in this volume.

sistent, and arbitrary human protections and could benefit from a reframing of humans' relationship with wetlands, and from recognition and support of the inherent rights of these ecosystems – including the right to exist and the right to be free from pollution and degradation.[80] Recognising and implementing the rights of wetlands provides a complementary, biocentric approach to wetland conservation and protection that could further the goals of the Ramsar Convention. The Convention has already started to take note of the possibilities of such an approach and could play an important role in championing the rights of wetlands moving forward.

While also promoting the pursuit of protective frameworks for wetlands, the Ramsar Convention has become a crucial part of the wider ecosystem of international environmental law, engaging closely with other natural resource treaties, formulating unique and far-reaching policies, and developing institutional innovations of its own. The second key theme explored in Part II of this book considers the more contextual role of the Ramsar Convention, as a collaborative partner to other likeminded institutions and its ability – in modern social media parlance – to act as an 'influencer' to ensure that wetland conservation may be integrated into the activities of a wider constellation of disparate international bodies. Biodiversity treaties have been increasingly required to simultaneously celebrate their specialist thematic niches and justify their running costs – and the financial and logistical implications of implementing their commitments assiduously by their respective parties – by successfully synergising agendas with the array of environmental regimes with which they share regulatory space. The contributions in Part II thus highlight the existential anxieties of a treaty reaching middle age and its dilemma as to whether to maintain the concerted focus of its youth or continue to mature into a more well-rounded global citizen concerned with a disparate range of overlapping issues. This has indeed proved to be a challenging tightrope to traverse, with commentators both rueing the Convention's titular focus on wetland birds as having infused the Ramsar regime with 'an emphasis which may not have been wholly to its advantage',[81] and lamenting the 'institutional drift' that has unmoored the Convention from its original mandate and 'airbrushed' waterfowl from its core agenda,[82] forcing it to become a more generalised and amorphous entity navigating an increasingly cluttered multilateral landscape.

---

[80] See 'Universal Declaration of the Rights of Wetlands', at: https://www.rightsofwetlands.org/.

[81] MJ Bowman, 'The Ramsar Convention Comes of Age' (1995) 42 *Netherlands International Law Review* 1, 7.

[82] P Bridgewater and RE Kim, '50 Years on, w(h)ither the Ramsar convention? A case of institutional drift' (2021) 30 *Biodiversity and Conservation* 3919, 3921.

The current and future place of the Ramsar Convention within what Bowman terms 'the ecology of treaty regime communities' remains a complex and multifaceted question. Arguably, the Ramsar Convention has performed best as a specialised partner – especially to the Convention on the Conservation of Migratory Species of Wild Animals (CMS)[83] and its key avifauna subsidiary instrument, AEWA.[84] As Lewis outlines,[85] the Ramsar Convention's ostensible distinction was predicated upon the regime being part of a wider constellation of supplementary treaties, with a series of contiguous remits. The Ramsar Convention has accordingly enjoyed deep roots of cooperation with the CMS, especially in the early years of the current century – even if synergies are now more ad hoc than deliberate and have a somewhat mixed record of achievement. Strong collaborations have been apparent from a scientific perspective, especially regarding emergency situations that affect a multitude of regulatory structures, such as sporadic but serious outbreaks of avian influenza.

The drive towards greater synergy has ebbed and flowed as an operative priority within international biodiversity regimes.[86] As Fajardo elaborates,[87] inter-treaty cooperation serves a vital purpose – not only in avoiding the pervasive risk of duplication and conflict associated with 'treaty congestion', but also in facilitating the smooth implementation of a suite of interconnected multilateral commitments on the ground. Such connections go beyond mere administrative efficiencies (although Lewis laments the continuing problems of uneven COP schedules, which often impede agile collaboration given the decision-making structures of MEAs), and closer relationships between treaties can facilitate the cross-pollination of strategies and guiding principles. In this respect, Fajardo expresses a degree of cautious optimism that a recent resetting of inter-treaty collaborative priorities may have value for regimes – such as the Ramsar Convention – that discharge their unique mandates within the wider context of an array of different actors. Further encouragement for effective collaboration can be seen in the context of international water law. As McIntyre

---

[83] Convention on the Conservation of Migratory Species of Wild Animals (CMS), adopted 23 June 1979, entered into force 1 November 1983, 1651 UNTS 67.

[84] AEWA, note 49.

[85] Chapter 7 in this volume.

[86] See further KN Scott, 'International Environmental Governance: Managing Fragmentation through International Connection' (2011) 6 *Melbourne Journal of International Law* 177; R Caddell, '"Only Connect"? Regime Interaction and Global Biodiversity Conservation' in M Bowman, P Davies, and E Goodwin (eds), *Research Handbook on Biodiversity and Law* (Edward Elgar Publishing, Cheltenham, 2016) 437.

[87] Chapter 8 in this volume.

explores,[88] the aquatic component of the Ramsar Convention's waterfowl focus also provides valuable scope for the Ramsar Convention to contribute effectively to the normative requirements of the ecosystem approach required under international water law. In this respect, McIntyre considers that the relations between the wetland regime and international water treaties buck the general trend of fragmentation and represent an intriguing example of greater multilateral convergence. Lewis, Fajardo, and McIntyre each point to the vital roles played by the machinery for dialogue between institutions, which can work to the mutual and individual benefit of each regime and ultimately may dictate the success or otherwise of such activities.

The success and effectiveness of every biodiversity treaty regime also depend on the overarching threat presented by climate change. As Cliquet explains,[89] the relationship between wetlands and climate change is a double-edged sword: on the one hand, wetlands are important for climate mitigation and adaptation, while on the other hand, wetlands face additional pressure because of changes in precipitation and sea-level rise. Climate change was not on the agenda when the Ramsar Convention was concluded, and in 2002, the COP's initial attempt to consider the interplay between wetlands and climate change proved contentious, with consensus being reached only after Resolution VIII.3 was whittled down from 19 pages to four, excising an annex developed by the STRP on wetlands and climate impacts, adaptation, and mitigation. Despite the halting beginnings, the Ramsar Convention framework has increasingly recognised the role of wetlands in climate change mitigation and adaptation – as has the broader international climate regime. Whereas the role of ecosystems in climate mitigation predominantly focused at first on forests, more recently greater attention has been given to other ecosystems – notably wetlands such as peatlands. The Paris Agreement calls for the conservation and restoration of sinks and reservoirs of greenhouses gases, and correspondingly, Ramsar IOPs have promoted wetland restoration as a 'nature-based solution'. Cliquet reports that wetlands are increasingly mentioned in Nationally Determined Contributions under the Paris Agreement but questions whether the current legal framework under the Ramsar Convention is sufficient to protect and restore remaining peatlands, mangroves, and other wetlands.

Elsewhere, the fingerprints of the Ramsar Convention are arguably more subtle, yet still significant. As Caddell elaborates,[90] aspects of the Convention and its terminology have taken root within the legislation of national and supranational regimes, albeit not always in a frictionless manner. One notable

---

[88] Chapter 9 in this volume.
[89] Chapter 11 in this volume.
[90] Chapter 12 in this volume.

example is the drive towards prohibiting lead shot within wetlands – a flagship policy of AEWA and one more recently embraced after much political wrangling by the European Union. This has seen the verbatim adoption of core definitions of the Ramsar Convention, which have been subject to a variety of creative challenges and interpretations by the judicial authorities of the European Union and are likely to be replicated on a national level as the respective Member States seek to transpose these requirements. Similarly, the Convention was a guiding light in the formation of the celebrated EU Birds Directive and formed the basis of its important but controversial protections for wetlands and avifauna, and has also provided significant interpretive guidance in implementing these supranational protections.

4   CONCLUSION

In some ways, the two lines of enquiry in this book highlight a potential tension within the Ramsar Convention: does the Convention's relationship with other international regimes detract from its core mission? That is, to the extent that the energy and resources of the Secretariat and other members of the Ramsar community focus on processes outside of the Ramsar Convention, has the Convention strayed too far from its institutional moorings? Although the first half century of the Ramsar Convention ushered in a greater appreciation for wetlands and attempts to protect wetlands and their ecosystem services through international designations and national laws and policies, much work remains to be done regarding the implementation of the wise use of wetlands and the maintenance of the ecological character of Ramsar Sites. To remain relevant in an ever-expanding multilateral universe, is it beneficial – even necessary – for wetland conservation for the Ramsar Convention to engage not only with the traditional biodiversity agreements but also with the water and climate sectors? No doubt the participants of the original MAR Conference would be heartened that the Ramsar Convention came into being and persists after 50 years, and might marvel at how it has evolved, despite disappointment that the progressive loss of wetlands has yet to be stemmed. It has been said that when an organism is confronted with change, it has three options: move, adapt, or die. The Ramsar Convention has nowhere to move (except perhaps under the UN umbrella, which is exceedingly unlikely), and its objectives are too important to allow it to die. Thus, it must adapt to changing natural and legal environments, and this book seeks to contribute to that effort as the Ramsar community navigates the next 50 years.

PART I

The Ramsar Convention – development and interpretation

1. The Ramsar Convention at 50: Cultural narratives, enlightenment, and the Wetland Conservation Project

Michael Bowman

1 INTRODUCTION

By common repute, the acid test of all truly monumental events in recent human history is that everyone can recall precisely what they were doing at the moment of their occurrence. Judged by that exacting standard, the conclusion of the Ramsar Wetlands Convention[1] could scarcely maintain a serious claim to such celebrity status, since – leaving aside any surviving 'founding fathers' of the instrument – it seems unlikely that anyone old enough to have been alive at the time will currently be able to recollect anything at all of their activities on 2 February 1971. Indeed, a great many people are probably still unaware of the Convention's very existence, even on the occasion of its fiftieth anniversary.

Yet the events typically selected to exemplify this 'universal recall' criterion (eg, the moon landings, the Kennedy assassinations, or 9/11) suggest that it is in reality applicable only to developments of an overtly seismic kind, rather

---

[1] Convention on Wetlands of International Importance especially as Waterfowl Habitat, adopted 2 February 1971, entered into force 21 December 1975, 996 UNTS 245 (amended 1982 & 1987). All texts, plus a wealth of additional materials, are available via the Convention's website at www.ramsar.org/. For discussion, see, eg, GVT Matthews, *The Ramsar Convention on Wetlands: Its History and Development* (Ramsar Bureau, Gland, 1993); Ramsar Convention Secretariat, *An Introduction to the Ramsar Convention on Wetlands* (Ramsar Handbooks, 5th edition, Ramsar Convention Secretariat, Gland, 2016); MJ Bowman, PGG Davies, and CJ Redgwell, *Lyster's International Wildlife Law* (2nd edition, Cambridge University Press, Cambridge, 2010) (hereinafter '*Lyster*'), Chapter 13; RC Gardner (ed), 'Special Issue: The Fortieth Anniversary of the Ramsar Convention on Wetlands' (2011) 14 *Journal of International Wildlife Law & Policy* 173–310. This chapter is intended to give a picture of progress on the occasion of the Convention's 50th birthday.

than to those of a more subtly seminal character. Occurrences of the latter ilk, indeed, seem by their very nature more likely to slip into existence unheralded or even unnoticed, revealing their true significance only gradually and through lessons spoken softly over time to those alert enough to pay attention. Eventually, their essential message may come to infuse our general discourse, even if their identity as its original source becomes contested or forgotten. Today, in the light of precisely such a sustained drip-feed of educative enrichment, the moment has surely arrived to acknowledge Ramsar's entitlement to be numbered among our most genuinely visionary and pioneering legal instruments, having helped – in its own quiet way – to bring radical refocus and reform not only to our approach to the conservation of wetlands and of natural ecosystems more generally, but also to our very conception of the global legal order and its inherent structural viability as we struggle to meet the challenges of the twenty-first century.

2 RAMSAR'S EARLY HISTORY

Impact of this kind could scarcely have been predicted, however, during the Convention's fraught early years. Indeed, a particularly inauspicious development had occurred in 1980, when the original World Conservation Strategy had explicitly referenced Ramsar in the context of a statement that 'weak' conventions were dangerous and to be avoided, since they created the impression that something was being done about conservation when in fact it was not.[2] Fortunately, there were no illusions even within Ramsar circles regarding the efficacy of the legal regime it had established; and at the first official wetlands conference convened under its aegis, in Cagliari, Sardinia, in 1980, several critical structural and procedural weaknesses were identified in the text as originally adopted.[3] Given that their remediation would require formal revision of the Convention, one key omission was the absence of any

---

[2] See International Union for Conservation of Nature (IUCN) et al, *World Conservation Strategy: Living Resource Conservation for Sustainable Development* (IUCN, Gland, 1980), Section 15.4–5.

[3] They are listed in Ramsar Convention, *Recommendation 1.8: Proposed amendments to the Convention* (1980).

amendment procedure allowing such changes to be made with the minimum of logistic complexity. Other problems identified included:

- the laxity and imprecision of the arrangements for ongoing institutional review of implementation – certainly by comparison with other globally oriented conservation treaties concluded in its aftermath;[4] and
- a relatively unusual feature of the testimonium clause, stipulating the primacy of the English language version in the event of any divergence among the authentic texts,[5] which had effectively deterred French-speaking countries from participation.

Accordingly, an extraordinary Conference of the Parties (COP) was convened in Paris during 1982, which took the first steps towards addressing these problems through the adoption of a subsidiary Protocol,[6] which established a formal procedure for future amendment of the Convention and reworded the testimonium clause to make all language texts equally authentic.[7] This Protocol did not enter force itself until 1986, meaning that a full 15 years had elapsed from the Convention's adoption before its most serious innate weaknesses could be resolved. Meanwhile, a second ordinary conference had taken place in Groningen, the Netherlands, during 1984, at which a taskforce was established to consider further remedial measures.[8] These resulted in agreement at the third meeting (held in Regina, Canada during 1987) to utilise the new amendment procedures for modifications to Articles 6 and 7,[9] which (once again borrowing from other conservation treaties concluded since its own adoption) essentially specified a formal timetable, nomenclature,

---

[4] In particular, the Convention on International Trade in Endangered Species of Wild Fauna and Flora (CITES), adopted 3 March 1973, entered into force 1 July 1975, 993 UNTS 243, and the Convention on the Conservation of Migratory Species of Wild Animals (CMS), adopted 23 June 1979, entered into force 1 November 1983, (1980) 19 ILM 15.

[5] Ie, English, French, German, and Russian; though it seems that not all of these were actually in existence at the time of the treaty's adoption. The negotiating text had been in English, which is doubtless why it was selected to prevail. Official translations into other United Nations (UN) languages were made subsequently.

[6] Paris Protocol of Amendment, adopted 3 December 1982, entered into force 1 October 1986 (1983) 22 ILM 698.

[7] See ibid, Arts 1 and 2.

[8] Ramsar Convention, *Summary Report of the Plenary Session* (1984), Sixth Session, Agenda Item 10.

[9] Misc 6 (1990), Cm 983; Matthews, note 1, 105.

and voting procedure for future meetings.[10] By the end of the decade, the Convention was finally up and running in earnest.[11]

Yet this institutional invigoration merely served to bring home the magnitude of the political and practical problems that would have to be overcome if the Convention were truly to make its mark. In particular, the principles and objectives embraced by this regime would ultimately require a complete reappraisal of our perceptions of the natural world and humankind's true place within it. Throughout the post-war era, the defence of fundamental human interests had generally been seen to be the province of the international human rights regime; but for all the valuable progress that this had accomplished, it was becoming increasingly apparent just how limited its practical import was by comparison with this emerging, much less well-known, legal regime that aimed to protect the very biosphere through which absolutely all earthly life was ultimately generated and sustained.

Indeed, it seemed on reflection entirely perverse that, despite according due recognition to the importance of the *social community* through which our individual and collective humanity was effectively realised, early human rights instruments had uniformly neglected the much more fundamental *ecological community* without which no life of any complexity could possibly flourish or survive at all. To appreciate these issues more fully, it is crucial to understand the general cultural background and political worldview prevailing at the time of the Convention's negotiation.

3 THE CULTURAL BACKDROP TO RAMSAR'S ADOPTION

Indeed, close examination of the general political, philosophical, and regulatory ethos that prevailed during the period when the Ramsar initiative was originally conceived[12] would surely provoke surprise that it was ever contemplated at all, let alone actually brought to fruition. In particular, the reconceived global order which emerged following the Second World War remained essentially cast in the old Westphalian mould, predicated upon the notion of a community of discrete, territorially defined nation states, each juridically sovereign within its own political boundaries. Although these had

---

[10] See note 4 above.

[11] See further MJ Bowman, 'The Ramsar Convention Comes of Age' (1995) 42 *Netherlands International Law Review* 1–52; and, on the amendment process specifically, Matthews, note 1, 73–80; MJ Bowman, 'The Multilateral Treaty Amendment Process – A Case Study' (1995) 44 *ICLQ* 540–559.

[12] For further information, see *Lyster*, note 1, 8–11.

been haphazardly superimposed upon, and without significant regard to, the natural ecological features and processes that ultimately underpinned human existential prospects, the risks inherent in these incongruities at a time of rapidly accelerating industrial development were scarcely even recognised in political circles, let alone addressed. Despite some unmistakable warnings of impending ecological turmoil – such as the North American Dust Bowl experience during the 1930s[13] – environmental issues received no explicit attention whatsoever in the United Nations (UN) Charter and none of the specialist technical agencies envisaged under its terms to serve the new community's vital interests had environmental protection as a primary goal.[14] Indeed, it took a formal resolution of the UN Economic and Social Council (ECOSOC) to establish that this issue fell legitimately within UN purview at all.[15]

Perhaps the sole significant glimmer of light in this regard emanated from the creation in 1948 of the International Union for the Conservation (originally 'Protection') of Nature (IUCN) under the auspices of the United Nations Educational, Scientific and Cultural Organization (UNESCO).[16] It was through IUCN that the wetland conservation project itself was subsequently launched by a scientific non-governmental organisation (NGO), the International Waterfowl and Wetlands Research Bureau (IWRB),[17] less than two decades into the post-war era – by which time the principal geopolitical preoccupations had become the Cold War and the process of decolonisation.[18]

---

[13] RA Reis, *The Dust Bowl* (Chelsea House, New York, 2008). For broader contemporary discussion of conservation issues, see SS Hayden, *The International Protection of Wildlife* (Columbia University Press, New York, 1942); L Mumford, *The Condition of Man* (Harcourt, Brace & Co, New York, 1944).

[14] It does not seem to feature in the preambles to these instruments, despite its criticality to the 'welfare' of humankind, which is a recurrent concern. Article I(2)(c) of the Food and Agriculture Organization (FAO) Constitution, 145 BFSP 910, does refer to 'the conservation of natural resources' as one of many topics which the FAO is mandated to encourage.

[15] ECOSOC Resolution 32(IV) (1947), which also convened the United Nations Conference on the Conservation and Utilisation of Resources – essentially just a forum for the exchange of scientific ideas – in 1949.

[16] IUPN first met in 1950. See generally R Boardman, *International Organization and the Conservation of Nature* (Macmillan, London, 1981), Chapter 2.

[17] The wetlands initiative was first proposed at the Project MAR Conference in 1962, following a proposal to IUCN by Luc Hoffman of IWRB. Matthews, note 1, 7.

[18] Regarding the latter, see the *Declaration on the Granting of Independence to Colonial Territories and Peoples*, UNGA Res 1514 (XV) (14 December 1960).

This latter development in particular, however vital a reform, was bound to complicate the realisation of any global conservation initiative, since an expanded constituency of international actors – many newly established and certain to be wary of perceived incursions into their recently acquired sovereign authority – hardly constituted the most promising setting for any such endeavour.

Nevertheless, the sovereignty of individual nation states was never intended to be absolute in the new world order, the largely untrammelled statist orientation of the past being now constrained by two highly significant innovations:

- the unprecedented procedures and powers designed to permit a centralised, community response to any unlawful use of force;[19] and
- the recognition of universal entitlement to basic human rights, the finer details of which were still in the process of elaboration.[20]

Yet although environmental protection should by this time surely have been recognised as a prerequisite to the achievement of both global security and human rights, these interconnections were essentially overlooked, for a complex array of reasons.

A key contributory factor here was the extent to which the academic disciplines which had exerted the strongest influence in shaping the new world order – namely, international relations, philosophy, economics, and social science – had become fossilised in cultural paradigms that were wholly outmoded and starkly at odds with the emerging findings of modern science. In the main, moreover, these paradigms were directly traceable to that particular historical phase of human intellectual development known as the 'Enlightenment',[21] when Europe's natural philosophers first became convinced that the universe could be rendered intelligible purely through the application of human ration-

---

[19] UN Charter, 1 UNTS xvi, Chapters VI, VII.

[20] UN Charter, Arts 1, 55, 56; 1948 Universal Declaration of Human Rights (UDHR), UNGA Res 217A(III).

[21] While typically associated with the eighteenth century, this phase can loosely be viewed as covering the entire period from the publication of Francis Bacon's *Novum Organum* in 1620 (where the reductive method in science was first expounded) to the death of Kant in 1804. Arguably, indeed, it still endures today: for a fuller sense of how the 'Enlightenment Project has cast Western civilization under its long shadow over the past two hundred years', see N Geras and R Wokler (eds), *The Enlightenment and Modernity* (Macmillan Press, Basingstoke, 2000); ML Davies (ed), *Thinking about the Enlightenment: Modernity and Its Ramifications* (Routledge, London, 2016).

ality and experience, and without the need for divine guidance or direction.[22] To their credit, they duly acknowledged that any such understanding must be securely grounded in the formal findings of science; but, much more fancifully, they imagined that the only pertinent body of science then available to them[23] – the mathematically derived Newtonian principles of motion and mechanics, through which everything from the cosmic peregrinations of the planets to the movement of balls across a billiard table could seemingly be calculated with precise accuracy – was essentially sufficient for that purpose.[24] Thus, the 'mechanical' (or 'mechanistic') worldview was born,[25] whereby the entire natural order might be viewed as a kind of giant machine – merely a more sophisticated version of the Great Clock of Strasbourg.

Needless to say, this perspective was hopelessly simplistic, generating a host of misconceptions,[26] a sample of which demands brief attention here. First, the idea that *living things* in particular could be adequately explained in such terms was woefully misguided, even by the standards of the day,[27] generating banal and crudely deterministic notions of animal behaviour that long maintained a stranglehold, even in scientific circles.[28] At the same time, however, the majority of these luminaries unsurprisingly baulked at the reduction of *their own kind* to the level of mere machinery incarnate: typically, they preferred to endorse the concept of 'reason' or 'rationality' as the magic ingredient

---

[22] That is not to say that they necessarily abandoned religious belief itself, for many remained extremely devout.

[23] Most other 'scientific' work – especially in the life sciences – was essentially descriptive or assertive, rather than genuinely explicatory.

[24] Newton himself cannot escape all blame for this, as he solemnly declared his principles to form 'the foundation of all philosophy'. See Rule III of his 'Rules of Reasoning in Philosophy', translated and extracted in I Kramnick (ed), *The Portable Enlightenment Reader* (Penguin Books, London, 1995) 46.

[25] For discussion, see DC Goodman and J Hedley Brooke (eds), *Towards a Mechanistic Philosophy* (Open University Press, Milton Keynes, 1974); P Ball, *Critical Mass: How One Thing Leads to Another* (Arrow Books, London, 2004), Chapters 1, 2; E Dolnick, *The Clockwork Universe: Isaac Newton, the Royal Society and the Birth of the Modern World* (Harper, London, 2011).

[26] This is not to deny that it also produced some valuable insights in the short term, especially regarding human physiology.

[27] Thus, the mechanists were opposed (albeit to little beneficial effect) by the 'vitalists'. See E Mayr, *This is Biology: The Science of the Living World* (Belknap Press, Cambridge, MA, 1997), Chapter 1.

[28] Key to this was the concept of 'instinct' – an amorphous, rag-bag term used indiscriminately to substitute for any meaningful explanation of animal behaviour. Another manifestation of such thinking was the simplistic 'stimulus-response' analysis of the Behaviourist tradition in psychology.

that liberated humankind from purely mechanistic causation and thereby set us apart from the rest of creation,[29] despite being incapable of providing any coherent or convincing account of exactly what it was or how it came to be.[30] A common misapprehension was to assume that it (along with other cognitive attributes such as thought, intelligence, self-consciousness, and even basic sentience) was somehow a product of language, despite this being another faculty of which they had no clear or compelling explanation.[31]

During the early nineteenth century, French positivist philosopher and 'father of sociology' Auguste Comte endeavoured to temper these follies by insisting on the importance of physics, chemistry, and biology as the successive intellectual foundations for the ontological study of humankind.[32] However, he was effectively confounded by the fact that the fundamental principles of biology still had not actually been formulated by the time of his death in 1857; furthermore, in his attempts to choose between rival emerging hypotheses, he repeatedly backed the wrong horse.[33] In the event, it was not until the 1940s that a successful scientific synthesis of the Darwinian concept of natural selection and the Mendelian findings on genetic inheritance was finally accomplished;[34] and even later before that synthesis could be integrated with the ecological insights and principles previously propounded by the likes

---

[29] This was, indeed, a pervasive theme from Descartes to Kant and beyond (though the likes of Hume did concede some degree of simple reason in animals).

[30] For one forlorn attempt by Kant, see HS Reiss (ed), *Kant: Political Writings* (2nd enlarged edition, Cambridge University Press, Cambridge, 1991) 42.

[31] This fallacy has exhibited a dogged durability in philosophical circles, where blindness to scientific findings has remained strong. For corrective discussion, see Stephen Pinker, *The Language Instinct: The New Science of Language and Mind* (Penguin, London, 1994); Robin Dunbar, *Grooming, Gossip and the Evolution of Language* (Faber and Faber, London, 1996); Peter Carruthers and Jill Boucher (eds), *Language and Thought: Interdisciplinary Themes* (Cambridge University Press, Cambridge, 1998).

[32] See generally G Lenzer (ed), *Auguste Comte and Positivism: The Essential Writings* (Routledge, London, 1997).

[33] Thus, he fervently embraced such misconceptions as the essential fixity of species boundaries, rather than speciation as a continuous process; the inheritance of *acquired* characteristics; the virtues of phrenology, rather than psychology, as the key to understanding human thinking; and the Enlightenment belief in nature's unrelenting progress towards ever-greater complexity and perfection, as most obviously reflected in humankind itself.

[34] See especially J Huxley, *Evolution: The Modern Synthesis* (George Allen & Unwin, London, 1942); E Mayr, *Systematics and the Origin of Species* (Harvard University Press, Cambridge, MA, 1942).

of Humboldt and Haeckel,[35] so as to form an essentially irrefutable basis for national, regional, and global policymaking regarding the natural world. UNESCO's Man and the Biosphere Conference in 1968 proved an important milestone in the consolidation of understanding here,[36] although by that time the Ramsar negotiations themselves were already well advanced.

A second source of Enlightenment misdirection – evident at least since the writings of Locke[37] – was the idea that no significant value could be attributed to raw nature of itself, since it was only when natural resources were transformed by human ingenuity into items of more immediate practical utility that value truly emerged. This perversely anthropocentric nostrum was in turn founded largely upon the twin beliefs that value stemmed primarily from rarity, and that natural resources themselves were essentially in endless supply.[38] It was on the basis of such delusions that Enlightenment authors stridently advocated the logging of forests and the draining of wetlands in the cause of advancing human convenience and wealth,[39] little realising how profoundly damaging to human interests society's enthusiastic accession to such demands would inevitably prove over time.

The follies inherent in these key tenets of Enlightenment thinking seem all too obvious now, the assumption of nature's inexhaustibility being manifestly ill-founded in simple factual terms;[40] while the obsessive preoccupation with untutored human preference has proved doubly instrumental in ensuring its falsity. In purely pragmatic terms, systemic neglect of the bio-preferences of species other than our own has inevitably ensured the diminution of primary natural productivity, leading inexorably to the disruption of wider planetary life-support systems. In a deeper, more philosophical sense, by blinding us

---

[35] For instructive discussion of their contributions, see A Wulf, *The Invention of Nature: The Adventures of Alexander von Humboldt, The Lost Hero of Science* (John Murray, London, 2016), especially Chapter 22.

[36] See Boardman, note 16, Chapter 4.

[37] See especially John Locke, *Two Treatises on Government: Civil Government* (Awnsham Churchill, London, 1690), Book II, Chapter V. Amazingly, such delusions persist even now. See GH Smith, 'John Locke: Money and Private Property,' 20 November 2015, at: https://www.libertarianism.org/columns/john-locke-money-private-property.

[38] Locke, note 37.

[39] See Kant, *Critique of Judgment* (1790; Oxford World Classics edition, N Walker ed, Oxford, 2007), Part II [67]; David Hume, 'On the Populousness of Ancient Nations', included in Volume I of *Essays and Treatises on Several Subjects* (new edition, T Cadell, London, 1793) 432.

[40] Indeed, many scientists believe that our planet is currently undergoing a sixth major extinction crisis.

to the need for meaningful recognition of the more fundamental, and essentially *biological*, dimension of 'value' – which alone makes the very faculty of preference both possible and important in the first place – this uncritical self-centredness has also served to distort and diminish our very concept of value by confining attention exclusively to its *secondary* aspects.[41]

After all, human preferences can only be shown to merit such scrupulous and exclusive respect by reference to some more fundamental, overriding 'good' embodied intrinsically in humans as such (ie, *peculiarly* and *universally*): no philosophical system worthy of the name can credibly consign this to mere unsubstantiated assumption. While Kant endeavoured to fill this gap by reference to his concept of 'human dignity' – which demanded that humans (and they alone) never be treated solely as means to an end, but always as ends in themselves[42] – he could never conceal the absence of any remotely plausible objective criterion for affording *absolutely all* human beings this entitlement while simultaneously denying it to *absolutely all* other entities.[43]

Although intellectually defensible routes undoubtedly exist through which the concepts of dignity and value (in its primary sense) can be rendered meaningful and viable,[44] the international community understandably had little energy or inclination for such arcane endeavours in the shattered aftermath of the Second World War, and the ramshackle Enlightenment conceptions were effectively imported wholesale and unreconstructed into the international legal order – to serve, indeed, as its very foundations.[45] In a chilling demonstration of their inherent unworldliness, US philosopher Mortimer Adler airily

---

[41] See further RJ McShea and DW McShea, 'Biology and Value Theory' in J Maienschein and M Ruse, *Biology and the Foundation of Ethics* (Cambridge University Press, Cambridge, 1999); M Fosci and T West, 'In Whose Interest? Instrumental and Intrinsic Value in Biodiversity Law' in MJ Bowman, PGG Davies, and EJ Goodwin, *Research Handbook on Biodiversity and Law* (Edward Elgar Publishing, Cheltenham, 2016).

[42] See, eg, Kant, *The Metaphysics of Morals* (1797; Cambridge University Press edition, M Gregor ed, Cambridge, 1996) 186–87.

[43] Kant's own attempt to advance 'reason' as the universal distinguishing feature would obviously have failed on empirical grounds even had he been able to provide an intelligible explanation of it.

[44] See MJ Bowman, 'Animals, Humans and the International Legal Order: Towards an Integrated Bioethical Perspective' in W Scholtz (ed), *Animal Welfare and International Environmental Law: From Conservation to Compassion* (Edward Elgar Publishing, Cheltenham, 2019).

[45] On human rights specifically, see J Morsink, *The Universal Declaration of Human Rights: Origins, Drafting and Intent* (University of Pennsylvania Press, Philadelphia, 2000), Chapter 8.

proclaimed to the Commonwealth Club in 1952 that, in ontological terms, chimpanzees were more closely akin to rocks and stones than they were to human beings.[46]

Even more troubling manifestations of the deficiencies of Enlightenment thinking became evident in relation to the international economy, where the anomalies highlighted above became entangled with misconceptions derived from Adam Smith's scientifically naive commendation of industrial mechanisation and free trade as the sure routes to human advancement. His principal errors in this instance stemmed from the failure to appreciate that Newton's laws, taken in isolation, were entirely insufficient to explain the workings *even of machines themselves* (let alone those of more complex entities). In particular, their functioning depends no less crucially upon the principles of *thermo-dynamics* – and especially the second law,[47] which predicates the deposit of *waste* into the surrounding environment as an inescapable outcome of sub-optimal mechanical efficiency.[48] Along with the *energy* inevitably frittered away in this process in the form of unwanted heat, light, and noise, the realities of industrial process are such that substantial quantities of waste *matter* are also generated – much of which may prove toxic, ecologically disruptive, or otherwise detrimental to us (as in the case of the greenhouse gases generated by fossil fuel consumption). In some cases, even the primary product itself may embody such hazards, which naturally become grossly intensified when the 'throwaway' culture of modern society is applied to items fabricated with a view to their robustness or indestructibility (as topically exemplified in the case of single-use plastics).[49]

As regards 'free trade', the expression is, of course, merely a convenient catchphrase for the activity of moving things around in pursuit of profit – a process which cannot rationally be regarded as either good or bad *of itself*, but only by reference to detailed consideration of (all) its practical effects. Since these are guaranteed to include the generation of yet further waste as a consequence of the transportation process, the damaging aspects of industrial

---

[46] MJ Adler, 'The Dignity of Man and the 21st Century', 10 October 1952, at: https://www.cooperative-individualism.org/adler-mortimer_dignity-of-man-and-the-21st-century-1952.htm.

[47] Stipulating that the entropy of any closed system can only increase over time. See J Daintith (ed), *Oxford Dictionary of Physics* (6th edition, Oxford University Press, Oxford, 2009) 546–47.

[48] Thus, the most crucial part of any machine is arguably the world which surrounds it. P Atkins, *Conjuring the Universe: The Origins of the Laws of Nature* (Oxford University Press, Oxford, 2018) 90–91.

[49] For up-to-the-minute scientific assessment of this accelerating crisis, see J Howgego, 'Waste Not ... Want Not?' *New Scientist* (12 February 2022) 38–47.

process itself are inevitably compounded. All these harms, moreover, occur even when the processes of manufacture and transport unfold exactly as planned; should something, by contrast, chance to go disastrously wrong (eg, think Chernobyl, Bhopal, Exxon Valdez, Fukushima, Deepwater Horizon), the damage can only be massively magnified.

Under any credible system of economic accounting, these costs – which are typically borne by the community at large – should plainly be factored into the equation to counterbalance whatever benefits might be derived from the transaction in question by the parties themselves, especially since they may very well outweigh them completely. The traditional failure of economists to ensure any such assessment[50] leaves the routine, uncritical description of these traded items as 'goods' heavy with unintended irony; while their further blind attachment to such ill-conceived notions as 'gross domestic product' has merely exacerbated the discipline's inherently despoliatory impacts.[51]

The failure of eighteenth-century philosophers and early economists to perceive the shortcomings inherent in their worldview was doubtless attributable to the very primitive state of development of the natural sciences at their times of writing. Yet this very naivety has served to precipitate one of the greatest follies in the entire history of human ideas: the attempt to propound precise regulatory principles for the *material exploitation* of the *oikos*, or human natural estate (through *oeconomics*), in advance of any adequately informed attempt to investigate its *fundamental nature and functioning* (through *oecologie*).

Even then, the most deleterious impacts of Enlightenment misconception could still have been avoided in the longer term if only the disciplines most severely blighted by it – namely, philosophy, politics, and above all, economics – had shown the willingness to progressively reformulate the essentials of their thinking so as to more effectively come to terms with underlying scientific reality. Yet reform of so radical a kind is difficult to stimulate authoritatively except from within, and the tiny handful of perceptive internal sceptics

---

[50] This stems in part from the very constricted conception of 'wealth' employed by early economists: originally the word simply meant 'welfare' or 'wellbeing' in the broadest sense (as still reflected today in the concept of 'commonwealth'). D Pilling, *The Growth Delusion: The Wealth and Well-being of Nations* (Bloomsbury Publishing, London, 2019) 189–202, 296–97. Both senses have been evident, however, since at least the thirteenth century. TF Hoad (ed), *Oxford Concise Dictionary of English Etymology* (Oxford University Press, Oxford, 2003).

[51] See further Pilling, note 50. For confirmation that even the more 'progressive' voices within the discipline of economics still lag well behind the 'asking rate' in the field of biodiversity, see D Helm and C Hepburn (eds), *Nature in the Balance: The Economics of Biodiversity* (Oxford University Press, Oxford, 2014).

in the field of economics have required time to build up their firepower.[52] It is accordingly fortunate that a second – and potentially equally productive – remedial strategy can be found in the simple promulgation of an alternative cultural narrative, the elucidatory power and attractiveness of which might hopefully prevail over time by highlighting these weaknesses more indirectly and incidentally. History will show that Ramsar itself has proved a vital instrument and rallying point for this latter kind of educative campaign.

4 RAMSAR'S EDUCATIONAL ASPECT

While a casual glance at the text of Ramsar would be unlikely to identify education as a central aspect of its mission, keener-eyed readers would doubtless notice that Articles 4(3) and (5) respectively require the parties to encourage research and data exchange on wetlands and their wildlife; and to promote training in the fields of wetland research, management, and wardening. These awareness-raising processes are, moreover, undoubtedly critical to advancing the Convention's fundamental aspiration for more effective wetland conservation and wise use on a global scale. The detailed guidance to governments embodied in the *Ramsar Wise Use Handbooks* obviously constitutes the Convention's primary educational resource in this regard.[53]

4.1 Enhancing Public Awareness

Yet Article 4 addresses only instruction of a technically specialist character and is silent on the equally crucial need for educational endeavours aimed at the general public, to counteract the enduringly unfavourable impression of bogs, swamps, and other wetland types that has traditionally dominated human consciousness. Here, Ramsar might seem at first sight to compare unfavourably

---

[52] For noteworthy examples, see S Keen, *Debunking Economics: The Naked Emperor Dethroned?* (2001; revised edition, Zed Books, London, 2011); GA Akerlof and RJ Shiller, *Animal Spirits: How Human Psychology Drives the Economy, and Why It Matters for Global Capitalism* (Princeton University Press, Princeton, 2009); Ha-Joon Chang, *23 Things They Don't Tell You About Capitalism* (Penguin Books, London, 2010). On the adverse implications of economic analysis for environmental protection specifically, see Pilling, note 50, Chapter 11; A Gillespie, *International Environmental Law, Policy and Ethics* (2nd edition, Oxford University Press, Oxford, 2015), Chapter IV.

[53] See https://www.ramsar.org/resources/the-handbooks; and further CM Finlayson et al (eds), *The Wetland Book: Structure and Function, Management, and Methods* (Springer, Dordrecht, 2018).

with other conservation conventions of the period:[54] even the African regional Convention of 1968[55] (for all its other weaknesses) required parties to ensure their peoples' appreciation of human dependence upon natural resources and understanding of the need for their rational utilisation. Methods specified included the incorporation of such issues into 'educational programmes at all levels' and 'information campaigns capable of acquainting the public with, and winning it over to, the idea of conservation',[56] making 'maximum use of the educational value of conservation areas' themselves for that purpose.[57] Yet these vital aspects of public education are in fact embraced by the Ramsar regime, even if a little effort may be required in order to uncover them.

Thus, the Convention's central obligation under Article 3.1, mandating the formulation and implementation of planning 'so as to *promote* the conservation ... and wise use' of wetlands, should be read as extremely wide-ranging, embracing not only the physical establishment of scientifically informed measures for wetland protection *in situ,* but also the fostering of public awareness regarding the importance of such programmes as a matter of principle; indeed, one of the commonest contemporary applications of the word 'promote' lies precisely in the realm of advertising and public relations. Thus, the awareness-fostering aspect of Ramsar's mission finds clear if rudimentary reflection in its preamble, which affirms specifically the 'interdependence of man and his environment', the 'fundamental ecological functions' of wetlands, and the irreplaceable values they consequently embody, in 'economic, cultural, scientific and recreational' terms.

The detailed elaboration of such matters, however, lies less in the bare text itself than in the gradual accretion of practice around it, as realised collectively in the programmes and pronouncements of treaty organs and individually in the arrangements for implementation adopted by states parties. This practice is in fact extremely rich, commencing with a call at the fourth COP for the development of comprehensive strategies for raising awareness of wetland values, both within the educational system itself and through more informal channels, with particular emphasis upon the use of nature reserves themselves for instructional purposes.[58] This process is now well advanced, with the

---

[54] See, eg, the Convention concerning the Protection of the World Cultural and Natural Heritage, adopted 16 November 1972, entered into force 17 December 1975, (1972) UNJYB 89, especially Articles 4, 5, 27, 28.

[55] The 1968 African Convention on the Conservation of Nature and Natural Resources, adopted 15 September 1968, entered into force 16 June 1969, 1001 UNTS 3.

[56] Ibid, Art XIII(1).

[57] Ibid, Art XIII(2).

[58] Ramsar Convention, *Recommendation 4.5: Education and training* (1990).

expanding network of wetland reserves managed in the United Kingdom by the regional water authorities, the Wildfowl and Wetlands Trust, and similar agencies constituting a noteworthy national example.

Further progress occurred at the seventh COP through the adoption of a formal outreach programme, which has subsequently evolved into a more comprehensive, ongoing regime for communication, capacity building, education, participation, and awareness (CEPA).[59] Within the Convention's current Strategic Plan,[60] CEPA constitutes a key medium for mainstreaming wetland conservation and wise use;[61] and *Ramsar Handbook 6* is devoted entirely to this topic.[62] CEPA is not the exclusive vehicle for education, however, and many other elements of the Convention's operation – including the Ramsar Advisory Missions – have been identified as contributing indirectly to this broad function.[63] The entire promotional/educational initiative is, moreover, also now encapsulated symbolically in the commemoration each February of World Wetland Day, on the anniversary of the Convention's adoption.[64]

Yet however sophisticated and productive such programmes and processes have turned out to be, they in truth reflect only part of Ramsar's overall educative impact to date, which has actually percolated much more widely across the international regulatory order, and through processes of a less overtly didactic and more exemplary or demonstrational character.

4.2 Ramsar and the Broader Educative Canvas

This wider impact is reflected in the way Ramsar has helped to transform our traditional perceptions not only of the natural world generally – extending far beyond its specific wetland remit – but also of the precise *methodologies* through which nature should best be regulated. Perhaps the pre-eminent contribution of the Ramsar project to the advancement of human interests, through the improved management of both its geophysical and its juridical estates,

---

[59] See https://www.ramsar.org/activity/the-cepa-programme.

[60] Ramsar Convention, *Resolution XII.2: The Ramsar Strategic Plan 2016–2024* (2015), as reviewed by *Resolution XIII.5: Review of the Fourth Strategic Plan of the Ramsar Convention* (2018).

[61] More precisely, this features as Target 16 of 19 in total and seeks to further Goal 4, regarding enhancement of the Convention's implementation.

[62] Ramsar Convention Secretariat, *Handbook 6: Wetland CEPA* (Ramsar Handbooks, 4th edition, Ramsar Convention Secretariat, Gland, 2010).

[63] See Ramsar Convention, *Resolution XIII.11: Ramsar Advisory Missions* (2018), preamble.

[64] For the most recent measure concerning this event, see Ramsar Convention, *Resolution XIII.1: World Wetlands Day* (2018).

has stemmed from its overdue infusion of a genuinely organic, ecologically informed, and holistic perspective into the barren landscape of abstracted, atomistic, mechanistic myopia within which the treaty-making process had previously for the most part been conducted.[65] This transformation has been manifest, moreover, throughout every dimension of the broad juridical forcefield which treaty regimes characteristically create for the purposes of their own functioning – namely, the normative, the instrumental, and the societal.

It will be recalled here that two distinct ways of conceptualising the juridical phenomenon of the treaty have been proposed: as *obligation* on the one hand and as *instrument* on the other, with the latter falling within the purview of the law of treaties, strictly so called, and the former being regulated by the law of state responsibility.[66] Yet there is also a third, essentially *societal*, dimension to treaty regimes, comprising the particular *juridical community* within which the constituent norms of the instrument are elaborated, interpreted, and applied. This aspect has not traditionally been a prime focus of attention for international lawyers as such, but it is certainly no less important to the overall functioning of the regime.

4.2.1 Expanding the focus of ecological normativity

Regarding the treaty-as-obligation, the most obvious advance lies in Ramsar's seemingly unprecedented identification of *ecosystems as such* as its focal point of normative conservation concern,[67] coupled with the exceptionally broad range of wetland types encompassed within that remit.[68] While it was not the first treaty historically to make reference to ecosystems or ecological considerations,[69] Ramsar seemingly was revolutionary in making these elements anything other than a purely peripheral focus of attention. Indeed, despite the

---

[65] On the distinction between 'organic' and 'mechanistic' styles of treaty-making, see MJ Bowman, 'The Interplay of Concept, Context and Content in the Modern Law of Treaties: Final Reflections' in MJ Bowman and D Kritsiotis (eds), *Conceptual and Contextual Perspectives on the Modern Law of Treaties* (Cambridge University Press, Cambridge, 2018), especially Section 2.2.

[66] S Rosenne, *Developments in the Law of Treaties 1945–1986* (Cambridge University Press, Cambridge, 1989).

[67] See further Matthews, note 1, Chapter 1.

[68] See Article 1, and the more detailed wetland classification system tabulated in Ramsar Convention Secretariat, *Handbook 17: Designating Ramsar Sites* (Ramsar Handbooks, 4th edition, 2010) 80–83.

[69] For an earlier example, see the 1968 African Convention, note 55, Arts III(4)(a) and (c), IV(a), X(i), XII, XIV(ii). Such considerations had also occasionally featured in fisheries conventions.

recurrent preoccupation within the text with 'waterfowl', it is noteworthy that very few of the substantive duties thereby established actually relate to them specifically,[70] being focused for the most part on wetlands themselves (albeit sometimes in conjunction with waterfowl, or with wetland flora and fauna more generally).[71] Reinforcing that holistic biological perspective, the central concern is clearly with maintenance of the fundamental *ecological character* of these particular landscape features.[72] Since habitat diminution and degradation have long been recognised as pre-eminent threats to nature conservation,[73] this was a particularly vital breakthrough.

Furthermore, although there was equally clear recognition of both wetlands and their wildlife as *resources*, and of the consequent inevitability of their exploitation, it was to be *natural ecology* rather than *human utility* which provided the Convention's primary intellectual focus and motivational force. That is, even if it were the 'economic' and 'recreational' values of wetlands that in reality motivated us most strongly to secure their protection, it is still upon *ecological* considerations that our conservation programmes should ultimately be grounded, since these are the inescapable natural prerequisites to those programmes' ultimate success, and hence to delivery of the wetland services we actually value most highly. By contrast, undue preoccupation with the financial and recreational returns themselves, to the neglect or subordination of basic ecology, risked merely defeating our own aims – certainly in the longer term.[74]

Ramsar's perspicacity in this respect was, moreover, greatly strengthened once the 'wise use' of wetlands was formally redefined in terms of mainte-

---

[70] The principal examples are Articles 2.6 and 4.4, each of only ancillary importance.

[71] Regarding the latter categories, see Articles 4.1, 4.3, and 5. Note also that only two of the nine current criteria for designating wetlands for 'the List' relate to waterfowl specifically (the same as for fish).

[72] See Convention on Wetlands, note 1, Art 3.2.

[73] Thus, RB Premack, *Essentials of Conservation Biology* (Sinauer Associates, Sunderland, Mass, 1993) 111–12, identifies the destruction, fragmentation, and degradation of habitat as three of the six primary drivers of biodiversity diminution at human hand (alongside direct overexploitation, the introduction of exotic species, and the increased spread of disease). Accordingly, habitat considerations now underpin every aspect of modern conservation policy. See generally Convention on Biological Diversity Secretariat, *Global Biodiversity Outlook 5* (UNEP/CBD, Montreal, 2020).

[74] See generally J Kay, *Obliquity: Why Our Goals are Best Achieved Indirectly* (Profile Books, London, 2010).

nance of their ecological character:[75] by making such perpetuation not merely an objective of the regime, but rather the formal litmus test of compliance with the parties' basic conservation obligations, the COP not only significantly enhanced the cohesion between the Convention's two major substantive provisions – Articles 3.1 and 3.2 – but also constructively recrystallised its overall mission. In addition, the regime's gradually growing emphasis on positive cooperative action to assist individual compliance – whether through financial assistance or technical support – has significantly strengthened its overtly ecological perspective.[76]

This cooperative ethos has, moreover, contributed to the entrenchment of a more genuinely *transnational* perspective upon ecosystem protection within the regime. This is reflected most obviously in the recognition that wetlands may sometimes extend across national boundaries and that 'water systems' may likewise be 'shared', demanding consultation and cooperation in their management.[77] More fundamentally, however, explicit acceptance of the idea that – even absent such transboundary extension – many wetland sites must be deemed of genuinely *international* importance and thereby merit conservation commitment from the global political community *as a whole*[78] represented a crucial development in the legal notion of *common concern*, which has subsequently attracted such widespread support as arguably to justify its recognition as a general principle of international law.[79] This must be counted as a highly important achievement, given the predominantly atomistic perception of the international community that previously prevailed.

Indubitably, there are grounds for lamenting the limited scale of the recovery in wetland health, abundance, and diversity that has been achieved under the Convention so far;[80] but allowance must be made here for the trajectory of

---

[75] See Ramsar Convention, *Resolution IX.1, Annex A: A Conceptual Framework for the wise use of wetlands and the maintenance of their ecological character* (2005).

[76] See further on these aspects *Lyster*, note 1, 435–48.

[77] Convention on Wetlands, note 1, Art 5. This particular perception, admittedly, was already enshrined in international watercourse law.

[78] Albeit without calling into question the 'exclusive sovereign rights' of states over sites within their territories. Ibid, Art 2.3.

[79] On the distinction between common concern (or common interest) and other collectivist legal conceptions, such as common property and common heritage, see *Lyster*, note 1, 48–52; D French, 'Common Concern, Common Heritage and other Global(-ising) Concepts: Rhetorical Devices, Legal Principles or a Fundamental Challenge?' in Bowman, Davies, and Goodwin, note 41.

[80] See, eg, N Dudley (ed), *Global Wetland Outlook: Special Edition 2021* (Ramsar Convention Secretariat, Gland, 2021); Ting Xu et al, 'Wetlands of

unrelenting, precipitate decline which it was originally designed to tackle, as well as for the pervasive governmental mindset of ignorance and ineptitude that not only brought it about originally but also serves as a formidable impediment to progress even today. In particular, the traditional culture of almost all major factions of government has remained firmly in thrall to Enlightenment misdirection, in all its many forms. This underscores the importance of the Convention's systematic promulgation of practical guidance on the detailed ramifications of 'wise use',[81] which seeks to extend the reach of ecological thinking all the way from individual wetland sites to the broadest policy deliberations at the national level. Yet the scale of the obstacles to be overcome in this regard is such that the achievement of meaningful progress cannot realistically be measured in temporal units smaller than decades.

Within the environmental field itself, however, traction has – happily – proved a little easier to secure. Here, one key aspect of the reorientation of legal attention that Ramsar has helped to bring about – namely, away from individual species exclusively and towards the wider ecological complex of which they form part – has been extremely influential in subsequent global conservation efforts. Most notably, the Convention on Biological Diversity (CBD)[82] astutely endeavours to address the conservation issue at multiple levels of biological organisation,[83] with the 'ecosystem approach' gradually assuming pride of place in its overall schematic;[84] and the UN, of course, very recently designated the period 2021–30 as the Decade of Ecosystem Restoration.[85]

While the punchy and pragmatic tone of the documentation surrounding this latest initiative leaves it generally unforthcoming regarding questions of underlying philosophy, it cannot avoid a passing affirmation that its overall objective entails a thoroughgoing restoration of 'the relationship between humans and nature' – a mission required 'for the health and well-being of

---

International Importance: Status, Threats and Future Protection' (2019) 16 *International Journal of Environmental Research and Public Health* 1818; V Reis et al, 'A Global Assessment of Inland Wetland Conservation Status' (2017) 67 *Bioscience* 523–33.

[81] Through the *Ramsar Handbooks*, note 53 and accompanying text.

[82] Convention on Biological Diversity (CBD), adopted 5 June 1992, entered into force 29 December 1993, 31 ILM 822.

[83] See the very definition of 'biological diversity' in Article 2.

[84] See especially CBD Decisions II/8 and V/6.

[85] See generally https://www.decadeonrestoration.org and United Nations Environment Programme (UNEP), *Ecosystem Restoration for People, Nature and Climate* (UNEP, Nairobi, 2021).

all life on Earth and that of future generations'.[86] This fleeting acknowledgement of nature as an intended beneficiary of conservation action in its own right is arguably tantamount to recognition (alongside all its other, purely anthropocentric, virtues) of the *intrinsic value* of nature – albeit a surprisingly muted one, given that the notion receives explicit endorsement globally in the CBD,[87] having previously found reflection in certain regional and soft law instruments.[88]

This coyness perhaps reflects the extent to which the dark shadow of eighteenth-century thinking still looms over international political deliberations generally. While it should by now be totally unnecessary to recapitulate the essential intellectual inescapability of the concept of nature's intrinsic value and the intellectual vacuity of the objections to it that have in the past been raised by mainstream philosophers and economists,[89] it must be admitted that an element of uncertainty remains regarding its applicability to *entire ecosystems*, by contrast to the individual organisms that represent its surest instantiation.[90] In particular, it has been argued that ecosystems lack the 'integrated unity' to be expected of a genuine system[91] and also suffer from much greater indeterminacy of physical delimitation. They might also seem strictly to lack the inbuilt capacity for existential preference which is the hallmark of entities that exhibit a genuine 'good of their own'.

Yet these objections might simply stem from the difficulty we naturally experience in distancing ourselves from ourselves as the appropriate paradigm for existential identity. Ecosystems can readily be acknowledged as simply a higher level of biological organisation, and objective criteria can certainly be devised in accordance with which it can be determined whether they are systemically hale, ailing, or failing at any given moment. At the very least, the applicability of intrinsic value in this context must be regarded as debatable; and there seems now to be a growing tendency – even in international legal instruments – towards the extension of recognition to ecosystems as genuine 'ends-in-themselves'.[92] This approach, moreover, clearly offers great prag-

---

[86] UNEP, note 85, Section 5.1 (emphasis added).
[87] CBD, note 82, preamble, first recital.
[88] For discussion, see *Lyster*, note 1, Chapter 3.
[89] For detailed discussion, see ibid, including the recommended further reading; and Fosci and West, note 41.
[90] See further *Lyster*, note 1, Chapter 3, Section 3.
[91] R Attfield, *The Ethics of Environmental Concern* (2nd edition, University of Georgia Press, Athens, Georgia, 1991), Chapter 8.
[92] In addition to the CBD itself, see the 1991 Environmental Protocol to the 1959 Antarctic Treaty, adopted 4 October 1991, entered into force 14 January 1998 (1991) 30 ILM 1461, Art 3(1); Protocol on Forestry to the 1992 Treaty of the

matic virtue in view of the unquestioned practical necessity of focusing our conservation efforts at the ecosystem level especially.

A final advantage here might arguably lie in the possibility of treating the intrinsic value of ecosystems as the foundation of some kind of formal legal *personhood*, as has already been accomplished for human beings within the international legal order. Although – even accepting in principle the intrinsic value of individual organisms – the recognition of every single one of them as a *legal person or rights holder* is likely to be perceived by states as an unworkable or even preposterous stratagem, the objections might prove less formidable in relation to ecosystems. Thus, while many might still scorn the notion that *trees* should have standing for legal purposes,[93] the idea that entire *forests* should do so might appear distinctly less unpalatable. Indeed, there is a growing incidence within domestic legal systems of the recognition of natural features – especially rivers – as the possessors of some form of legal status in their own right.[94] This approach has, moreover, already attracted cautious commendation in international policy documents[95] and is clearly worthy of more detailed exploration.[96]

Given the early date of Ramsar's adoption, it is unsurprising that it made no explicit reference to the notions of conservation for nature's sake, intrinsic value, or the juridical personality of natural ecosystems. And yet, through its very focus upon ecosystems as such, and the maintenance of their essential characteristics, the Convention helped to sow the intellectual seeds for the emergence of all these ideas. Furthermore, through the subsequent exposition of detailed COP guidance on the typology and inventory of wetland sites, the precise indicia and implications of ecological change, and the crucial regulatory processes of impact assessment, site management and wetland restoration, it has provided the substantive technical wherewithal to translate such ideas into practical realisation. Few forms of instruction could possibly have been more important to human wellbeing.

Southern African Development Community, adopted 3 October 2002, entered into force 17 July 2009, preamble.

[93] For the *locus classicus* on this point, see C Stone, 'Should Trees have Standing? Towards Legal Rights for Natural Objects' (1972) 45 *Southern California Law Review* 450.

[94] For a rapidly growing array of examples, see www.harmonywithnatureun.org/rightsOfNature/.

[95] Eg, 'The Future We Want' (Outcome Document of the UN Conference on Sustainable Development, Rio de Janeiro, 20–22 June 2012) at www.un.org/disabilities/documents/rio20_outcome_document_complete.pdf [39], [40].

[96] For an invaluable primer, see DR Boyd, *The Rights of Nature: A Legal Revolution that Could Save the World* (ECW Press, Toronto, 2017).

4.2.2 Institutional biology of treaty regimes

Yet Ramsar's imaginative infusion of organic thinking into the international legal order was by no means limited to its purely normative dimension, being evident also – if only metaphorically – at the structural, *instrumental* level. For it was this Convention that ushered in the modern era of 'living instruments' for environmental protection – that is, treaty regimes which were themselves conceived essentially in the image of biological entities rather than mere mechanisms.[97] Underpinning this approach is the manifest need for continuous operational adjustment to the evolving challenges to be addressed under the regime in question, which typically concern wide-ranging, technically complex subject matter that is inherently in a state of perpetual flux and/or highly contentious politically. Given the reality of the daily struggle for survival faced by all living things and the adaptive flexibility that this requires, organisms offer a far more appropriate model for treatymakers than do mere machines, which are typically designed and constructed to offer a predetermined solution to a specific, fixed challenge.

Until the 1970s, the 'organic' style of treaty arrangement in international law generally was largely confined to those that created formal international organisations, commonly imbued with legal personality on their own account and equipped with internal institutional organs charged with the ongoing performance of designated tasks. These constituted a category that Rosenne judged to be so fundamentally distinct from 'ordinary' legal agreements that he remained forever unconvinced of the feasibility of their regulation under a single, unified law-of-treaties regime.[98] The International Law Commission plainly disagreed, however, designing the Vienna Convention on the Law of Treaties to be of universal application to legally binding inter-state written agreements of every kind.[99]

The wisdom of this approach was promptly confirmed by the successive appearances of Ramsar and other, similarly structured, wildlife conservation regimes, which effectively staked out an ontological middle ground between the two previously familiar treaty types so as to produce a virtual continuum.[100] This new hybrid style of treaty arrangement[101] required at the absolute

---

[97] See note 65.

[98] Rosenne, note 66, 252.

[99] Vienna Convention on the Law of Treaties, adopted 23 May 1969, entered into force 27 January 1980, 1155 UNTS 331, Arts 1, 2(1)(a).

[100] See further Bowman, note 65, Section 2.3.1.

[101] See further RR Churchill and G Ulfstein, 'Autonomous Institutional Arrangements in Multilateral Environmental Agreements: A Little-Noticed Phenomenon in International Law' (2000) 94 *American Journal of International Law* 623–59; MJ Bowman, 'Beyond the "Keystone" COPs: The Ecology of

minimum the establishment of a regular cycle of meetings of the contracting parties (since they collectively possess the exclusive juridical authority to shape the treaty's future development),[102] together with a small cadre of administrators (the 'Secretariat', or 'Bureau', as it was originally known under Ramsar) dedicated solely and specifically to the fine-detailed administration of the international aspects of the relevant programme under the parties' overall direction.[103]

Experience rapidly demonstrated, however, that the operational gap between this (typically tri-annual) convocation of participating governments and the day-to-day activities of the secretariat was excessive, requiring to be bridged by a representative sub-set of the former (most often labelled the 'Standing Committee'), whose essential role was to supervise the latter to ensure the proper implementation of COP policy. Equally, the unusually intense technical demands of treaty operations in this field dictated the need for formal, in-house scientific input to all these other organs (though a wide array of options was available regarding how this might best be delivered).[104] In the case of Ramsar, these latter elements were factored into the system retrospectively and through COP resolutions, taking account of experience under other treaty regimes. Thus, upon its creation in 1993, the Scientific and Technical Review Panel (STRP) became the final key component of the Ramsar organography,[105] following the prior establishment of a Standing Committee constituted primarily on a regional basis.[106]

Institutional Governance in Conservation Treaty Regimes' in M Fitzmaurice and D French (eds), *International Environmental Law and Governance* (Brill Nijhoff, Leiden, 2015).

[102] On the role of COPs specifically, see A Wiersema, 'The New International Law Makers? Conferences of the Parties to Multilateral Environmental Agreements' (2009) 31 *Michigan Journal of International Law* 231–87; PGG Davies, 'Non-Compliance – A Pivotal or Secondary Function of COP Governance?' in Fitzmaurice and French, note 101.

[103] Convention on Wetlands, note 1, Arts 6–8.

[104] For further detail, see Bowman, note 101, Section 3.1.4.

[105] Ramsar Convention, *Resolution 5.5, Establishment of a Scientific and Technical Review Panel* (1993); see currently Ramsar Convention, *Resolution XII.5: New framework for delivery of scientific and technical advice and guidance on the Convention* (2015).

[106] Ramsar Convention, *Resolution 3.3: Establishment of a Standing Committee* (1987); see currently *Resolution XIV.2: Responsibilities, roles and composition of the Standing Committee and regional categorization of countries under the Convention on Wetlands* (2022).

These four key institutional organs could then be supplemented, on either a permanent or *ad hoc* basis, by whatever additional bodies might be deemed necessary – and all without the creation of a conventional inter-governmental organisation imbued with formal legal personality in its own right. This compromise solution to organic management arguably achieved the best of both prior regulatory worlds: it avoided the greater bureaucracy, formality, expense, and accommodation requirements of a large-scale permanent organisation, while nevertheless securing the technical competence, administrative flexibility, and adaptive potential which such entities typically provide.

4.2.3   The ecology of treaty regime communities

The final, specifically *societal*, aspect of the regime follows inexorably from the organic nature of the treaty instrument, as discussed above, and the creation in particular of the permanent institutions for which such agreements characteristically provide. To a degree, no doubt, this societal dimension is present in every international treaty, since the states which become party are thereby unavoidably brought together into a form of normatively grounded political association. Yet this relationship, being essentially of only an abstract and almost ethereal character, is unlikely of itself to prove strongly conducive to the advancement of the Convention's objectives – especially where these are not seen as being closely aligned to the obvious and immediate political interests of states as such (as has commonly been true of conservation regimes in particular).[107]

Where, by contrast, legal instruments of the organic kind are concerned, a radical transformation is wrought not merely in the dynamic but also in the very fabric of this normative community, which automatically assumes flesh-and-blood form simply by virtue of the fact that the various institutional 'organs' of the living instrument have to be constituted, sustained and populated, creating a social community in an absolutely literal sense. This community should, moreover, be naturally disposed towards commitment to the Convention's objectives, since the bureaucratic agency which lies at the system's heart is constitutionally dedicated to that cause exclusively, having typically no other loyalty or function.[108] In addition, since this body becomes effectively the first port of call for each state party in conducting its

---

[107] A situation that gives rise to Simon Lyster's noted category of 'sleeping treaties'.

[108] For the functions of the Ramsar Bureau (Secretariat), see Article 8.2. Given that this body was to be provided by IUCN, a special ring-fenced unit was established within that organisation in order to ensure its complete independence and commitment to the treaty.

treaty-related business, the normally fragile nature of the normative network is radically transformed and strengthened by links to the centre of its web. An additional central hub is provided by the STRP through the delivery of the services specified by its work plan; while a vital element of scientific objectivity is ensured through the fact that its members, however nominated, serve as independent experts rather than national/organisational representatives.[109]

This entrenchment of purposive commitment is further strengthened by the fact that each state party attending COP meetings must be represented physically by a delegation of individuals – which, in Ramsar's case, should include persons with expertise regarding wetlands or waterfowl gained from scientific, administrative, or other experience.[110] The opportunity for profitable, ongoing interaction with similarly qualified representatives of other nations consequently arises as an inherent feature of the regime, encouraging the sharing of information and experience, the identification of best practice, and the development of collaborative programmes for conservation.

Such exchanges will doubtless have represented a highly salutary experience for all those involved – and one which, in the early 1970s, must have seemed rather novel. For, while international convocations of scientists themselves have been relatively commonplace since the mid-nineteenth century at least,[111] in this instance the technocrats in question form part of a national delegation exercising official powers of a specifically legal and political nature. Insofar as the government scientists and specialist administrators from whose ranks these individuals are drawn will have had prior experience of such deliberations at the purely national level, it will probably have been as the isolated and beleaguered junior partner in a policy debate dominated by other – very differently trained and motivated – agencies of government. By stark contrast, the demands and opportunities associated with these new, organically inspired treaty arrangements serve significantly to enhance the prospect of properly informed and more constructive debate and decision-making with their counterparts from other nations, which will (hopefully) be as untrammelled as they (ideally) are by such troublesome intellectual baggage.

---

[109] Resolution XII.5, note 105, Annex [15(ii)], [25].

[110] Convention on Wetlands, note 1, Art 7. For exploration of this crucial but widely neglected aspect of treaty arrangements, see EJ Goodwin, 'Delegate Preparation and Participation in Conferences of the Parties to Environmental Treaties' in Fitzmaurice and French, note 101; and EJ Goodwin, 'State Delegations and the Influence of COP Decisions' (2019) 31 *Journal of Environmental Law* 235.

[111] On the origins of the process, see M Crosland, 'The Congress on Definitive Metric Standards, 1798–99: The First International Scientific Conference?' (1969) 60 *Isis* 226–31.

Such interactions are, moreover, capable of producing outcomes of a kind not readily predictable by reference to the established conventions of international relations scholarship, given their undue preoccupation with conflicts and accommodations between 'the powers'. It is clear, for example, that national delegations have on occasion enlisted the support of the Ramsar community to mitigate or forestall environmentally damaging projects within *their own* country that are driven by government agencies of a very different stripe.[112] It helps, of course, that the Ramsar regime also offers governments the prospect of financial and technical support – however modest – to balance the effects of such encroachments into their policy-making processes.[113]

Yet for all that, the members of national delegations always remain representatives of their respective states and are ultimately constrained by the dictates of governmental policies insofar as they have been expressly formulated. Accordingly, the natural potential of COP deliberations to secure the enhancement of ecological awareness and the specification of sound conservation policy for wetlands can be greatly increased by input from entities that are inherently unconstrained by such bonds and free to speak from a more independent, pro-conservation standpoint. Most obviously, environmental NGOs have a crucial role to play in this regard.

While it may therefore seem surprising that no explicit provision for the participation of NGOs appeared in the original text of Ramsar, it must be remembered that the relevant institutional reference in the pre-amendment wording was not actually to Conferences *of the Parties* as such, but rather to 'Conferences on the Conservation of Wetlands and Waterfowl',[114] which manifestly left open the specifics of entitlement to participate. Given the pre-eminent role of NGOs throughout the Ramsar negotiations, their continued participation could scarcely have been in question; indeed, it was IWRB that took the initiative to organise the inaugural Ramsar meeting at Heiligenhafen,

---

[112] See, eg, the Mauritanian delegation's expression of concern regarding the planned routing of the Inter-Maghreb Highway through the Banc d'Arguin wetland site. Ramsar Convention, *Doc WG C 51 (Rev): Summary Report of Workshop A* (1993), and Recommendation C5.1, at: https://www.ramsar.org/sites/default/files/documents/library/cop5_proc_workshop_a_annexes_efs.pdf; Wendy Strahm et al, *Mission de Suivi Réactif: Parc National du Banc d'Arguin (Mauritanie), 6–13 January 2014* (IUCN/UNESCO, March 2014), especially Section 3.2 and Figure 4.

[113] In the Mauritanian case, a major international project to support traditional local fishing methods, with financial backing latterly provided (primarily from Europe) by Western Africa's first biodiversity trust fund. See Justin Woolford, 'Banc d'Arguin – conservation, development, and finance at the crossroads', 7 October 2020, at: https://mava-foundation.org/blog-5-icons-insight-banc-darguin/.

[114] Convention on Wetlands, note 1, Art 6.1.

even before the Convention's formal entry into force.[115] Furthermore, in addition to the 30-plus states represented, as either parties or observers, at its first formal meeting at Cagliari in 1980, and the five inter-governmental organisations also present in the latter capacity, the official record discloses that the International Council for Bird Protection, IUCN, IWRB, the World Wide Fund for Nature (WWF) and the International Council for Game and Wildlife Conservation not only attended, but also were in several cases active participants in the discussions.[116] Participation arrangements have subsequently been consolidated through formal Rules of Procedure[117] adopted to govern 'meetings of the Conference of the Contracting Parties', as they are now officially designated by virtue of the amended Article 6.1.

A notable feature of Ramsar operations has, moreover, been the unusual warmth with which the participation of non-state entities has been embraced, resulting in an arguably unique 'family feeling' within the regime.[118] No doubt this ethos has been facilitated by various factors, including the essential nature of the Convention's substantive business (which is less overtly emotive in character than in treaty regimes addressing, say, the direct exploitation of animals); the relatively unthreatening nature of the substantive provisions themselves; the frontline involvement of NGOs from the moment of the Convention's conception; and the highly exceptional circumstance that one of their number – IUCN – has provided the vital Secretariat services from the outset.[119] A crucial later development was the decision in 1999 to create a special status of 'International Organization Partners', which is open to entities (whether non-governmental or inter-governmental in character) which maintain a (near-) global programme of relevant activities and possess a proven capacity and willingness to advance Ramsar's mission through technical and scientific support and assistance with policy development.[120] Those recognised

---

[115] Matthews, note 1, 48–49.

[116] See the Ramsar Convention, *Report of the Conference* (1980).

[117] For the current version, see Ramsar Convention, *Rules of Procedure* (2022), and on NGOs specifically, Rule 7.

[118] See the observation to that specific effect of the Colombian delegate at the Brisbane COP, Ramsar Convention, *Notes of the First Plenary Session* (1996) [89].

[119] See Convention on Wetlands, note 1, Art 8.1; following a recent review, it was decided to retain this arrangement, which was originally envisaged to be only temporary.

[120] Ramsar Convention, *Resolution VII.3: Partnerships with international organizations* (1999). For the 2018 Memorandum of Cooperation between the Secretariat and the IOPs, see *'Partners for Wetlands' Memorandum of Cooperation* (2018).

initially[121] were IUCN itself, Birdlife International,[122] Wetlands International[123] and WWF International, with later additions being the International Water Management Institute and the Wildfowl and Wetlands Trust.[124] These bodies have access to all Ramsar meetings, where they serve not merely as observers but as formal advisers.

The Ramsar community in its widest sense, moreover, embraces links with numerous other key players, including river basin authorities; the secretariats of other multilateral environmental agreements; UN agencies in the fields of conservation, trade and development, tourism, and humanitarian affairs; scientific and technical organisations; and even elements of the business sector[125] – not to mention community groups, Indigenous peoples, and other local stakeholders.[126] The inclusiveness of this social network should itself serve to enhance the strength, vibrancy, productivity, resilience, and durability of the Ramsar normative regime – not least because it contrives to replicate the essential features and processes of a thriving biological community,[127] within which diversity, structured integration, interdependence, adaptability, and social learning are typically crucial.[128] Further, the four key treaty organs fulfil the critical role of 'keystone' species, binding the entire system together. The enhanced tensile strength and creative potential of the network created by such regimes stem precisely from the combination of shared commitment to the object and purpose of the Convention with the considerable divergence of personal motivations and constitutional standpoints from which it is approached.

---

[121] Ibid, Resolution VII.3 (1999).

[122] Originally the International Council for Bird Protection.

[123] Formed by the amalgamation of IWRB with similar organisations from beyond Europe.

[124] See Ramsar Convention, *Resolution IX.16: The Convention's International Organization Partners (IOPs)* (2005); and Ramsar Convention, *Resolution XII.3: Enhancing the languages of the Convention and its visibility and stature, and increasing synergies with other multilateral environmental agreements and other international institutions* (2015) [59].

[125] See generally Ramsar Convention Secretariat, *Handbook 5: Partnerships* (Ramsar Handbooks, 4th edition, Ramsar Convention Secretariat, Gland, 2010), and https://www.ramsar.org/about/partnerships.

[126] Ramsar Convention Secretariat, *Handbook 7: Participatory skills* (Ramsar Handbooks, 4th edition, Ramsar Convention Secretariat, Gland, 2010).

[127] As to which, see Premack, note 73, 34–51.

[128] See, eg, RM May, *Stability and Complexity in Model Ecosystems* (new edition, Princeton University Press, Princeton, 2001); RM May, S Levin, and G Sugihara, 'Complex Systems: Ecology for Bankers' (2008) 451 *Nature* 893.

The specifically *legal* character of the bonds that unite this community reminds us that a final form of invaluable input is that obtainable from legal scholars themselves. Encouragingly, the Ramsar regime has displayed a commendable openness to such contributions – whether through the internal commissioning of reports and opinions on normative questions,[129] balanced receptiveness to analysis and prescriptions from wholly independent sources,[130] or the formal incorporation of legal expertise into the partnership system.[131] Such input has indeed proved crucial to the realisation of certain reforms within the regime, such as the modified definition of 'wise use', affirming the continued significance and vibrancy of legal scholarship in the substantive development of public international law, just as the International Court of Justice Statute envisages.[132]

5 RAMSAR AT 50

While it is, of course, essential not to present too rose-tinted a picture of the political ramifications and practical results of the features described above, it remains impossible to disregard the nature and scale of the impact that such developments have wrought on the overall dynamics of international relations. Like no other subject, international environmental law highlights the counterproductive eccentricity of a world artificially disintegrated into a jigsaw of territorially discrete geopolitical units, fashioned entirely in response to the vagaries of recent human history, and with scant regard for the inherent ecological unity of the biosphere as a whole or the need to nurture its fundamental processes. The introduction into this field of the 'living instruments' of the 1970s, with Ramsar in the van, represented at the very least a small counter-

---

[129] In addition to Matthews, note 1, see, eg, V Koester, *The Ramsar Convention on the Conservation of Wetlands: A Legal Analysis of the Adoption and Implementation of the Convention in Denmark* (IUCN/Ramsar Bureau, Gland, 1989); C de Klemm and I Créteaux, *The Legal Development of the Ramsar Convention on Wetlands of International Importance Especially as Waterfowl Habitat* (Ramsar Bureau, Gland, 1995).

[130] Several external studies have been formally incorporated into the Ramsar website and are accessible via the Documents library ('Full search') at: https://www.ramsar.org/search.

[131] Thus, the list of cooperating technical organisations listed in *Handbook 5* includes Stetson University College of Law, in reflection of the splendid work of Royal Gardner and his various collaborators.

[132] See Statute of the International Court of Justice, Art 38(1)(d), at: https://www.icj-cij.org/statute.

balance to this dead weight of ecologically dysfunctional fragmentation and the scientific ignorance and naivety that had served to compound its effects.

The extent to which their collective message has progressively worked its way into the thinking of the world's political agencies is strikingly reflected in the address given by UN Secretary-General Antonio Guterres at Columbia University in December 2020, launching the Decade for Ecosystem Restoration.[133] Put simply, he declared, 'the state of the planet is broken'. Humanity has long been 'waging war on nature' – an enterprise which can only be regarded as 'suicidal', because 'nature always strikes back'. The unmistakeable evidence of this folly is expressed in widespread species extinction and habitat loss and degradation, the build-up of greenhouse gases and the prevalence of air and water pollution, which every year kill over 9 million people. Extreme weather events, zoonotic contagion, social displacement, economic and political instability, and international conflict are merely the symptoms of this underlying ecological degradation; while human activities are indisputably at the root of this 'descent towards chaos'. Accordingly, 'Making peace with nature is the defining task of the 21st century. It must be the top, top priority for everyone everywhere.'

The two key elements of this peace restoration process, moreover, consist in the protection and restoration of natural ecosystems and the radical reform of existing economic and financial practice – in essence, the wholehearted embracing of the Ramsar metanarrative at the expense of its hopelessly perverse and outmoded Enlightenment counterpart.

While the precise ramifications of these twin projects are spelled out in some detail in the statement, and in relation to the current crises regarding both biodiversity and climate change, it is more useful to concentrate here on another critical recent development regarding the financial aspects in particular. This concerns the publication of a major report from the field of economics, timed fittingly (albeit probably by pure coincidence) to celebrate Ramsar's 50th anniversary on 2 February 2021. This global review on the economics of biodiversity – headed by Cambridge University Professor of Economics Sir Partha Dasgupta[134] – aspires to elucidate afresh the notion of sustainable development through the unification of enlightened microeconomic reasoning with its macroeconomic counterpart, thereby enabling us to understand 'why

---

[133] United Nations, *The UN Secretary-General Speaks on the State of the Planet*, at: https://www.un.org/en/climatechange/un-secretary-general-speaks-state-planet.

[134] P Dasgupta, *The Economics of Biodiversity: The Dasgupta Review* (HM Treasury, London, 2021), at: https://www.gov.uk/government/publications/final-report-the-economics-of-biodiversity-the-dasgupta-review.

and how in recent decades we have disrupted Nature's processes to the detriment of our own and our descendants' lives and what we can do to change direction'.[135] Despite the report's rather flustered and unconvincing attempt to gloss over the high degree of responsibility borne by economists themselves,[136] it does at least lay the groundwork for a less destructive future for the discipline, highlighted by its promise of a much richer conception of 'wealth' than Adam Smith himself had ever been able to conceive.[137]

The report's headline messages accordingly begin with the long overdue recognition that humankind is ecologically 'part of nature, not separate from it'; and that Nature itself is not merely an asset but manifestly the most precious that we possess, upon which our economies, livelihoods, and well-being inherently depend. Indeed, it constitutes more than merely an economic good, the 'intrinsic worth' of which is increasingly recognised.[138] Biological diversity is, moreover, absolutely central to the productivity, resilience, and adaptability that Nature affords us, with the result that we allow it to decline at our peril. Yet estimates reveal that between 1992 and 2014, increases in 'produced' and 'human' capital were achieved only at the expense of a 40% reduction per person in 'natural' capital,[139] suggesting that we would now require the natural resources of 1.6 Planet Earths to maintain current human lifestyles. Biodiversity is declining more rapidly than at any time in human history, with species extinctions currently occurring at some 100–1000 times the natural background rate and many ecosystems already degraded beyond repair or poised on the brink of catastrophe.

The seriousness of this situation requires urgent, radical, and transformative action, geared towards ensuring not only that our demands upon Nature do not exceed its supply, but also that Nature's supply itself is significantly increased over current levels. This plainly necessitates changes in our formal measures of economic success, through the recognition that the concept of gross domestic product (while still deemed necessary 'for short-run macroeconomic analysis and management') fails totally to account for the depreciation of assets, especially natural wealth, and has therefore been operating as an active

---

[135] Ibid, Preface, 5.

[136] Ibid, 4, footnote 5.

[137] Ibid, Preface, 5.

[138] His own evident intuitive sympathy for the idea of intrinsic value might profitably have been bolstered by an explicit acknowledgement that the concept has been formally endorsed by 196 governments, through their participation in the Biodiversity Convention!

[139] These various forms of capital are defined to embrace such assets as 'roads, machines, buildings, factories and ports' ('produced'); 'health and education' (human); and Nature itself, as reflected in biological diversity ('natural').

encouragement to unsustainable behaviour. Natural capital should therefore be introduced into national accounting systems and accounting frameworks further developed to meet this need. This in turn will require a transformation in our systems and institutions, especially in the realms of education and finance, so that knowledge can be more effectively developed and shared across all geopolitical levels – global, regional, national, and local – to allow for collaborative planning and participation.

Natural ecosystems that represent global goods are particularly in need of supranational protection, whether through financial support for governmental efforts to conserve those falling within national boundaries or the imposition of rents or charges on the use of those located beyond them, and a total prohibition on utilisation where necessary. The current situation, in which 'financial flows devoted to enhancing natural assets are small and are dwarfed by subsidies and other financial flows that harm' them, requires radical overhaul. All impacts and dependencies upon Nature must be fully accounted for in business, along with full risk disclosure. Public education programmes are essential to enable urbanised populations to reconnect with Nature, both to improve health and wellbeing directly and to ensure that informed choices and consumer pressure for increased sustainability are facilitated.

These findings are encouraging signs of a long overdue movement within the discipline towards genuine willingness finally to engage with earthly reality, rather than the crude and clumsy caricatures of it with which it has for so long been preoccupied. Almost every one of the key ideas embraced is, of course, already firmly embedded in the jurisprudence of international environmental law, and especially the Ramsar Convention, but only tenuously grasped as yet by the global legal regimes governing finance, energy, trade, transport, and human rights. The report itself may not necessarily provide all of the answers currently required from the perspective of economics itself, but it does at least finally begin to identify and address some of the relevant questions. Its authority is greatly strengthened by its realisation as a large-scale collaborative effort, commissioned in 2019 by then British Chancellor of the Exchequer Philip Hammond, with a view to guiding governmental policy for the future. Furthermore, it has attracted a host of positive endorsements from the realms of conservation, business, finance, politics, and academia.[140] Finally, and most importantly, it seems already to have become enshrined in the evolving political agenda at the global level, including in the documentation for the Decade of Ecosystem Restoration itself.[141]

---

[140] For a sample of these responses, see the website cited at note 134 above.

[141] Indeed, the Review features prominently in the introduction to the UNEP report cited at note 85.

It is doubtless too early to determine whether and to what extent the economic follies and fantasies of the past will finally be laid to rest as a consequence of these developments. Should they somehow escape this fate, we will be forced to bury in their place any lingering conceit that 'reason and conscience' – as the Universal Declaration of Human Rights insists – represent the defining traits of humankind,[142] and meekly acknowledge the less attractive reality of a collective mindset dominated by ignorance, insensibility, and myopic self-indulgence. Whatever the eventual outcome, however, few human initiatives could surely mount a more persuasive claim than the Ramsar Convention to have charted a viable regulatory path for steering us towards the ecological light and out of the long, dark shadow of the Enlightenment.

---

[142] UDHR, note 20, Art 1.

2. The Ramsar Convention and the concept of consensus
Royal C. Gardner[1]

1 INTRODUCTION

A defining characteristic of the Ramsar Convention is its strong tradition of consensus. Although the Convention's text and Rules of Procedure allow for decision-making by majority vote,[2] the Ramsar practice 'has always been a consensus'.[3] The tradition of consensus is frequently invoked at Ramsar Conferences of the Parties (COPs) to encourage contracting parties to resolve their differences over the text of draft resolutions.[4] Indeed, there had never been a vote on the merits of COP resolution until COP14 in 2022 regarding Russian aggression in Ukraine (discussed in Chapter 3).[5]

---

[1] Thanks are expressed to Stetson University College of Law for a scholarship grant that supported this work and to my former Scientific and Technical Review Panel colleagues for all they have done for wetland conservation and wise use.

[2] Eg, Convention on Wetlands of International Importance especially as Waterfowl Habitat, adopted 2 February 1971, entered into force 21 December 1975, 996 UNTS 245 (amended 1982 & 1987), Art 6.5 (two-thirds majority to adopt budget), Art 7.2 (simple majority to adopt COP resolutions); Ramsar Convention, COP13, *Rules of Procedure*, Rules 38–45.

[3] GVT Matthews, *The Ramsar Convention on Wetlands: its History and Development* (Ramsar Convention Bureau, Gland, 1993) 76.

[4] As early as COP4, the COP Chair stated that 'Ramsar has always operated on a basis of consensus.' Ramsar Convention, COP4, *Proceedings of the Fourth Meeting of the Conference of the Contacting Parties* (1990) 29.

[5] Two votes, however, were conducted in 1984 at COP2 on proposed recommendations. The first was whether to move forward with Recommendation 2.2, which concerned amendments to the Convention. That vote failed (seven in favour of 'abandoning' the recommendation, ten against, and eight abstentions), and Recommendation 2.2 was adopted by consensus. Ramsar Convention, COP2, [*Untitled Conference Report*] (1984) 29–30. The second involved a recommendation to establish an interim Standing Committee. At Finland's suggestion, discus-

This Chapter examines the history and impact of the Ramsar tradition of consensus. It begins by defining the concept and noting how it has been applied in other treaty regimes, with particular attention to a legal opinion by the United Nations (UN) Office of Legal Affairs that was requested in light of a controversy that grew out of Convention on Biological Diversity (CBD) COP6 (and later spilled over into Ramsar COP8). The Ramsar COP has achieved consensus primarily through negotiation, but it has also employed several other mechanisms. Most controversially, on two occasions, procedural votes or straw polls were conducted to attempt to resolve disputes involving structural arrangements: the regional placement of Israel (COP7) and the institutional hosting of the Secretariat (COP11). More typically, if contracting parties cannot agree on acceptable language through negotiation, a concerned contracting party will not block a resolution's adoption by consensus but will instead enter a reservation to the resolution or issue a formal statement to be included in the Conference Report. These reservations and statements for the record – which have dealt with a variety of concerns, from the management of transboundary waters to climate change – allow for disagreements to be voiced in a manner that nevertheless permits consensus. In rare cases, where there has been significant opposition to a draft resolution, a sponsoring party has simply withdrawn the resolution. A common theme here is a concern that the draft resolution's subject matter goes beyond the Ramsar Convention's competencies and mandate.

The Chapter also considers the effect of the tradition of consensus on efforts to accomplish the Convention's objectives. While the tradition of consensus can create an atmosphere of collegiality and greater acceptance of resolutions, it also leads to the text of resolutions being diluted to the lowest common denominator. As indicated by the results of the 2018 *Global Wetland Outlook*,[6] the gentle exhortations and recommendations contained in resolutions do not appear to readily contribute to 'stem[ming] the progressive encroachment on and the loss of wetlands'.[7]

Yet there is an opportunity for the Convention to leverage its tradition of consensus. In the context of the Intergovernmental Science-Policy Platform on Biodiversity and Ecosystem Services (IPBES) assessments, the summaries

---

sion was suspended and the matter was put to a vote. The recommendation 'was not accepted' by a vote of four in favour, 13 against, and seven abstentions. Ibid, 30–31.

[6] RC Gardner and CM Finlayson, *Global Wetland Outlook: State of the world's wetlands and their services to people 2018* (Ramsar Convention Secretariat, Gland).

[7] Convention on Wetlands, note 2, Preamble.

for policymakers are reviewed line by line and approved by IPBES members by consensus.[8] Future editions of the *Global Wetland Outlook* should include a summary for policymakers that should similarly be approved by consensus by the Ramsar COP. Such an exercise would require contracting parties to confront more directly the status of the world's wetlands, the drivers of wetland loss and degradation, and possible response actions – but in a manner keeping with the Ramsar spirit of consensus.

2 CONSENSUS DECISION-MAKING

Consent is a foundation of international law.[9] States are bound only by those conventions that they consent to join. A fundamental tenet of international law is that a treaty may not 'create either obligations or rights for a third State without its consent'.[10] Unsurprisingly, sovereign states are reluctant to enter into arrangements through which they cede decision-making authority. Conventions for which COP decisions are made by majority vote generally have opt-out provisions: a state on the losing end of a vote may issue a reservation[11] or decline to accept a conservation measure.[12] In many conventions, including the Ramsar Convention, consensus decision-making is the common

---

[8] Intergovernmental Science-Policy Platform on Biodiversity and Ecosystem Services, *Decision IPBES-3/3: Procedures for the preparation of Platform deliverables* (2015) 3, at: https://ipbes.net/sites/default/files/downloads/Decision_IPBES_3_3_EN_0.pdf.

[9] A D'Amato and K Engel (eds), *International Environmental Law Anthology* (Anderson Publishing Co., Cincinnati, OH, 1996) 45. Of course, some international legal obligations – such as those based on customary international law or *jus cogens* – may apply to a state even without its consent. D Shelton, 'Normative Hierarchy in International Law' (2006) 100 *American Journal of International Law* 291, 302 (noting that the source of peremptory norms has been 'attributed to state consent' but also to 'natural law, necessity, international public order, and the development of constitutional principles').

[10] Vienna Convention on the Law of Treaties, adopted 23 May 1969, entered into force 27 January 1980, 1155 UNTS 332, Art 34.

[11] Eg, Convention on International Trade in Endangered Species of Wild Fauna and Flora, adopted 3 March 1973, entered into force 1 July 1975, 993 UNTS 243, Art XV (reservation procedure).

[12] Eg, International Convention for the Regulation of Whaling, adopted 12 February 1946, entered into force 10 November 1948, 161 UNTS 72, Art V (objection procedure).

practice – an approach that is consistent with consent but avoids the need for voting.[13]

'Consensus' in this context is defined as 'the practice of adoption of resolutions or decisions by general agreement without resort to voting in the absence of any formal objection that would stand in the way of a decision being declared adopted in that manner'.[14] The benefits of consensus decision-making are manifold: it is an efficient means to ensure the buy-in of all states, including significant actors;[15] it promotes compromise and a sense of community;[16] and it allows the full participation of all states, even those that may not be politically powerful.[17] One downside is, of course, that a few states – or even a single state – can make a formal objection to prevent a resolution or decision from being adopted by consensus. For example, the Copenhagen Accord under the UN Framework Convention on Climate Change (UNFCCC) was derailed when a handful of states objected.[18]

The meaning of 'consensus' played a central role in the consideration of the Guiding Principles for the Prevention, Introduction and Mitigation of Impacts of Alien Species that Threaten Ecosystems, Habitats or Species ('Guiding Principles') at CBD COP6.[19] The Guiding Principles endorsed the use of the precautionary approach in dealing with invasive alien species and stated that those 'responsible for the introduction of invasive alien species should bear the cost of control measures and biological diversity restoration'.[20] Australia expressed its concern that some of the Guiding Principles 'present[ed] a strong and unacceptable risk of increased trade protectionism' and formally objected

---

[13] Eg, L Goldsworthy, 'Consensus Decision-making in CCAMLR: Achilles' Heel or Fundamental to Its Success?' (2022) 22 *International Environmental Agreements* 411.

[14] Letter from H Corell, Under-Secretary-General for Legal Affairs, United Nations, to H Zedan, Executive Secretary, Convention on Biological Diversity (6 June 2002) 1 (hereinafter 'UN Legal Opinion').

[15] J Brunée, 'COPing with Consent: Law-Making Under Multilateral Environmental Agreements' (2002) 15 *Leiden Journal of International Law* 1, 10.

[16] Ibid, 40–41.

[17] Ibid, 41.

[18] L Rajamani, 'The Cancun Climate Agreements: Reading the Text, Subtext and Tea Leaves' (2011) 60 *International and Comparative Law Quarterly* 499, 499–500.

[19] The Guiding Principles are an annex to Convention on Biological Diversity, *Decision VI/23: Alien species that threaten ecosystems, habitats or species* (2002).

[20] Ibid, Guiding Principle 12 (responsibility attaches when national laws and regulations are not complied with).

to their adoption.[21] Despite Australia's formal objection, the COP Chair nevertheless declared that 'substantial consensus' existed and ruled that the decision that included the Guiding Principles was adopted.[22] Several parties – including Argentina, Canada, and Spain – made reservations regarding the process of adoption by consensus although a formal objection had been made.[23]

Post-COP, the CBD Executive Secretary requested a UN legal opinion on the matter. After defining 'consensus' (as above), the UN legal opinion noted that 'consensus' does not mean that all states are 'in favour of every element of the resolution or decision'.[24] States that do not wish to block adoption by consensus can register their disagreement in a number of ways – by making 'reservations, declarations, statements of interpretation and/or statements of position'.[25] When there is a formal objection, however, a decision cannot be adopted by consensus.[26] Accordingly, the legal opinion characterised the action of the CBD COP Chair as 'contrary to the established practice', with 'serious flaws in the procedure'.[27] Nevertheless, the legal opinion found that because Australia had failed to properly object once the decision had been adopted, the CBD COP decision and the Guiding Principles could stand as adopted.[28] Australia vigorously disagreed with that conclusion.[29]

The contretemps was reignited at Ramsar COP8 when the contracting parties considered a draft resolution on invasive species and wetlands. EU Member States sought to amend the draft to include two specific references to the CBD COP decision and the Guiding Principles.[30] Australia objected and *Earth Negotiations Bulletin* reported that '[s]ome of discord from ... CBD COP6 on the invasive species issue spilled over into the Ramsar Contact Group considering the matter'.[31] Eventually, the International Union for Conservation of Nature (IUCN) endeavoured to broker a compromise, suggesting that the

---

[21] RC Gardner, 'Perspectives on Wetlands and Biodiversity: International Law, Iraqi Marshlands, and Incentives for Restoration' (2004) 15 *Colorado Journal of International Environmental Law and Policy* 1, 4.

[22] TR Young, 'Brief Thoughts on COP-6' (2002) 32 *Environmental Policy and Law* 133, 135.

[23] UN Legal Opinion, note 14, 2.

[24] Ibid, 1.

[25] Ibid.

[26] Ibid.

[27] Ibid, 2.

[28] Ibid.

[29] Ramsar Convention, COP8, *Conference Report* (2002) [91].

[30] Gardner, note 21, 5–6. CBD COP6 was held in April 2002, with Ramsar COP8 following in November 2002.

[31] Ibid, 6.

Ramsar resolution generally (and vaguely) refers to 'any relevant guidelines adopted under other conventions', thus avoiding express mention of the CBD's Guiding Principles.[32] The EU Member States accepted IUCN's compromise language, as long as the phrase 'or guiding principles' – in lower case – was inserted after 'relevant guidelines'.[33] With no explicit reference to the Guiding Principles (upper case), Australia joined the consensus. Both the EU Member States and Australia entered statements for the record that reiterated their positions on the status of the CBD decision and the Guiding Principles.[34] Both statements also invoked the spirit of 'conciliation and consensus' as the reason for their decision to support the compromise language.[35]

A single party that blocks consensus can be accused of bad faith and 'hijacking' a COP.[36] Yet declaring consensus in the face of a formal objection also poses risks. At the UNFCCC COP that followed the disappointment in Copenhagen, Bolivia formally objected to the Cancun Agreements.[37] Despite the objection, the COP Chair 'declared consensus ... to thunderous applause'.[38] The result called into question the definition of 'consensus', with an Indian delegate characterising it as 'terror by applause'.[39]

Ramsar has never adopted a resolution using a 'substantial consensus' or 'consensus minus one' approach. Ramsar hews to the traditional definition of 'consensus', where one party can force a vote. But if a sole party objects to a particular resolution, that party does not necessarily stand alone. The objecting state may have silent partners which also object but which do not wish to be viewed as uncooperative by blocking consensus. A contracting party may also rely on others to take the lead in negotiating language or to provide support for amendments.

I observed this in 2002, at Ramsar COP8, when the contracting parties considered a wetland and climate change resolution for the first time.[40] After the draft resolution was introduced in plenary, a contact group was established to try to resolve concerns. At an early contact group meeting, a US Department of State representative stated that the United States had no objection to the draft resolution. A delegate from Japan, who was sitting in front of the US

---

[32] Ibid, 6–7.
[33] Ramsar Convention, note 29, [90].
[34] Ibid, [90–91].
[35] Ibid.
[36] Young, note 22, 135.
[37] Rajamani, note 18, 515.
[38] Ibid.
[39] Ibid, 516.
[40] See Ramsar Convention, *Resolution VIII.3: Climate change and wetlands: impacts, adaptation, and mitigation* (2002). See also Chapter 11 of this volume.

delegation, wheeled around and asked, 'Are you sure?' The US delegate had indeed stated the US position correctly,[41] although – as discussed below – the US position evolved over the course of the two-week COP.

BOX 2.1 CONSENSUS AT EXTRAORDINARY CONFERENCES OF THE CONTRACTING PARTIES

Ramsar has held three extraordinary COPs, all of which operated based on consensus. The first in 1982 considered Article 10 bis, which set forth a procedure for amending the Convention. Article 10 bis provides that a COP may adopt an amendment based on a two-thirds majority vote and any such amendment will enter into force once two-thirds of the contracting parties have deposited an instrument of acceptance. The COP adopted the text by consensus.

At the second extraordinary COP in 1987 (held in conjunction with COP3), the contracting parties considered amendments to the Convention such as the timing of COPs and budgetary matters. Article 6.5 states that the budget requires a two-thirds majority for approval and Article 6.6 requires unanimity with respect to each contracting party's financial contribution. Although Article 10 bis permitted such amendments to be adopted by a two-thirds majority, the text was again adopted by consensus, with only the United States making a reservation. (And although Article 6.5 authorises budget approval by voting, subsequent COPs have always reached consensus on that issue.)

The third extraordinary COP occurred in 2021 and was held online due to the COVID-19 pandemic. COP14 had been scheduled to take place in Wuhan, China, in October 2021 but was postponed until November 2022. A principal business item was the adoption of the budget, as required by Article 6.5. Reaching consensus was not difficult; however, meeting the quorum requirements was more so. A quorum required 115 contracting parties, but only 105 had submitted correct credentials. Ten other credentials were incorrect: seven were not signed by an official at the appropriate level and three had not been translated into one of the three Convention languages. Accordingly, the extraordinary COP was unable to make a decision and was adjourned until the following week. When the Extraordinary COP resumed, the credentials issues had been resolved and the COP was able to

---

[41] Instructions for the US delegation (on file with the author) stated that the United States had 'no objection with the guidelines and general information provided in DR-3'.

approve the budget by consensus.

In theory, an online COP can allow greater participation as there are no travel expenses with having a larger delegation (eg, Zambia had four delegates at the third extraordinary COP). In practice, however, connectivity challenges in some countries can inhibit active participation. (Even Switzerland had technical issues at one point.) Accordingly, many contracting parties expressed concern that the online meeting should not set a precedent. Ultimately, it was agreed (by consensus) to include in the Conference Report a statement that the parties recognised that the online modalities 'respond to the current extraordinary circumstance as related to the COVID-19 pandemic and do not set a precedent in normal circumstances for the organization of similar meetings under the Convention in the future'.

Sources: The Final Act of the Conference to Conclude a Protocol to the Convention on Wetlands of International Importance especially as Waterfowl Habitat (1982); Amendments of the Convention Adopted by the Extraordinary Conference (1987); Conference Report of the Third Extraordinary Meeting of the Conference of the Contracting Parties (2021); GVT Matthews, *The Ramsar Convention on Wetlands: its History and Development* (Ramsar Convention Bureau, Gland, 1993).

3   VOTING ON WHETHER TO HAVE A VOTE

While prior to 2022 the Ramsar COP had never held a vote on the merits of any resolution, it has on two occasions conducted procedural votes – in part to discourage any individual contracting party from being the first to demand a vote on the merits. Both involved structural issues within the Convention and did not directly concern the conservation and wise use of wetlands. The motivations behind each proposal, however, were dramatically different: the first at COP7 was an attempt to introduce larger geopolitical disputes into Ramsar processes, while the second at COP11 was part of an effort to increase the visibility of Ramsar.

3.1   Placement of Israel

The Ramsar Convention can be viewed as an example of 'environmental peacebuilding' – where 'cooperation around mutual interests in shared natural resources' can help to avoid and resolve conflict.[42] The Ramsar Convention

---

[42] T Ide et al, 'The past and future(s) of environmental peacebuilding' (2021) 97 *International Affairs* 1, 2. See also PJ Griffin and SH Ali, 'Managing transboundary wetlands: the Ramsar Convention as a means of ecological diplomacy' (2014) 4 *Journal of Environmental Studies and Sciences* 230, 231 (discussing

was developed and negotiated by representatives from Western Europe and the Soviet Union in the 1960s and 1970s during the Cold War.[43] As Matthews noted, the participation of a Soviet delegate and expert at the Second European Meeting on Wildfowl Conservation in the Netherlands in 1966 was quite 'politically remarkable' given the dynamics at the time.[44] The Soviet delegation remained engaged throughout the negotiation process and was one of the original parties present in Iran in 1971.[45]

Discussions at Ramsar COPs have generally avoided debates focused on geopolitical matters. To be sure, there have been exceptions – typically involving territorial disputes, such Argentina's objections to certain UK observer organisations because they had a connection to the Falklands/Malvinas Islands[46] and Ukraine's expressed concern over the status of Ramsar Sites in Russia-occupied Crimea.[47] At COP3, Iran wanted it noted that Iraq (then a non-party) had damaged a Ramsar Site with chemical weapons, but after Jordan intervened to dispute the assertion, the Chair advised that the COP was 'an inappropriate forum ... and that only ecological comment would be accepted'.[48] Indeed, Ramsar COPs traditionally focus on matters related to wetland conservation and the Convention's administration rather than the wider geopolitical context.

Since 1999 at COP7, the Ramsar Convention has been formally organised into six biogeographical regions: Africa, Asia, Europe, Latin America and the Caribbean (initially referred to as the Neotropics), North America, and Oceania.[49] These regional categories are reflected in the Convention's bodies and processes. The Secretariat's staff includes Senior Advisors and Junior Professionals based on regions,[50] and the Scientific and Technical Review

---

how 'the Ramsar Convention can enhance its role in utilizing conservation as a peace-building tool').

[43] Matthews, note 3, 13–28.
[44] Ibid, 14.
[45] Ibid, 25.
[46] Ramsar Convention, COP13, *Conference Report* (2108) Annexes 2, 3.
[47] Ibid, Annexes 4, 5.
[48] Ramsar Convention, COP3, [*Untitled Conference Report*] (1987) 44. Iran placed the 'Shadegan Marshes & mudflats of Khor-al Amaya & Khor Musa' Ramsar Site on the Montreux Record in 1993. Source: https://rsis.ramsar.org/ris/41?language=en.
[49] See Ramsar Convention, *Resolution VII.1: Regional Categorization of countries under the Convention, and composition, roles and responsibilities of the Standing Committee, including tasks of Standing Committee members* (1999).
[50] Source: The Convention on Wetlands, The Secretariat, https://www.ramsar.org/about/the-secretariat.

Panel (STRP), the Convention's scientific advisory body, is expected to have regional representation.[51] Membership on the Standing Committee is based proportionally on the number of contracting parties in each region.[52] Furthermore, contracting parties meet at pre-COP regional meetings to discuss priorities, identify possible positions on COP issues, and exchange information.[53] At COPs, the schedule usually allows time for regional meetings each morning.[54]

The regionalisation was not without controversy, as Middle East disputes emerged at Ramsar COP7 when Iran and Syria objected to Israel's placement in the Asia region. They asserted that, rather than being based solely on environmental grounds or geographic location, a contracting party's regional placement should also depend on the 'concerns' of other contracting parties in the region to ensure a cooperative atmosphere.[55] Syria proposed an amendment that would have effectively excluded Israel from the Asia region, but the COP President ruled that it could not be considered.[56] Syria pressed for a vote on its amendment and the COP President decided to hold a preliminary vote on whether to consider it.[57]

Two votes actually took place. First, Syria requested that the vote on whether to consider its proposal be held by secret ballot. That request was denied, with only 37 votes in favour of a secret ballot, 53 votes against and ten abstentions.[58] The plenary then rejected the request to consider Syria's proposal, with 22 votes in favour of its consideration, 46 votes against and 35 abstentions.[59] The COP eventually adopted the resolution on regionalisation by consensus, but it included a provision that allowed contracting parties to participate in a nearby alternative region at their own request.[60]

---

[51] Ramsar Convention, *Resolution XII.5: New framework for delivery of scientific and technical advice and guidance on the Convention* (2015) Annex 1 [7, 25].

[52] Ramsar Convention, *Resolution XIII.4: Responsibilities, roles and composition of the Standing Committee and regional categorization of countries under the Convention* (2018) Annex 1 [4].

[53] Eg, The Convention on Wetlands, Regional Pre-COP meeting 2018: Asia, https://www.ramsar.org/event/regional-pre-cop-meeting-2018-asia.

[54] Eg, Ramsar Convention COP13, *Doc.3.2 Rev.1: Provisional working programme* (2018).

[55] Ramsar Convention, COP7, [*Untitled Conference Report*] (1999) [47].

[56] Ibid, [88].

[57] Ibid, [92].

[58] Ibid, [93].

[59] Ibid, [95]. Three parties elected not to participate.

[60] Ramsar Convention, note 49, [5].

After the resolution's adoption, Israel notified the COP that although it remained a member of the Asia region, it requested 'to participate temporarily within the alternative region of Europe'.[61] This temporary arrangement continued until 2018 when COP13 – at Israel's request – formally placed Israel in the European region.[62] Because COP13 was held in the United Arab Emirates, Israel was not present for the adoption of the resolution. Ultimately, Iran and Syria's objective was accomplished – but through consensus.

3.2 Hosting of the Secretariat

The only other occasion when procedural votes took place was in Bucharest at COP11 over the hosting of the Secretariat, the Convention's administrative body. The roots of this controversy can be traced back to the Convention's founding.

Article 8 of the Ramsar Convention established a 'Bureau' to carry out certain administrative functions.[63] Specifically, the Bureau is to organise the COP and to maintain the List of Wetlands of International Importance.[64] It is also to receive notifications from the contracting parties about the status of Ramsar Sites: when a Ramsar Site's boundaries are increased or reduced (including when a Site is deleted from the List) and when human-induced negative changes occur in a Site's ecological character.[65] The Bureau is expected to inform the contracting parties about the status of Ramsar Sites so that it may be discussed at the COP.[66] Relatedly, the Bureau is then to inform the relevant contracting parties about any COP recommendations regarding those Ramsar Sites.[67] The Convention assigns these 'continuing bureau duties' to IUCN until two-thirds of the contracting parties decide otherwise.[68] IUCN continues to

---

[61] Ramsar Convention, note 55, [104].

[62] Ramsar Convention, *Resolution XIII.4: Responsibilities, roles and composition of the Standing Committee and regional categorization of countries under the Convention* (2018) Annex 2.

[63] Article 2 refers to the 'bureau established by Article 8'. Unlike Article 6, which expressly 'established' a COP, Article 8 does not use the term 'establish' and appears to set up the Bureau implicitly by assigning it specific tasks.

[64] Convention on Wetlands, note 2, Art 8.2.

[65] Ibid, Arts 3.2 and 8.2. See further Chapter 3 of this volume.

[66] Convention on Wetlands, note 2, Art 8.2.

[67] Ibid.

[68] Ibid, Art 8.1.

serve as the physical host of the Ramsar Bureau, which has been known as the 'Secretariat' since a COP decision (by consensus) in 2005.[69]

IUCN's express inclusion in the Convention text reflects the important role that non-governmental organisations (NGOs) have played since the Ramsar Convention's inception.[70] It also highlights that the Ramsar Convention remains outside the UN system. Indeed, the conclusion and signing of the Convention in February 1971 pre-date the establishment of the UN Environment Programme, which administers and/or supports many secretariats of multilateral environmental agreements (MEAs).[71]

Although the Convention text established a Bureau (Secretariat), it did not resolve technical matters regarding the legal personality of that administrative body. An entity may have legal personality under national and/or international law.[72] An entity with legal personality under national law (eg, a corporation) possesses rights and is subject to duties under that domestic legal scheme.[73] The decision whether to afford an entity domestic legal personality is a matter for an individual state.[74] In contrast, international legal personality is more complex. An entity with international legal personality may engage in treaty-making, possess diplomatic immunity, and have standing in international tribunals and organisations such as the UN.[75] An MEA's secretariat may be recognised to have international legal personality through treaty provisions or COP decisions.[76]

---

[69] Ramsar Convention, *Resolution IX.10: Use of the term and status of the 'Ramsar Secretariat'* (2005).

[70] See Matthews, note 3, Chapter 3.

[71] 'Secretariats were established within the United Nations Environment Programme (UNEP) for the Basel Convention, CBD, CITES, CMS, Rotterdam Convention and Stockholm Convention.' Ramsar Convention, COP10, *Doc 35: Report on the Legal Personality of the Ramsar Secretariat* (2008) 37[102]. See also P Kishore, 'A Comparative Analysis of Secretariats Created Under Select Treaty Regimes' (2011) 45 *International Lawyer* 1051, 1054 (noting that UNEP hosts the Ozone Secretariat).

[72] BH Desai, *Multilateral Environmental Agreements: Legal Status of the Secretariats* (Cambridge University Press, New York, 2010) 138.

[73] Ramsar Convention, *SC55 Doc 9: Review of the legal status of Ramsar Regional Initiatives and the implications for the Convention* (2018) [34–35].

[74] See Desai, note 72, 138 (noting that the domestic legal capacity of international organisations is 'generally enshrined in the headquarters agreement').

[75] Ramsar Convention, COP10, *Doc 35: Report on the Legal Personality of the Ramsar Secretariat* (2008) 3[14].

[76] Ibid, 18 [28.g].

The Ramsar Secretariat's legal status is ambiguous at best. It appears to lack a firm foundation for both domestic and international legal personality.[77] IUCN, on the other hand, does have legal personality under Swiss law.[78] Thus, in 1990, the COP stated that 'the Secretary General shall be responsible' to the COP (and to the Standing Committee intersessionally) for all Convention matters, 'except for those requiring the exercise of legal personality on behalf of the Convention (e.g. establishment of the separate bank account, formal personnel and contract administration, etc.)'.[79] Matters requiring legal personality 'shall rest with the Director General of IUCN'.[80] IUCN, however, formally delegated that authority to the Secretariat through a memorandum of agreement.[81]

Over the years, questions about the Ramsar Secretariat's legal status and its relationship with IUCN raised logistical concerns.[82] Because it has no domestic legal personality, the Ramsar Secretariat occasionally found it impossible to enter into binding contracts as 'the Ramsar Secretariat'.[83] Because the Secretariat lacked status as an international organisation, its staff encountered '[f]requent difficulty in obtaining travel visas'.[84] Some contracting parties baulked at submitting annual contributions to IUCN, stating that they could not send funds to an NGO.[85] And many contracting parties expressed concern about Ramsar's relatively low profile. Its legal status made it difficult for the Secretariat to formally engage at the UN and some contracting parties suggested that Ramsar would have a higher profile within their countries if it were formally part of the UN system.[86] As part of the UN, the Ramsar Convention would also be obliged to operate in all six UN languages.

---

[77] See ibid, 2–3 [11] (Ramsar Secretariat's domestic personality depends on delegation from IUCN), 24 [52] (suggesting that the Ramsar Secretariat is imbued with 'some aspects of international legal personality' due to COP decisions); Desai, note 72, 153 (contending that the Ramsar Secretariat 'has no legal standing').

[78] Ramsar Convention, note 75, 3 [13].

[79] Ramsar Convention, *Resolution 4.15: Resolution on Secretariat matters* (1990) [1(e)].

[80] Ibid.

[81] Ibid, [2].

[82] Ramsar Convention, *DOC. SC36-15: Legal Status of the Ramsar Convention Secretariat* (2008) 2–3.

[83] Ibid, 3.

[84] Ibid.

[85] Ibid.

[86] Ibid.

After years of study and discussions regarding institutional hosting arrangements, the issue came to a head in 2012, with two starkly competing proposals: Alternative 1 would maintain IUCN as host of the Secretariat, while Alternative 2 would move the Secretariat to the United Nations Environment Programme (UNEP).[87] A sharp division existed between those contracting parties that wanted the Secretariat to move to UNEP to increase Ramsar's visibility and those (including most of the countries whose contributions provided the Convention's budget) that were concerned that the UN pay scale and language requirements would result in higher financial assessments and/or fewer services to the contracting parties.[88]

The Alternate President who was chairing the COP at the time suggested that a 'straw poll' would be useful in deciding how to move forward.[89] When some contracting parties objected, noting that the COP Rules of Procedure did not contemplate such indicative votes, the Alternate President decided that there should be a vote on whether to hold the indicative vote.[90]

Many contracting parties were not pleased with this process.[91] However, 61 contracting parties were in favour of an indicative vote, with 44 opposed and ten abstentions.[92] The indicative vote was then conducted: 61 contracting parties favoured remaining with IUCN, 26 contracting parties favoured moving to UNEP and 18 contracting parties abstained.[93] Based on this outcome, the IUCN alternative was used 'as a starting point for building consensus'.[94]

A 'Friends of the Chair' group was established to try to accommodate competing views.[95] After two revisions of the draft resolution and a great deal of discussion in plenary, the resolution was finally adopted by consensus.[96] On the main issue, Resolution XI.1 expressed confidence in IUCN and decided that it should continue to host the Secretariat.[97] The resolution sought to address the visibility issue by calling on the Standing Committee to consider

---

[87] Ramsar Convention, COP11, *Conference Report* (2012) [137].
[88] Earth Negotiations Bulletin, *Report of main proceedings for 8 July 2012*, at: https://enb.iisd.org/events/11th-meeting-conference-parties-ramsar-convention-cop11/report-main-proceedings-8-july-2012.
[89] Ramsar Convention, note 87, [140].
[90] Ibid, [142]–[144].
[91] Ibid, [165], [219].
[92] Ibid, [146].
[93] Ibid, [148]–[149].
[94] Ibid, [218].
[95] Ibid.
[96] Ibid, [480].
[97] Ramsar Convention, *Resolution XI.1: Institutional hosting of the Ramsar Secretariat* (2012) [12].

a range of actions, including accommodating Arabic, Chinese, and Russian as working languages into the Convention; increasing high-level political engagement by holding ministerial segments at COPs; and enhancing synergies with other MEAs and UNEP.[98] The issue of the Secretariat's legal status remained unresolved.

Of course, consensus did not mean that all the contracting parties agreed. For example, South Africa wanted its position in favour of moving the Secretariat to UNEP noted on the record but agreed to the resolution 'in the spirit of consensus'.[99] Similarly, after the resolution's adoption, Costa Rica made a political declaration on behalf of eight other Latin American and Caribbean states, stating that the issue should be revisited in light of the upcoming 'Rio+20' UN Summit on Sustainable Development.[100]

Most of the efforts to increase the visibility and stature of the Ramsar Convention remain works in progress. While no new official languages have been introduced, COP13 had Arabic interpretation (as it was held in the United Arab Emirates) and endorsed the goal of adding Arabic as an official language.[101] The Biodiversity Liaison Group provides a forum for the secretariats to coordinate and exchange information, although a similar mechanism for the MEAs' scientific advisory bodies has foundered.[102] And while the Ramsar Convention has not yet achieved status as a UN observer,[103] the UN General Assembly has officially recognised 2 February as World Wetlands Day, commemorating the adoption of the Ramsar Convention in 1971.[104] Moreover, the Ramsar Secretariat has been named a co-custodian with UNEP for Sustainable Development Goal Indicator 6.6.1, which tracks the change in extent of water-related ecosystems.[105]

---

[98] Ibid, [17].

[99] Ramsar Convention, note 87, [265].

[100] Ibid, [482].

[101] Ramsar Convention, *Resolution XIII.6: Language strategy for the Convention* (2018) Annex 1.

[102] RC Gardner and A Grobicki, 'Synergies between the Convention on Wetlands of International Importance and other multilateral environmental agreements: possibilities and pitfalls' in Balakrishna Pisupati, *Understanding synergies and mainstreaming among the biodiversity related conventions* (UN Environment, Nairobi, 2016) 57–58, 64–65.

[103] See Ramsar Convention, *SC59 Doc.14: Report of the Observer Status Working Group* (2021).

[104] UNGA, *World Wetlands Day*, Resolution 75/317 (30 August 2021).

[105] United Nations Department of Economic and Social Affairs, 'Wetland inventories to support Contracting Parties to achieve Indicator 6.6.1', at: https://

4 RESERVATIONS

The Vienna Convention on the Law of Treaties recognises that, under certain circumstances, a party may make a reservation to a treaty provision – a unilateral statement that the treaty provision does not apply to that party.[106] No contracting party has ever made a reservation to the Ramsar Convention itself.[107] On occasion, however, a contracting party has made a reservation with respect to a COP resolution.

The first Ramsar reservations involved the administrative operation of the Convention. At COP3, Resolution 3.3 created a Standing Committee comprised of a representative from each region, plus the prior and upcoming COP host country.[108] In Iran's view, the Standing Committee's membership 'should reflect that Asia was the biggest continent', and Iran made a reservation with respect to the composition of the Standing Committee.[109] While the COP has never endorsed representation based on geographical size, the Standing Committee's composition is now based proportionally on the number of contracting parties within each region.[110] Africa, as the region with the most contracting parties, currently has the greatest numerical representation within the Standing Committee.[111]

Also at COP3, Austria expressed concerns about Resolution 3.4, which encouraged the provisional implementation of the Regina Amendments (adopted at the second extraordinary COP held during the intervals of COP3), pending their entry into force.[112] The Regina Amendments dealt with, among other things, financial and budgetary matters. Austria noted that its domestic constitutional procedures did not allow for the provisional application of trea-

---

sdgs.un.org/partnerships/wetland-inventories-support-contracting-parties-achieve-indicator-661.

[106] Vienna Convention, note 10, Art 19.

[107] Source: https://treaties.un.org/pages/showDetails.aspx?objid=0800000280104c20.

[108] Ramsar Convention, *Resolution 3.3: Establishment of a Standing Committee* (1987) [2].

[109] Ramsar Convention, note 48, 37.

[110] Ramsar Convention, note 62, Annex 1 [4].

[111] Source: The Convention on Wetlands, Current Standing Committee (2023–2025): Members, https://www.ramsar.org/current-standing-committee-2023-2025-members.

[112] Ramsar Convention, note 48, 20–21.

ties and therefore entered a reservation to the resolution, but did not block its adoption by consensus.[113] Austria made a similar reservation at COP4.[114]

More than a decade passed without a Ramsar reservation, as none occurred at the next three successive COPs.[115] COP8 saw the return of reservations, in part related to the scope of Ramsar's mandate. For example, Brazil stated that Ramsar was not the appropriate forum to consider issues related to cultural values, which should instead fall under the auspices of the United Nations Educational, Scientific and Cultural Organization.[116] Brazil also issued a reservation to a resolution on wetlands and agriculture, contending that the World Trade Organization was the appropriate venue.[117]

The uptick in reservations was also attributable to Turkey, which acceded to the Convention in 1994. As Table 2.1 illustrates, Turkey is responsible for the majority of all reservations. Indeed, at COP12 and COP13, Turkey was the only contracting party to enter a reservation. Turkey's reservations have centred on transboundary water resources.

Turkey resolutely defends its control over transboundary rivers within its territory, questioning the concept of 'shared' water resources.[118] Its hydroelectric and irrigation projects have been a source of tension in the region.[119] Turkey is not a party to the Convention on the Law of the Non-navigational Uses of International Watercourses and places a reservation on any mention – even indirect – to that treaty.[120] Similarly, Turkey challenged references to the

---

[113] Ibid, 34.

[114] Ramsar Convention, note 4, 45.

[115] At COP7, Chile suggested a goal of 50% of all Ramsar Sites having management plans was more realistic than a goal of 75% and requested that 'its reservations be recorded'. Ramsar Convention, note 55, [129]. In this context, it is likely that 'reservations' refers to 'concerns', rather than a formal reservation to the resolution. Chile's concerns proved prescient: at the end of 2021, the Ramsar Sites Information Service indicated that only 49% of Ramsar Sites had a management plan available. See https://rsis.ramsar.org/.

[116] Ramsar Convention, note 29, [92]. See further Chapter 10 of this volume.

[117] Ramsar Convention, note 29, [94].

[118] See, eg, Ramsar Convention, *Resolution X.19: Wetlands and river basin management: consolidated scientific and technical guidance* (2008) [9] and footnote 1 (explaining that not all parties accept the definition of shared or transboundary river basins).

[119] See, eg, Ramsar Convention, COP10, *Conference Report* (2008) [91–92] (discussing desiccation of the Hawizeh Marsh in Iraq).

[120] Convention on the Law of the Non-navigational Uses of International Watercourses, adopted 21 May 1997, entered into force 17 August 2014, (1997) 36 ILM 700. In a similar vein at COP10, the United States placed a reservation on

Table 2.1 Ramsar Conference Reports: Reservations on Adopted Resolutions

COP (Year)	Resolution number and title	Contracting party
COP3 (1987)	3.3: Establishment of a Standing Committee	Iran
	3.4: Provisional Implementation of the Amendments to the Convention	Austria
COP4 (1990)	Annex to DOC. C.4.13: Resolution on financial and budgetary matters	Austria
COP8 (2002)	VIII.1: Guidelines for the allocation and management of water for maintaining the ecological functions of wetlands VIII.2: The Report of the World Commission on Dams (WCD) and its relevance to the Ramsar Convention VIII.25: The Ramsar Strategic Plan 2003–2008	Turkey
	VIII.19: Guiding principles for taking into account the cultural values of wetlands for the effective management of sites VIII.34: Agriculture, wetlands and water resource management	Brazil
COP9 (2005)	IX.1 Additional scientific and technical guidance for implementing the Ramsar wise use concept IX.3 Engagement of the Ramsar Convention on Wetlands in ongoing multilateral processes dealing with water	Turkey
COP10 (2008)	X.24: Climate change and wetlands X.26: Wetlands and extractive industries X.31: Enhancing biodiversity in rice paddies as wetland systems	United States
COP11 (2012)	XI.8: Streamlining procedures for describing Ramsar Sites at the time of designation and subsequent updates XI.20: Promoting sustainable investment by the public and private sectors to ensure the maintenance of the benefits people and nature gain from wetlands	Turkey
	XI.19: Adjustments to the terms of Resolution VII.1 on the composition, roles, and responsibilities of the Standing Committee and regional categorization of countries under the Convention	Iran

COP (Year)	Resolution number and title	Contracting party
COP12 (2015)	XII.12: Call to action to ensure and protect the water requirements of wetlands for the present and the future	Turkey
COP13 (2018)	XIII.8: Future implementation of scientific and technical aspects of the Convention for 2019–2021 XIII.11: Ramsar Advisory Missions XIII.22: Wetlands in West Asia	Turkey

Sources: Ramsar Conference Reports (https://www.ramsar.org/about/proceedings-of-past-cops).

Report of the World Commission on Dams, which it characterised as having 'no worldwide acceptance and ... [as] subject to criticism of many countries'.[121] More broadly, Turkey has repeatedly asserted that transboundary water resource management is beyond the Ramsar Convention's mandate and therefore is 'irrelevant to the context and obligations of the Ramsar Convention'.[122] Despite its strong views, however, Turkey has never formally objected to block consensus.

5   STATEMENTS FOR THE RECORD

Rather than making a reservation to a resolution, contracting parties will also occasionally request that statements for the record be included in the Conference Report.[123] These statements emphasise that party's understanding or interpretation of a particular provision.

For example, at COP9, in the context of a resolution on synergies with other international organisations dealing with biological diversity, the United States 'wish[ed] to strongly note for the record' that it defined the term 'harmonization' to mean 'the reduction of duplication in national reporting'.[124] As a non-party to the CBD, the United States was apparently concerned that harmonisation of reporting requirements would somehow entangle it in CBD

---

the use of language from the CBD, to which it is not a party. Ramsar Convention, note 119, [301].

[121] Ramsar Convention, note 29, [83].

[122] Ibid. See further Chapter 9 of this volume.

[123] Statements for the record have been made for a range of issues, including some where another party has made a reservation (eg, Egypt at COP7 regarding regional groupings and Colombia and Malaysia at COP8 regarding cultural values).

[124] Ramsar Convention, COP9, *Report of the Meeting* (2005) [348].

matters. Nevertheless, 'in the interest of consensus in the Ramsar Convention', the United States 'reluctantly' accepted the inclusion of 'harmonization'.[125]

A statement might also stress a party's view that the decision does not set a precedent for the future. At COP11, Denmark (on behalf of the EU Member States present and Croatia) noted that it was 'of great importance for us that the budget is agreed by consensus', but expressed regret that the current global economic situation resulted in a flat budget.[126] The statement characterised the 0% increase as 'an exception' and 'anticipate[d] that this decision does not set a precedent for the budgets in the subsequent years of the Ramsar Convention and other international environmental agreements and conventions'.[127] COP12, COP13, and the third extraordinary COP in 2021 all approved budgets with no increase in the core budget, however.[128]

Some statements for the record have focused on site designation issues. At COP9, India wanted to note that guidance for addressing Ramsar Sites that no longer met the designation criteria should not be prescriptive and 'impinge on the sovereign rights of a Contracting Party'.[129] Also at COP9, Brazil was worried about the credibility of the List of Wetlands of International Importance, expressing 'great concern' about designating artificial wetlands as Ramsar Sites in cases where they had replaced natural wetlands.[130]

Climate change resolutions have proven to be particularly contentious,[131] although consensus has been achieved through negotiation and the use of statements for the record. At COP8, the first wetlands and climate change resolution was cut down from 19 pages to four as an annex developed by the STRP on wetlands and climate impacts, adaptation, and mitigation was excised.[132] Initially, the United States had no objection to the draft resolution's guidelines and general information contained in the annex. The US position shifted over the course of the COP, however. After the truncated resolution was adopted, the United States made a statement 'to interject a word of caution', emphasising that 'climate change is a difficult subject and the stakes are great' and calling into question the precision of the supporting information papers.[133]

---

[125] Ibid.
[126] Ramsar Convention, note 87, [363].
[127] Ibid.
[128] The core budget has remained flat at CHF 5.081 million during this time period.
[129] Ramsar Convention, note 124, [220].
[130] Ibid, [330].
[131] See further Chapter 11 of this volume.
[132] See Ramsar Convention, note 40.
[133] Ramsar Convention, note 29, [89].

At COP10, negotiations over a climate change resolution went to the last minutes of the COP. In its intervention after the draft resolution was introduced, Brazil made a statement for the record, noting the importance of common but differentiated responsibilities.[134] Brazil also considered it premature to come to conclusions about wetlands and climate change mitigation.[135] After numerous contact group meetings, the clock was winding down as a third revision of the draft resolution was negotiated in plenary. The COP Alternate President requested interested parties to move to the hallway to try to reach consensus. The impromptu scrum was able to reach compromise language and Brazil joined the consensus without a further statement for the record.[136]

When another climate change resolution was considered at COP11, Brazil made a statement for the record that expressed its belief that the final text did not adequately reflect its views.[137] While not obstructing the resolution's adoption by the COP, Brazil did emphasise 'its position that Resolution XI.14 does not, in any way, impinge on the work of the UNFCCC, which is the sole multilateral forum mandated to address issues regarding climate change'.[138]

COP13 saw several other climate change-related resolutions adopted.[139] The most controversial was Resolution XIII.14 on promoting conservation, restoration and sustainable management of coastal blue-carbon ecosystems – in part because not all parties agreed on the definition of 'blue carbon'.[140] The matter was resolved by including a footnote that acknowledged that 'not all Contracting Parties endorse [the resolution's] definition or recognise the Ramsar Convention as the competent forum to address mitigation reporting and accounting arrangements'.[141] With that caveat, no reservations or statements for the record were made and the resolution was adopted by consensus.

Interestingly, statements made for the record do not always make it into the Conference Report. At COP11, when the draft resolution on guidelines

---

[134] Ramsar Convention, note 119, [154].

[135] Ibid.

[136] Ibid, [371].

[137] Ramsar Convention, note 87, [470].

[138] Ibid.

[139] Five resolutions had 'climate change' or 'carbon' in their titles, and several other resolutions referred to climate change. Source: https://www.ramsar.org/cop13-resolutions.

[140] Ramsar Convention, *Resolution XIII.14: Promoting conservation, restoration and sustainable management of coastal blue-carbon ecosystems* (2018). The resolution defined 'blue carbon' as '[t]he carbon captured by living organisms in coastal (e.g. mangroves, saltmarshes and seagrasses) and marine ecosystems and stored in biomass and sediments'.

[141] Ibid, footnote 1.

for avoiding, mitigating, and compensating wetland losses was introduced in plenary, Mexico requested that all references to the 'no net loss' concept be deleted.[142] STRP members that had drafted the guidelines (and other parties) preferred that the 'no net loss' references remain in the text because COP10 had given the STRP the task to report on lessons learned from 'no net loss' and the guidelines did not advocate any position on the 'no net loss' approach.[143] STRP members conferred with a member of the Mexican delegation who agreed that 'no net loss' could remain in the resolution because Mexico had stated for the record (for the Conference Report) that Mexican law does not permit the 'no net loss' concept.[144] The resolution was eventually adopted by consensus,[145] but the Conference Report did not include Mexico's statement. In early COPs (with fewer parties), Conference Reports provided detailed accounts of interventions. Now, however, when there are numerous interventions on a draft resolution, the Conference Report generally only lists which parties made a comment without describing the substance. If a party wishes its views to be included in the Conference Report, it must expressly so state (and should provide the precise text to the rapporteur).

6   WITHDRAWAL OF DRAFT RESOLUTIONS

When there is significant opposition to a draft resolution, the sponsoring party may simply withdraw it from consideration. At COP9, for example, Switzerland proposed a resolution to develop synergies between the Ramsar Convention and the Antarctic Convention,[146] which requested the respective secretariats to collaborate in identifying which Antarctic wetlands satisfied the Ramsar Site designation criteria.[147] A wetland may only be designated a Ramsar Site by the country that has sovereignty over that area and many parties expressed concern that such a joint exercise could undermine Article IV of the Antarctic Treaty, which holds all Antarctic territorial claims in abey-

---

[142] Earth Negotiations Bulletin, 'Report of main proceedings for 10 July 2012' at: https://enb.iisd.org/events/11th-meeting-conference-parties-ramsar-convention-cop11/report-main-proceedings-10-july-2012.

[143] Ramsar Convention, *Resolution X.10: Future implementation of scientific and technical aspects of the Convention* (2008) Annex 2 [9.1].

[144] Personal notes of the author.

[145] Ramsar Convention, note 87, 51.

[146] 1959 Antarctic Treaty, adopted 1 December 1959, entered into force 23 June 1961, 402 UNTS 71.

[147] W Burns et al, 'International Environmental Law' (2006) 40 *International Lawyer* 197, 203.

ance.[148] In a statement for the record, Switzerland declared that, '[i]n a spirit of compromise and openness', it would withdraw the draft resolution but hoped that the Ramsar Secretariat would nevertheless 'have the opportunity to exchange information' about the conservation of polar wetlands with the Antarctic Treaty.[149]

At COP13, Sweden again broached the Antarctic issue, proposing a draft resolution on wetlands in polar and subpolar regions.[150] The draft resolution encouraged parties to designate Ramsar Sites in polar and subpolar regions – except for Antarctica, where parties were encouraged to 'designate more protected areas through international agreements in the area covered by the Antarctic Treaty, where there are unprotected wetland biodiversity hotspots'.[151] Because a number of parties at a pre-COP Standing Committee meeting 'raised concerns that the draft resolution covered matters within the mandate of the Antarctic Treaty', all references to Antarctica were bracketed.[152] At COP13, consensus was achieved once all mentions of Antarctica were excised from the text and the resolution applied only to the Arctic and sub-Arctic.[153]

Another COP13 item that proved controversial was Senegal and the Central African Republic's draft resolution on wetlands, peace, and security.[154] The proposal, among other things, called on the STRP to provide guidance on mapping and assessing the vulnerability of 'wetland security hotspots'.[155] More than a dozen parties intervened at the COP to oppose the proposal on the

---

[148] Ibid.

[149] Ramsar Convention, note 124, [179]. The Antarctic Treaty Secretariat was an invited STRP observer organisation from 2009 to 2012, as was the Conservation of Arctic Flora and Fauna Working Group of the Arctic Council from 2013 to 2018. Neither participated in the work of the STRP and both were thus deleted from the list of COP-approved observer organisations.

[150] Ramsar Convention, COP13, *Doc 18.25: Draft resolution on wetlands in polar and subpolar regions* (2018).

[151] Ibid, [27].

[152] Ramsar Convention, *Report and Decisions of the 54th Meeting of the Standing Committee* (2018) [56].

[153] Ramsar Convention, *Resolution XIII.23: Wetlands in the Arctic and sub-Arctic* (2018).

[154] Ramsar Convention, COP13, *Doc 18.19: Draft resolution on the importance of wetlands [for] [in the context of] peace and [human security] [security]* (2018). Even the title was bracketed.

[155] Ibid, [19]. At COP13, Wetlands International organised a side event to discuss the 'bilateral relationship between human security [and] healthy wetlands'. Wetlands International Eastern Africa, Twitter Post, 19 October 2018, 4:37 a.m., at: https://twitter.com/WetlandsIntEA/status/1053203446600470528.

grounds that many portions of the text 'lay outside the scope and competence of the Convention'.[156] With consensus not forthcoming, the sponsoring parties withdrew the draft resolution rather than forcing a vote.[157]

7 CONSENSUS AS A STRENGTH AND A WEAKNESS

The Ramsar tradition of consensus has no doubt contributed to the 'Ramsar spirit' – the generally collegial atmosphere at COPs and other Ramsar meetings, where parties negotiate to resolve differences. Consensus requires all parties' views to be taken into account, thus in theory ensuring the full participation of all parties – even those that may be less politically powerful or influential. Furthermore, a resolution adopted by consensus suggests that all states (except for those that made a reservation or statement for the record) have committed to its provisions. In some jurisdictions, a resolution adopted by consensus is binding in domestic law.[158] Even if a resolution's terms are not binding, they may nevertheless have persuasive force as 'soft law' or through the principle of estoppel.[159]

Ramsar's emphasis on consensus is one of its strengths, but it can also be a weakness. From a procedural standpoint, as a practical matter, not all parties can participate in contact group discussions that are critical in arriving at agreed-upon resolution language. Given the number of resolutions at each COP, multiple contact groups may meet at the same time. Due to budget constraints, many parties have small delegations, consisting of just one or two delegates. The delegations are simply not large enough to cover every contact group and thus their concerns may not be addressed.[160] A shift to a voting model, however, would not necessarily result in all voices being heard and would disrupt Ramsar's culture for uncertain benefits.

Virtual meetings can offer an opportunity for greater participation, but – as was seen at the third extraordinary COP – parties are concerned that technical

---

[156] Ramsar Convention, note 46, [100].

[157] Ibid, [204].

[158] RC Gardner et al, 'Ramsar at the national level: Application and incorporation into domestic law' in P Gell, N Davidson, and CM Finlayson (eds), *Ramsar Wetlands: Values, Assessment, Management* (Elsevier, Amsterdam, 2023) 78–79 (discussing effect of Ramsar resolution under Dutch law). See also JM Verschuuren, 'Verdrag van Ramsar art. 3' (2008) 35 *Milieu en Recht* 28.

[159] Gardner, note 21, 7–8.

[160] Ibid, 8–9.

limitations and connectivity issues could make such participation illusory.[161] Furthermore, in-person meetings are critical to building a sense of community. The conversations in the corridors and shared meals (even if a limp sandwich at a side event) forge relationships that carry over into the implementation of the Convention.

A greater concern pertains to the substantive results of consensus. While consensus secures greater buy-in from parties, it is achieved by weakening provisions concerning wetland conservation and wise use. For example, COP13 considered a draft resolution on wetlands and agriculture.[162] Agriculture has long been a driver of wetland loss and degradation, in part because of subsidies that encouraged wetland drainage.[163] The draft resolution contained two references to subsidies: one encouraging parties to review their agricultural subsidies and assess their impacts on wetlands and the other calling for policymakers and decisionmakers to withdraw subsidies that harmed wetlands.[164] But the final version, approved by consensus, did not contain the term 'subsidy'.[165]

Consensus and COP resolutions are not ends to themselves. Rather, they are mechanisms that should contribute to the Convention's overall objective of 'stem[ming] the progressive encroachment on and the loss of wetlands'.[166] Yet it is uncertain to what extent consensus-driven COPs and their resolutions have delivered on that objective. The Wetlands Extent Trends Index found that between 1970 and 2015, the area of both natural inland and marine/coastal wetland area had declined by approximately 35%, where data was available.[167]

---

[161] Connectivity issues were also a challenge with online meetings of other international bodies, such as the Arctic Council, during the COVID-19 pandemic. M Cela and P Hansson, 'A Challenging Chairmanship in Turbulent Times' (2021) 11 The Polar Journal 43, 52.

[162] Ramsar Convention, COP13, *Doc 18.21: Draft resolution on agriculture in wetlands* (2018).

[163] Gardner and Finlayson, note 6, 67.

[164] Ramsar Convention, note 162, [23, 26].

[165] See Ramsar Convention, *Resolution XIII.19: Sustainable agriculture in wetlands* (2018). Interestingly, an initial draft of the Ramsar Convention contained two provisions that prohibited parties from subsidising wetland drainage and modification, but both were ultimately dropped. DA Stroud et al, 'Development of the Text of the Ramsar Convention: 1965–1971' (2022) 72 *Marine and Freshwater Research* 1, 11.

[166] Convention on Wetlands, note 2, Preamble.

[167] SE Darrah et al, 'Improvements to the Wetland Extent Trends (WET) index as a tool for monitoring natural and human-made wetlands' (2019) 99 *Ecological Indicators* 294, 295.

This wetland loss figure was highlighted in the 2018 and 2021 editions of the *Global Wetland Outlook*, the Ramsar Convention's flagship publication.[168]

What can be done, in the context of the COP, to reverse these negative trends? I suggest one small step regarding the *Global Wetland Outlook* that would be in keeping with the Ramsar tradition of consensus.

The production of the inaugural *Global Wetland Outlook*, which was a Ramsar-wide effort led by the STRP, was viewed as 'the single most important outcome' of COP13.[169] The lead coordinating authors presented the report's main findings at a special plenary session, which covered wetland status and trends, drivers of change, and response actions.[170] Yet an important link was missing: the Secretariat did not prepare a summary for policymakers.[171] The lead coordinating authors subsequently published a journal paper highlighting key takeaways for policymakers and decisionmakers, but it exists behind a paywall and is not readily accessible.[172] Future editions of the *Global Wetland Outlook* can remedy this situation by following the example of IPBES and engage the Ramsar COP by seeking its approval of the text of a summary for policymakers.

IPBES produces thematic as well as regional and global assessments on biodiversity and ecosystem services.[173] The assessment reports are 'accepted' by the Plenary (the equivalent of a COP), which signifies that the Plenary considers the material to present 'a comprehensive and balanced view of the subject matter'.[174] In contrast, the summaries for policymakers are 'approved' by the Plenary, which means 'that the material has been subject to detailed, line-by-line discussion and agreement by consensus at a session of the

---

[168] Gardner and Finlayson, note 6, 19; M Courouble et al, *Global Wetland Outlook: Special Edition 2021* (Ramsar Convention Secretariat, Gland) 21.

[169] Earth Negotiations Bulletin, 'Summary of the thirteenth meeting of the Conference of the Parties to the Ramsar Convention on Wetlands: 22–29 October 2018' at: https://enb.iisd.org/events/13th-meeting-conference-contracting-parties-ramsar-convention-wetlands-cop13/summary-report.

[170] Ibid.

[171] CM Finlayson and RC Gardner 'Ten key issues from the Global Wetland Outlook for decision makers' (2020) 72 *Marine and Freshwater Research* 301, 301.

[172] See CSIRO Publishing, Marine & Freshwater Research, https://www.publish.csiro.au/mf/MF20079.

[173] Source: IPBES, About, https://www.ipbes.net/about.

[174] IPBES, note 8, 3.

Plenary'.[175] The benefit of such a process is that IPBES members are more engaged and have ownership of the documents.[176]

The Ramsar COP should follow this approach, at least for a summary for policymakers. Requiring parties to approve such a document – by consensus – would more directly focus the COP's attention on the state of the world's wetlands, the drivers causing wetland loss and degradation, and the possible response actions to improve the situation. The summary for policymakers would contain recommended actions in a single document and would have greater visibility than any single resolution. In this way, Ramsar's tradition of consensus can be leveraged to better promote wetland conservation and wise use.

---

[175] Ibid.
[176] See M Kowarsch et al, 'Scientific assessments to facilitate deliberative policy learning' (2016) 2 *Palgrave Communications* 16092.

3. Ramsar Sites and their ecological character: A cornerstone of the Ramsar Convention on Wetlands

Royal C. Gardner, Marcela Bonells, and Katherine Pratt

1 INTRODUCTION

Concluded in 1971, the Ramsar Convention on Wetlands[1] formally introduced the concept of internationally designated sites located on the sovereign territory of a state.[2] Prior conventions contained provisions for protected areas, but typically as national sites, such as national parks. For example, the 1942 Convention on Nature Protection and Wildlife Preservation in the Western Hemisphere – hailed as 'visionary' in its scope by Bowman et al[3] – called on governments to 'explore at once the possibility of establishing in their territories national parks, national reserves, nature monuments, and strict wilderness reserves'.[4] Similarly, the 1968 African Convention on the Conservation of Nature and Natural Resources contemplated that its parties would 'maintain and extend' protected natural resource areas, such as strict natural reserves, national parks, or special reserves.[5] The 1964 Agreed Measures for the

---

[1] Convention on Wetlands of International Importance especially as Waterfowl Habitat, adopted 2 February 1971, entered into force 21 December 1975, 996 UNTS 247 (amended 1982 & 1987).
[2] Ibid, Art 2.
[3] M Bowman et al, *Lyster's International Wildlife Law* (Cambridge University Press, Cambridge, UK, 2010) 241.
[4] Convention on Nature Protection and Wildlife Preservation in the Western Hemisphere, adopted 27 October 1940, entered into force 30 April 1942, 161 UNTS 193, Art II.
[5] African Convention on the Conservation of Nature and Natural Resources (with annexed list of protected species), adopted 15 September 1968, entered into force 16 June 1969, 1001 UNTS 3, Art X.

Conservation of Antarctic Fauna and Flora provided for the designation of 'Specially Protected Areas ... to preserve their unique natural ecological system';[6] but because it was adopted pursuant to the Antarctic Treaty, such designations would not be located on any territory recognised as under the sovereign control of a state.[7] Thus, as Stroud et al observed, the drafters of the Ramsar Convention had 'no other models of what international site-based protection under a multilateral treaty may look like'.[8]

This Chapter discusses issues related to the Ramsar Convention's signal contribution to international environmental law and a cornerstone of the Convention itself: the designation of Wetlands of International Importance ('Ramsar Sites'). It examines the designation and listing process, the criteria for which initially focused on wetlands as habitat for waterbirds, but which have since expanded to include other ecologically based factors. The Chapter also notes the thorny and still unresolved matter of Ramsar Site designations on disputed territories, where Ramsar contracting parties disagree about which state has sovereignty. The claim that Ramsar Sites are the world's largest network of protected areas is also challenged.

Furthermore, the Chapter reviews the obligations that a contracting party assumes when it designates a wetland as a Ramsar Site. The chief responsibility is to maintain the Ramsar Site's ecological character – a core Ramsar concept. Over the years, the Ramsar Convention has developed different mechanisms to facilitate the response of contracting parties to adverse changes to a site's ecological character, but significant gaps remain. As we will see, many Ramsar Sites lack management plans. The Chapter also explores how the role of the Conference of the Parties (COP) has evolved with respect to weighing in on site-specific challenges, culminating with COP14's resolution regarding damage to Ukrainian Ramsar Sites as a result of Russian aggression. Finally, the Chapter discusses the creation of site-based recognitions beyond the designation of a single site, including Transboundary Ramsar Sites and Wetland City Accreditation. While the development of such positive incentives for wetland conservation is to be lauded, the need to improve efforts to conserve Ramsar Sites should not be overlooked. The Ramsar Convention was a forerunner with respect to internationally designated sites; however, when

---

[6] Agreed Measures for the Conservation of Antarctic Fauna and Flora, adopted 2–13 June 1964, entered into force 1 November 1982, 17 UTS 996, Art VIII.

[7] See Antarctic Treaty, adopted 1 December 1959, entered into force 23 June 1961, 402 UNTS 71, Art IV.

[8] DA Stroud et al, 'Development of the text of the Ramsar Convention: 1965–1971' (2022) 73 *Marine and Freshwater Research* 1107, 1113.

it comes to reporting and monitoring, it has since lagged behind global counterparts such as the World Heritage Convention[9] and the Biosphere Reserve Program,[10] which have periodic site review processes in place.

2 THE DESIGNATION AND LISTING PROCESS

When joining the Convention, a Ramsar contracting party must designate at least one site within its territory as a Wetland of International Importance.[11] The only guidance that the Convention text provides regarding site selection, in Article 2.2, is that the wetlands should be internationally significant based on 'ecology, botany, zoology, limnology or hydrology'.[12] Article 2.2 also states that the first additions to the list should be based on a wetland's international importance to waterfowl – a point emphasised in the formal title of the Convention. Over time, the COP has refined the criteria for designation. As a technical matter, when a contracting party designates a site, it submits a questionnaire with information about the site – the Ramsar Information Sheet (RIS) – to the Ramsar Secretariat, which, after a review, adds the site to the List.[13] As of November 2023, 172 contracting parties had designated nearly 2500 sites as Wetlands of International Importance.[14]

2.1 The Development of Listing Criteria

The 1962 Project MAR Conference – organised by the International Union for Conservation of Nature (IUCN), the International Council for Bird Preservation (now BirdLife International) and the International Wildfowl Research Bureau (now Wetlands International) – was instrumental in high-

---

[9] Convention for the Protection of the World Cultural and Natural Heritage, adopted 23 November 1972, entered into force 17 December 1975, 1037 UNTS 15; UNESCO World Heritage Convention, Operational Guidelines (2021) 169–76, 199–210 at: https://whc.unesco.org/en/guidelines/; see further Chapters 10 and 6 in this volume.

[10] Source: Man and the Biosphere Programme, Designation and Review Process, https://en.unesco.org/biosphere/designation.

[11] Convention on Wetlands, note 1, Art 2.1.

[12] Ibid, Art 2.2.

[13] RC Gardner and KD Connolly, 'The Ramsar Convention on Wetlands: Assessment of International Designations Within the United States' (2007) 37 *Environmental Law Review* 10089, 10090–91; Ramsar Convention Secretariat, *The Ramsar Convention Manual* (Ramsar Convention Secretariat, Gland, 6th ed, 2013) 54.

[14] Source: Ramsar Sites Information Service, https://rsis.ramsar.org/.

lighting the need for a coordinated international effort to protect wetlands.[15] The Conference requested that IUCN produce a list 'of European and North African wetlands of international importance', and stated that 'this list may be considered as a foundation for an international convention on wetlands'.[16] The MAR Conference decided that ornithological data would be the primary basis for listing, for several reasons. Such data was readily available; bird populations could be used as a surrogate for a site's other ecological values; and migratory birds illustrated the international aspect of wetlands.[17] This emphasis on waterbirds was apparent at COP1 when the contracting parties adopted three criteria for listing, with importance to waterfowl as the first.[18] Any wetland that regularly supported 10 000 ducks, geese, swans, or coots or 20 000 waders could qualify as a Ramsar Site.[19] In addition, a wetland could be a Ramsar Site if it regularly supported '1% of the individuals or 1% of the breeding pairs in a population of one species or subspecies of waterfowl'.[20] However, COP1 also approved criteria that did not necessarily apply directly to waterbirds: a wetland could be added to the List if it demonstrated importance to plants or animals (eg, supporting rare, vulnerable, or endangered species); or if the wetland type was representative or unique.[21]

Over time, the COP refined and added to the listing criteria. Efforts to expand the criteria to include a wetland's cultural value or socioeconomic ecosystem services have been rejected – in part because some contracting parties see non-ecological criteria as going beyond the categories identified in Article 2.2.[22] The current listing criteria have remained unchanged since 2005,[23]

---

[15] GVT Matthews, *The Ramsar Convention on Wetlands: its History and Development* (Ramsar Convention Secretariat, Gland, reissued, 2013) 29–30.

[16] PJS Olney et al, *Project Mar: The conservation and management of temperate marshes, bogs and other wetlands* (IUCN, 1965) 7, at: https://portals.iucn.org/library/sites/library/files/documents/NS-005.pdf. The MAR list identified 236 individual wetlands for consideration; thus far, it appears that approximately 78% have been designated as Ramsar Sites. Compare ibid with https://rsis.ramsar.org/ (spreadsheet on file with the authors).

[17] Ibid.

[18] DA Stroud and NC Davidson, 'Fifty years of criteria development for selecting wetlands of international importance' (2022) *73 Marine and Freshwater Research* 1134, 1136.

[19] Ibid, 1138.

[20] Ibid.

[21] Ibid, 1136.

[22] Ibid, 1136–37.

[23] Ramsar Convention, COP9, *Resolution IX.1 Annex B: Revised Strategic Framework and guidelines for the future development of the List of Wetlands of International Importance* (2005) 5.

although COP14 provided guidance on alternative mechanisms to determine whether the 1% population threshold for Criterion 6 has been satisfied.[24]

Although the criteria now encompass many more wetland characteristics beyond those related to waterbirds, the Convention's origins remain a strong influence in the listing process. For example, 70% (30 out of 43) of Ramsar Site listings in 2022 invoked the waterbird-specific Criterion 5 and/or 6. Furthermore, in the 13 designations that did not rely on waterbird-specific criteria, waterbirds were mentioned in relation to other criteria (eg, Criterion 2) in all but one.[25]

2.2 The Matter of Sovereignty

An early draft of the Convention provided that Ramsar Sites 'shall be the subject of the joint care of the Contracting Parties' and contemplated the need for a COP opinion prior to a contracting party delisting a Ramsar Site or reducing its area.[26] Even though these proposed COP opinions were supposed to be made 'without prejudice to the exclusive rights attached to ... sovereignty', ultimately such an approach was jettisoned as politically unpalatable.[27] The final text of the Convention makes no reference to joint care or management of Ramsar Sites and expressly emphasises that 'inclusion of a wetland in the List does not prejudice the exclusive sovereign rights of the Contracting Party in whose territory the wetland is situated'.[28]

A contracting party is almost entirely in control of the designation process. It is the contracting party in the first instance that decides to submit, through its administrative authority, the RIS that serves as the application for listing.[29] The Ramsar Secretariat's review is essentially limited to ensuring that the RIS is properly completed. This deference to contracting parties in the listing process contrasts with later international designation regimes, such as for World Heritage Sites or United Nations Educational, Scientific and Cultural Organization (UNESCO) Global Geoparks, where applications are subject to a more rigorous review.[30]

---

[24] Ramsar Convention, COP14, *Resolution XIV.18: Waterbird population estimates to support new and existing Ramsar Site designations under Ramsar Criterion 6 – use of alternative estimates* (2022) 2.

[25] Source: https://rsis.ramsar.org/ (spreadsheet on file with the authors).

[26] Stroud et al, note 8, 1113.

[27] Ibid.

[28] Convention on Wetlands, note 1, Art 2.3.

[29] Ramsar Convention Secretariat, note 13, 75.

[30] UNESCO, note 9, Chapter II; UNESCO International Geoscience and Geoparks Programme, 'Submit a UNESCO Global Geopark proposal' at: https://www.unesco.org/en/iggp/geoparks/proposals?hub=67817.

Controversy may arise when a contracting party designates a Ramsar Site in an area that is subject to a territorial dispute. For example, in 2001 the United Kingdom designated Bertha's Beach and Sea Lion Island – which are located on the Falkland Islands (known as the Islas Malvinas in Argentina) – as Ramsar Sites. Argentina considers the Islas Malvinas to be its territory and subject to its sovereignty. Accordingly, Argentina made a diplomatic protest to these designations, which is noted in the Sites' descriptions on the Ramsar Sites Information Service (RSIS) to further evidence that a territorial claim has not been abandoned.[31]

At COP14, Algeria submitted a draft resolution that called on the Secretariat to take the geographical coordinates provided by the United Nations Geospatial Network into account when listing Ramsar Sites.[32] The draft resolution also requested that the Secretariat keep only those Ramsar Sites that were 'in line with' the United Nations Geospatial Network on the List of Wetlands of International Importance – effectively asking the Secretariat to delist some Sites that were designated in areas where a contracting party's sovereignty was in question.[33] The COP was reluctant to approve a process to delist Ramsar Sites and deferred consideration of Algeria's proposal.[34] The COP did, however, request that the Secretariat prepare a technical report on all aspects of the listing process, which is expected to include the question of territorial sovereignty.[35]

2.3   The Largest Protected Area Network?

The nearly 2500 Ramsar Sites encompass more than 256 million hectares (2.56 million square kilometres (km^2)),[36] which almost equals the area of the land mass of Kazakhstan (2.7 million km^2). Ramsar Sites cover approximately

---

[31] Ramsar Sites Information Service, *Bertha's Beach* at: https://rsis.ramsar.org/ris/1103 and *Sea Lion Island* at: https://rsis.ramsar.org/ris/1104.

[32] Ramsar Convention, COP14, *Doc 18.16 Rev 1: Amended Draft resolution on the Ramsar List* (2022).

[33] Ibid.

[34] International Institute for Sustainable Development, 'Summary of the Fourteenth Meeting of the Conference of the Contracting Parties to the Ramsar Convention on Wetlands: 5–13 November 2022' (2022) 17:54 *Earth Negotiations Bulletin* 11, at: https://enb.iisd.org/sites/default/files/2022-11/enb1754e.pdf; Ramsar Convention, *Resolution XIV.13: The status of Sites in the List of Wetlands of International Importance* (2022) [22].

[35] Ramsar Convention, note 34, [21].

[36] Source: The Convention on Wetlands, Highlights, https://www.ramsar.org/.

13%–18% of all wetlands.[37] Greater precision is difficult because some Ramsar Sites include both wetland and non-wetland areas (eg, buffer zones or the larger catchment area), and knowledge gaps exist with respect to the extent and distribution of many wetland types.[38] Nevertheless, the number of Ramsar Sites designated and the total area covered demonstrate a serious commitment by the contracting parties to the cause of wetland conservation.[39] Figure 3.1 tracks the growth of Ramsar designations over time, while Figure 3.2 provides a breakdown of Ramsar designations by region. While Europe accounts for the greatest number of individual Sites, Africa has designated the greatest area by far.

It has been stated – and indeed continues to be detailed on the Ramsar Convention's website – that 'the Ramsar List is the world's largest network of protected areas'.[40] While this may have been true at one time, it is an incorrect characterisation today. In 2021, the total area of Natural World Heritage Sites was more than 303 million hectares, and this figure does not include cultural sites or mixed natural and cultural sites.[41] Moreover, there is a debate about whether Ramsar Sites, by virtue of their designation alone and regardless of any additional national legal protections, qualify as 'protected areas' under either the IUCN or Convention on Biological Diversity (CBD) definitions.[42] One view is that a Ramsar Site is indeed a protected area under international law – once a contracting party designates a site, that country has certain

---

[37] RC Gardner, CM Finlayson, and E Okuno, 'Global Wetland Outlook: Technical Note to Introduction' (Ramsar Convention Secretariat Gland, 2018) 6.
[38] Ibid.
[39] RC Gardner and CM Finlayson, *Global Wetland Outlook: State of the world's wetlands and their services to people 2018* (Ramsar Convention Secretariat, Gland, 2018) 13.
[40] Source: The Convention on Wetlands, Wetlands of International Importance, https://www.ramsar.org/our-work/wetlands-international-importance#:~:text=Today%2C%20the%20Ramsar%20List%20is,than%202.5%20million%20square%20kilometres.
[41] UNESCO World Heritage Convention, World Heritage List Statistics at: https://whc.unesco.org/en/list/stat/.
[42] The IUCN defines a 'protected area' as 'a clearly defined geographical space, recognised, dedicated and managed, through legal or other effective means, to achieve the long term conservation of nature with associated ecosystem services and cultural values.' Source: IUCN, Effective protected areas, https://www.iucn.org/our-work/topic/effective-protected-areas. The CBD defines a 'protected area' in Article 2 of the Convention as 'a geographically defined area which is designated or regulated and managed to achieve specific conservation objectives.' Convention on Biological Diversity, adopted 16 November 1972, entered into force 29 December 1993, 1037 UNTS 151, Art 2.

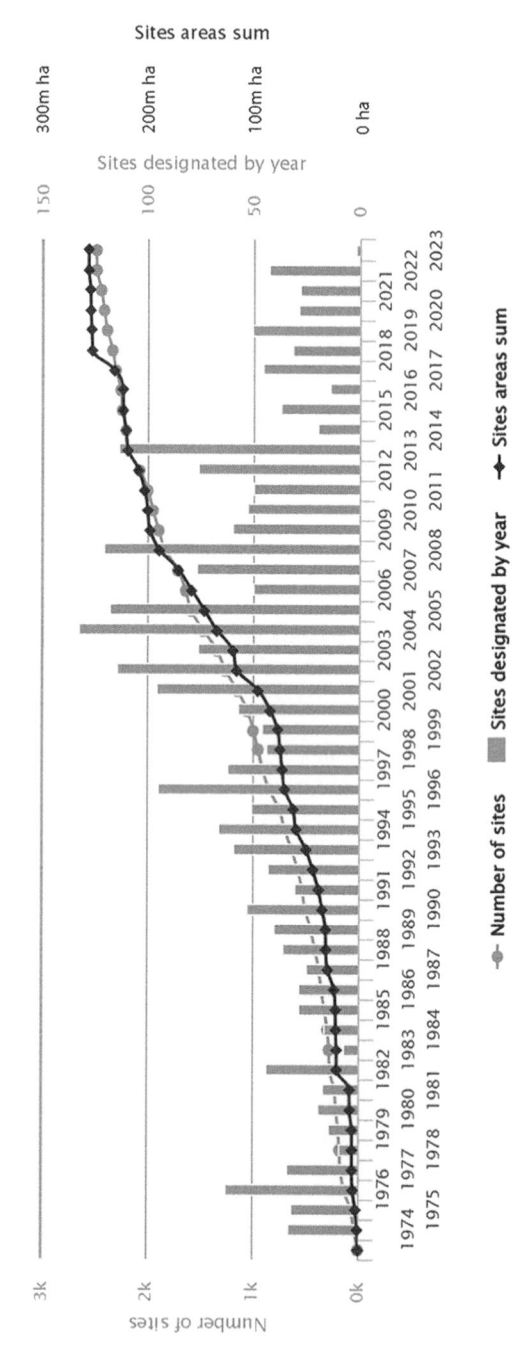

Figure 3.1 Sites number and area by year

Source: Ramsar Sites Information Service at: https://rsis.ramsar.org/.

Figure 3.2 Sites number and area by region

Source: Ramsar Sites Information Service at: https://rsis.ramsar.org/.

international obligations regarding the maintenance of the site's ecological character (discussed further below), regardless of whether the country takes any additional steps to protect the site through domestic law. But Ramsar Resolution VIII.13 makes a distinction between 'nationally relevant protected area status' and 'international conservation designations', such as Ramsar designations.[43] And others have suggested that Ramsar Sites (or portions thereof) lacking 'nationally relevant protected area status' may be 'other effective area based conservation measures' (OECMs).[44] Whatever one's position on the protected area debate, rather than claiming that Ramsar has the largest network, it would be more precise to say that '[t]he designation of sites through the Ramsar Convention has led to the most extensive list of sites focusing on wetland conservation'.[45]

Moreover, the extent of Ramsar designations contributes to global targets beyond Ramsar. For example, Target 11 of the CBD's Aichi Biodiversity Targets aimed for at least 17% of terrestrial and inland water and 10% of coastal and marine areas to be managed as protected areas or OECMs by 2020.[46] These were some of the few aspects of the Aichi Biodiversity Targets that were actually met in a timely fashion.[47] Whether these protected areas, including the Ramsar Sites, are effectively managed is a different matter.

3   OBLIGATIONS

The management of Ramsar Sites is tied to the Ramsar concept of 'ecological character'.

This obligation flows directly from the Convention's text. Article 3.1 states that contracting parties 'shall formulate and implement their planning so as to promote the conservation of the wetlands included in the List'.[48] If the 'eco-

---

[43] Ramsar Convention, *Resolution VIII.13: Enhancing the information on Wetlands of International Importance (Ramsar sites)* (2002) 17.

[44] H Jonas and D Laffoley, 'Exploring "Other Effective" Forms of Coastal and Marine Conservation', 7 September 2017, *IUCN News & Events* at: https://www.iucn.org/news/protected-areas/201709/exploring-%E2%80%98other-effective%E2%80%99-forms-coastal-and-marine-conservation.

[45] D Juffe-Bignoli et al, 'Achieving Aichi Biodiversity Target 11 to improve the performance of protected areas and conserve freshwater biodiversity' (2016) 26:S1 *Aquatic Conservation: Marine and Freshwater Ecosystems* 133, 135.

[46] Ibid.

[47] Secretariat of the Convention on Biological Diversity, *Global Biodiversity Outlook 5* (Convention on Biological Diversity, Montreal 2020) 82. See further Chapter 8 in this volume.

[48] Convention on Wetlands, note 1, Art 3.1.

logical character' of a Ramsar Site 'has changed, is changing or is likely to change as the result of technological developments, pollution or other human interference', Article 3.2 requires that the contracting party notify the Ramsar Secretariat 'without delay'.[49] A contracting party is therefore 'expected to establish management planning and monitoring mechanisms for Ramsar Sites, and to invoke appropriate response options' to resolve human-induced negative change.[50] In this sense, the Ramsar Convention places an affirmative duty on a contracting party to maintain the ecological character of its Ramsar Sites.[51]

3.1 The Concept of Ecological Character and Baselines

'Ecological character' is a cornerstone of the Convention's mission to conserve and wisely use wetlands, but this definition has been modified over time. COP7 defined 'ecological character' as 'the sum of the biological, physical, and chemical components of the wetland ecosystem, and their interactions, which maintain the wetland and its products, functions, and attributes'.[52] After the Millennium Ecosystem Assessment (MA), the Scientific and Technical Review Panel (STRP) – the Convention's scientific advisory body – recommended that the 'ecological character' definition be updated to reflect the MA's terminology, stating that it should be viewed as 'the combination of the ecosystem components, processes and services that characterise the wetland at a given point in time'.[53] The notion of 'ecosystem services' concerned some contracting parties,[54] as in some contexts 'services' may imply that 'people

---

[49] Ibid, Art 3.2.

[50] Ramsar Convention, *Resolution XI.9: An Integrated Framework and guidelines for avoiding, mitigating and compensating for wetland losses (2012)* Annex [19] (internal quotation marks omitted). See further Chapter 5 in this volume.

[51] See, eg, ibid; Ramsar Convention Secretariat, *Handbook 19: Addressing change in wetland ecological character* (Ramsar Handbooks, 4th edition, Ramsar Convention Secretariat, Gland, 2010) 8.

[52] Ramsar Convention, *Resolution VII.10: Wetland Risk Assessment Framework* (1999) [11].

[53] Ramsar Convention, COP9, *Doc 16 Information Paper: Rationale for proposals for A Conceptual Framework for the wise use of wetlands and the updating of wise use and ecological character definitions (COP9 DR1 Annex A)* (2005) [28].

[54] See Earth Negotiations Bulletin, 'Summary of the Ninth Meeting of the Conference of the Contracting Parties to the Ramsar Convention on Wetlands: 8–15 November 2005' 5, at: https://enb.iisd.org/events/9th-meeting-conference-parties-ramsar-convention-cop9/summary-report-8-15-november-2005.

must now pay for what were previously free benefits'.[55] Ultimately, COP9 revised the definition to mean 'the combination of the ecosystem components, processes and benefits/services that characterise the wetland at a given point in time'.[56] A footnote was added to the definition to clarify that the term 'ecosystem benefits' is consistent with the MA's concept of 'ecosystem services', which are 'the benefits that people receive from ecosystems'.[57]

At the time of designating a Ramsar Site, contracting parties provide a description of the Site's ecological character, through the RIS and an ecological character description sheet,[58] which serves as the baseline against which to measure any changes in ecological character.[59] This information is based on the status of the Site at the time of designation.[60] However, it 'provides a static snapshot' of the Site and does not necessarily capture fluctuations or changes in time.[61] This may pose difficulties in identifying an accurate baseline in the case of shifting conditions – that is, when ecosystems change to fundamentally different states, sometimes prompted by human activities, giving rise to novel ecosystems.[62] This can present challenges for protected areas, which are reliant on 'history to define their reference conditions for tolerance of change'.[63] Thus, as recommended by the STRP, it is important to include information on the natural range of variability of a Site in the ecological character description or RIS.[64] Management plans and monitoring of the Site will also help to detect any changes.[65]

---

[55] E Lugo, 'Ecosystem Services, the Millennium Ecosystem Assessment, and the Conceptual Difference Between Benefits Provided by Ecosystems and Benefits Provided by People' (2008) 23 *Journal of Land Use & Environmental Law* 243, 255.

[56] Ramsar Convention, *Resolution IX.1 Annex A: A Conceptual Framework for the wise use of wetlands and the maintenance of their ecological character* (2005) [15].

[57] Ibid, 5.

[58] See Ramsar Convention, *Resolution X.15: Describing the ecological character of wetlands, and data needs and formats for core inventory: harmonized scientific and technical guidance* (2008).

[59] D Pritchard, *Change in ecological character of wetland sites – a review of Ramsar guidance and mechanisms* (2014) (consultant report for the Ramsar Convention Secretariat) 2, at: https://www.ramsar.org/document/change-ecological-character-wetland-sites-ramsar-guidance-and-mechanisms-long-version.

[60] Ibid.

[61] Ibid, 13 (internal quotation marks omitted).

[62] Ibid, 23.

[63] Ibid.

[64] Ibid, 22–23.

[65] Ibid, 76–78.

3.2 Management Plans

Designation of a site as a protected area does not, by itself, lead to positive ecological outcomes. Management (or some type of protective measures) is a necessary precondition.[66] Somewhat surprisingly, based on RSIS data, only about half of all Ramsar Sites have a management plan. When disaggregated by region, data as of August 2023 shows that Africa (32.3%) and Asia (39%) have the lowest percentage of Ramsar Sites with a management plan, while Oceania (74.1%) has the highest percentage.

As Kingsford et al concluded, effective management remains a significant challenge in light of the fact that so many Ramsar Sites do not have management plans.[67] Ideally, the management of a Ramsar Site will include 'a vision of a desired future state, involving stakeholders, clear objectives, triggers, risk assessment supported by co-design and co-management actions, and transdisciplinary science, with evaluation, reporting and updating'.[68]

3.3 The Obligation to Report Change in Ecological Character

Article 3.2 of the Convention exhorts contracting parties to:

> arrange to be informed at the earliest possible time if the ecological character of any wetland in its territory and included in the List has changed, is changing or is likely to change as the result of technological developments, pollution or other human interference.[69]

This provision is precautionary in nature and requires not only the reporting of negative, human-induced changes that have occurred (past) and that are occurring (ongoing), but also any likely changes (foreseeable).[70] To meet this obligation, contracting parties must put in place mechanisms to detect, report, and respond to such changes. Resolution X.16 provides guidance to parties in this regard.[71] It is important to note that a change in ecological character does

---

[66] F Leverington et al, 'A Global Analysis of Protected Area Management Effectiveness' (2010) 46 *Environmental Management* 685, 685–86.

[67] RT Kingsford et al, 'Ramsar Wetlands of International Importance – Improving Conservation Outcomes' (2021) 9 *Frontiers in Environmental Science* 1, 3.

[68] Ibid.

[69] Convention on Wetlands, note 1, Art 3.2.

[70] Pritchard, note 59, 18.

[71] See Ramsar Convention, *Resolution X.16: A Framework for processes of detecting, reporting and responding to change in wetland ecological character* (2008).

not mean a change in the international importance of a Site or vice versa. These are different concepts: the former is an ecological concept and deals only with ecological character; while the latter is a political concept concerned with whether a Site meets the criteria for designation.[72]

3.3.1 Triggering the obligation to report

The reporting requirement of Article 3.2 has a number of key elements which are important for determining when it is triggered, including the nature (human versus natural), degree (actual versus likely), and significance of the change (trivial versus significant). The reporting obligation will be triggered only when a change is human induced – for example, as a result of pollution, development or other human activity.[73] Additionally, the change must negatively impact on any of the processes, services, or components outlined in the ecological description of the Site. There is no need to report natural evolutionary changes or positive human-induced changes.[74] The latter can be reported through other mechanisms, such as the RIS, which should be updated every six years; or National Reports, which should be submitted in advance of each COP.[75]

The COP has not yet addressed whether negative impacts as a result of climate change trigger the reporting obligation under Article 3.2. The STRP, however, discussed the issue in February 2010, noting that it was unlikely that the drafters of the Convention intended for Article 3.2 to cover indirect, human-induced negative changes.[76] Application of the reporting requirement would be very difficult, as virtually all Sites would be impacted given the global nature of climate change.[77] Instead, the STRP recommended that climate-related issues affecting Ramsar Sites should be addressed through Resolutions X.24 (*Climate change and wetlands*) and XI.14 (*Climate change and wetlands: implications for the Ramsar Convention on Wetlands*) until the COP decides otherwise.[78]

As noted above, the reporting requirement under Article 3.2 applies to changes that are likely to occur. While there is no guidance from the

---

[72] Pritchard, note 59, 16.
[73] Ibid, 65.
[74] Ramsar Convention, COP10, *Doc 27: Background and rationale to the Framework for processes of detecting, reporting and responding to change in wetland ecological character* (2008) [42]. There is no guidance to parties on how to identify human-induced changes from natural changes. Ibid, [40].
[75] Pritchard, note 59, 26–27.
[76] Ibid, 69.
[77] Ramsar Convention, note 74, [40]. On climate change and the Ramsar Convention, see further Chapter 11 in this volume.
[78] Pritchard, note 59, 69.

Ramsar Sites and their ecological character 93

Convention on what constitutes a likely change, some contracting parties have adopted their own standards of proof.[79] For instance, Australia's National Guidelines for Notifying Change in Ecological Character of Australian Ramsar Sites (Article 3.2) define a 'likely change' to be 'when there is evidence that a change in character will occur in the future or is imminent'.[80] Any notification of change in ecological character must be based on 'best available science'.[81]

There is a related reporting requirement for the Secretariat, pursuant to Article 8.2, which requires the Secretariat to make 'known to the Contracting Party concerned, the recommendations of the Conferences in respect of such alterations to the List or of changes in the character of wetlands included therein'. [82] This means that the Secretariat must report to parties (at the COP) any changes in the ecological character of Ramsar Sites (ie, Article 3.2 reports) and responses.[83] The COP, in turn, may offer its advice through a resolution or recommendation (discussed further below).[84]

3.3.2 Reporting procedure

In principle, the administrative authority[85] within each contracting party is responsible for reporting to the Secretariat. In practice, however, the

---

[79] Although there is no guidance on this matter, Ramsar Convention, *Resolution VII.10: Wetland Risk Assessment Framework* (1999), would be relevant – particularly the section on early warning indicators on pages 6–12.

[80] Australian Department of the Environment, Water, Heritage and the Arts, *National Guidelines for Notifying Change in Ecological Character of Australian Ramsar Sites (Article 3.2): Module 3 of the National Guidelines for Ramsar Wetlands – Implementing the Ramsar Convention in Australia* (Australian Government, Canberra, 2009) 6, at: https://www.awe.gov.au/sites/default/files/documents/module-3-change.pdf.

[81] Ibid, 7.

[82] Convention on Wetlands, note 1, Art 8.2(e).

[83] See, eg, Ramsar Convention, COP14, *Doc 10 Rev 1: Report of the Secretariat pursuant to Article 8.2 on the List of Wetlands of International Importance* (2022) Annexes 4a and 4b, 5–6.

[84] Convention on Wetlands, note 1, Art 8.2(e).

[85] The Administrative Authority is the government agency or ministry within a contracting party responsible for national implementation of the Convention. Ramsar Convention Secretariat, note 12, 39. While neither Article 3.2 nor related resolutions specify a particular format for reporting, reports should be official, submitted by the administrative authority, and in writing. In practice, however, parties use three mechanisms for reporting changes in ecological character. First, a party may formally notify the Secretariat in writing of any actual or likely changes. Second, a party may make the notification in an update to the RIS. Third, it may

Secretariat often receives reports from third parties, such as civil society or non-governmental organisations (NGOs).[86] In such cases, the Secretariat will follow up with the administrative authority for confirmation.[87] Recognising this fact, Resolution VIII.8 encourages contracting parties to put in place mechanisms to facilitate local and Indigenous communities and NGOs communicating such concerns to the administrative authorities.[88]

3.3.3 Tracking reported changes in ecological character

When the Secretariat receives an Article 3.2 report, it will open an Article 3.2 file in the RSIS for tracking purposes.[89] Only when the issues have been resolved will the file be closed. However, it may take years to close an Article 3.2 file, as some issues are slow to resolve. Historically, it has taken anywhere from one to 26 years to close an Article 3.2 file.[90]

By the time of COP14 (2022), 28% of parties had reported cases of actual and likely changes in ecological character, up from 21% for COP13 (2018).[91] As Table 3.1 demonstrates, the number of open files or unresolved cases significantly increased from 2012 to 2022, although from 2018 to 2022 the totals remained relatively constant.

After making an Article 3.2 report to the Secretariat, a contracting party sometimes fails to provide any follow-up information, suggesting a lack of response actions. At COP14, the Secretariat noted that for 97 sites with an open Article 3.2 file, contracting parties had neglected to provide any updates for more than five years, even though they 'are urged to report to each annual meeting of the Standing Committee the status of these Sites and any steps taken to address negative changes to their ecological character'.[92]

---

notify via the National Reporting format (Section 2.6) submitted to the COP every three years.

[86] Pritchard, note 59, 26.

[87] Ibid.

[88] Ramsar Convention, *Resolution VIII.8: Assessing and reporting the status and trends of wetlands, and the implementation of Article 3.2 of the Convention* (2002) [12].

[89] RC Gardner, E Okuno, and D Pritchard, 'Ramsar Convention governance and processes at the international level' in P Gell, N Davidson, and CM Finlayson (eds), *Ramsar Wetlands: Values, Assessment, Management* (Elsevier, Amsterdam, 2023) 48–49; Pritchard, note 59, 32.

[90] Ramsar Convention, COP13, *Doc. 11.1: Report of the Secretary General on the implementation of the Convention: Global implementation* (2018) [49].

[91] Ramsar Convention, COP14, *Doc. 9.1: Report of the Secretary General on the implementation of the Convention: Global implementation* (2022) [48].

[92] Ramsar Convention, note 91, [50].

Table 3.1   Open Article 3.2 Files

2012	97 open files (153 total, with 56 closed).
2015	123 open files (144 total, with 21 closed – plus another 64 reported by third parties but unconfirmed by contracting parties).
2018	145 open files (168 total, with 23 closed – plus another 60 reported by third parties but unconfirmed by contracting parties).
2022	149 open files (175 total, with 26 closed – plus another 56 reported by third parties but unconfirmed by contracting parties).

Sources: Ramsar COP11 Doc 8, COP12 Doc 7, COP13 Doc 12, COP14 Doc 10 Rev 1.

4   RESPONSES TO ADVERSE CHANGE IN THE ECOLOGICAL CHARACTER OF RAMSAR SITES

The Convention itself does not specify response actions to changes in a Ramsar Site's ecological character, except where a contracting party, for reasons of urgent national interest, decides to delist a Site or reduce its boundaries.[93] In those cases, as discussed in Chapter 5, the Convention calls on the contracting party to provide ecological compensation. However, processes for detecting, reporting, and responding to changes in ecological character are elaborated in Ramsar resolutions and guidance.[94] Response options include the contracting party listing the Ramsar Site on the Montreux Record, a registry of Ramsar Sites where adverse change in ecological change has occurred.[95] In addition, the contracting party may request the Secretariat to establish a Ramsar Advisory Mission (RAM) to visit the affected Ramsar Site to make recommendations.[96] Resolution IX.6 also calls upon a contracting party to make 'at least equivalent provision of [ecological] compensation' when there is unavoidable loss of ecological character at a Ramsar Site.[97] These processes and mechanisms underscore the seriousness of the obligation to maintain the ecological character of Ramsar Sites. Indeed, as noted below, in one instance, the International Court of Justice (ICJ) was requested to intervene in a dispute over the ecological character of Ramsar Sites.

---

[93]   Convention on Wetlands, note 1, Art 4.2.
[94]   See, eg, Ramsar Convention, note 50 and note 74; see also Ramsar Convention Secretariat, note 51.
[95]   Ramsar Convention, *Ramsar Information Paper no. 6: The Montreux Record and the Ramsar Advisory Missions* (2007).
[96]   Ibid.
[97]   Ramsar Convention, *Resolution IX.6: Guidance for addressing Ramsar sites or parts of sites which no longer meet the Criteria for designation* (2005) [15].

4.1 The Montreux Record

The Montreux Record is a list of Ramsar Sites where changes in ecological character have occurred, are occurring, or are likely to occur as a result of human activities.[98] The Record was created by resolution 'to identify priority sites for positive national and international conservation attention.'[99] The Record is a useful tool in the following circumstances:

> a) demonstrating national commitment to resolve the adverse changes would assist in their resolution; b) highlighting particularly serious cases would be beneficial at national and/or international level; c) positive national and international conservation attention would benefit the site; and/or d) inclusion on the Record would provide guidance in the allocation of resources available under financial mechanisms.[100]

The Montreux Record is a voluntary mechanism to help respond to changes in ecological character. A Site can only be listed at the request of a contracting party, and once the changes have been addressed, the Site can be removed from the Montreux Record through a special procedure.[101] While the Montreux Record is a mechanism to bring attention to Sites at risk, it is neither a punitive measure nor one intended to mandate compliance with the Convention. However, its use by the contracting parties has declined in recent years, suggesting that it is viewed with negative connotations. While, as of September 2023, 46 Ramsar Sites were listed on the Montreux Record,[102] only two had been added since 2017.[103]

---

[98] Ramsar Convention, note 95.

[99] Ibid; see Ramsar Convention, *Resolution 5.4: The Record of Ramsar sites where changes in ecological character have occurred, are occurring, or are likely to occur (Montreux Record)* (1993). See Ramsar Convention, *Resolution XII.6: The status of Sites in the Ramsar List of Wetlands of International Importance* (2015), for the most up-to-date guidance on adding Sites to the Record.

[100] Ramsar Convention, note 88, [19]; Ramsar Convention, note 95.

[101] Ramsar Convention, *Resolution VI.1: Working definitions of ecological character, guidelines for describing and maintaining the ecological character of listed sites, and guidelines for operation of the Montreux Record* (1996) [3.3].

[102] Ramsar Convention, COP14, *Doc 10 Rev 1: Report of the Secretariat pursuant to Article 8.2 on the List of Wetlands of International Importance* (2022) [21].

[103] Ramsar Convention, COP13, *Doc 12: Report of the Secretary General pursuant to Article 8.2 concerning the List of Wetlands of International Importance* (2018) [27].

4.2　Ramsar Advisory Missions

In the early years of the Convention, the COP discussed and made recommendations regarding specific Ramsar Sites where adverse ecological change was either threatened or actually occurring. For example, COP3 adopted a recommendation concerning the Azraq wetland in Jordan, calling for an environmental impact assessment (EIA) of the Amman Water Authority's pumping and suggesting that extractions be reduced by half.[104] COPs rarely focus on individual sites now, with such technical advice and guidance being provided through RAMs.[105]

RAMs are independent technical missions organised by the Secretariat at the request of a contracting party to help address issues at Sites where changes in ecological character have occurred, are occurring, or are likely to occur.[106] RAMs provide Recommendations through a report which, after being approved by the contracting party, is made publicly available on the Convention's website.[107] RAMs can only be sent to a contracting party's territory upon its request and it has discretion as to whether and how to adopt a RAM's recommendations.[108] As Stroud et al observe, RAMs have contributed to positive outcomes at Ramsar Sites, such as at Chilika Lagoon in India and the Ouse Washes in the UK.[109] COP13 recognised the RAM mechanism as 'a useful tool … to assist with implementation of the Convention' and encouraged its greater use.[110]

At COP14, the Plenary again turned to the status of specific Ramsar Sites in a resolution that condemned Russian aggression in Ukraine.[111] (Chapter 2

---

[104] Ramsar Convention, *Recommendation 3.8: Conservation of Azraq Ramsar site* (1987).

[105] Stroud et al, note 8, 1114.

[106] RC Gardner et al, 'Ramsar Advisory Missions: Technical Advice on Ramsar Sites' *Ramsar Briefing Note 8* (Ramsar Convention Secretariat, Gland, Switzerland, 2018) 2. Up to 2022, the RAM mechanism had been used 97 times. Ramsar Convention, *List of Ramsar Advisory Missions* (2022).

[107] RAM reports are available at https://www.ramsar.org/search?sort_bef_combine=search_date_DESC&f[]=document_type%3A2906.

[108] Gardner et al, note 106, 3–4.

[109] Stroud et al, note 8, 1114.

[110] Ramsar Convention, *Resolution XIII.11: Ramsar Advisory Missions* (2018) [9].

[111] Ramsar Convention, *Resolution XIV.20: The Ramsar Convention's response to environmental emergency in Ukraine relating to the damage of its Wetlands of International Importance (Ramsar Sites) stemming from the Russian Federation's aggression* (2022).

notes that the COP departed from its tradition of consensus in this regard.) The Resolution recognised 'the devastating impact of the Russian Federation's aggression' on Ukraine's environment, 'including the disruption of the ecological status of 16 Ramsar Sites and potential damage to another 15 Ramsar Sites within Ukraine'.[112] The Secretariat was requested to provide a damage and mitigation report to COP15.[113]

4.3 Ecological Character and the International Court of Justice

The Ramsar Convention played a significant role in the ICJ case *Certain Activities carried out by Nicaragua in the Border Area (Costa Rica v Nicaragua)*.[114] The dispute involved two Ramsar Sites: *Humedal Caribe Noreste* in Costa Rica and *Refugio de Vida Silvestre Río San Juan* in Nicaragua.[115] Costa Rica alleged that Nicaragua had violated its territorial sovereignty through a military presence in Costa Rica's Ramsar Site, and that a Nicaraguan dredging project was adversely affecting its ecological character.[116] Costa Rica also made an Article 3.2 report to the Secretariat and requested a RAM.[117] In turn, Nicaragua claimed that Costa Rica's road construction was negatively affecting Nicaragua's Ramsar Site and also made an Article 3.2 report, requested a RAM, and instituted an ICJ action.[118] The ICJ consolidated the two cases.

---

[112] Ibid, [12].

[113] Ibid, [18].

[114] See *Certain Activities Carried Out by Nicaragua in the Border Area (Costa Rica v Nicaragua), Judgment of 16 December 2015*, ICJ Reports 2015, 664.

[115] JK Cogan 'International Decisions: Certain Activities Carried Out by Nicaragua in the Border Area (Costa Rica v. Nicaragua); Construction of a Road in Costa Rica Along the San Juan River (Nicaragua v. Costa Rica)' (2016) 110 *American Journal of International Law* 320, 320–21.

[116] *Certain Activities Carried Out by Nicaragua in the Border Area (Costa Rica v Nicaragua), Application Instituting Proceedings*, ICJ Reports 2010, 18.

[117] Ramsar Convention Secretariat, *Notification: Ramsar Advisory Mission to the Humedal Caribe Noreste Ramsar Site, Costa Rica* (2010) at: https://www.ramsar.org/sites/default/files/documents/tmp/pdf/diplomatic_notes/DN2010-8E.pdf.

[118] T Jones and D Pritchard, *STRP Task 4.2 – Comprehensive review and Analysis of Ramsar Advisory Mission (RAM) reports* (2018) 14, at: https://www.ramsar.org/sites/default/files/documents/library/review_analysis_ram_reports_e.pdf; *Construction of a Road in Costa Rica Along the San Juan River (Nicaragua v Costa Rica) Application Instituting Proceedings*, ICJ Reports 2011.

In March 2011, the ICJ issued a provisional order requiring Nicaragua and Costa Rica to remove all their personnel from the disputed area.[119] However, Costa Rica was permitted to send civilian personnel responsible for environmental protection to the area, but only to avoid irreparable harm to the Ramsar Site and only in consultation with the Ramsar Secretariat.[120] Consequently, the Secretariat organised a RAM, which concluded that the site's ecological character was being affected and recommended that it be placed on the Montreux Record.[121] After Nicaragua continued excavation work in the area, in November 2013 the ICJ issued a new order again calling on Nicaragua to cease its dredging activities and permitting Costa Rican environmental officials to enter the area after consultation with the Ramsar Secretariat.[122] This dispute marked the first time that the ICJ, when issuing orders, relied on the expertise of the Secretariat, underscoring the Ramsar Convention's importance to wetland protection. In the end, the ICJ upheld Costa Rica's territorial claims.[123]

The case had broader significance for international environmental law in two other respects. First, the ICJ discussed at length the duty under customary international law to prepare EIAs when a proposed activity implicates transboundary harm, building on the *Pulp Mills* case.[124] Second, the ICJ required Nicaragua to pay Costa Rica to compensate for environmental harm caused – the first time that it issued an environmental compensation decision.[125]

---

[119] *Certain Activities Carried Out by Nicaragua in the Border Area (Costa Rica v Nicaragua), Order of 8 March 2011*, ICJ Reports 2011, 25.

[120] Ibid.

[121] *Certain Activities Carried Out by Nicaragua in the Border Area (Costa Rica v Nicaragua), Reply of Costa Rica on Compensation*, ICJ Reports 2017, 168–70.

[122] *Certain Activities Carried Out by Nicaragua in the Border Area (Costa Rica v Nicaragua), Construction of a Road in Costa Rica Along the San Juan River (Nicaragua v Costa Rica), Request Presented by Costa Rica for the Indication of New Provisional Measures, Order of November 2013*, ICJ Reports 2013, [59].

[123] *Certain Activities Carried Out by Nicaragua in the Border Area (Costa Rica v Nicaragua)*, note 114, [229].

[124] J Bendel and J Harrison, 'Determining the legal nature and content of EIAs in International Environmental Law: What does the ICJ decision in the joined Costa Rica v Nicaragua/Nicaragua v Costa Rica cases tell us?' (2017) 42 *Questions in International Law* 13, 14.

[125] IUCN, 'ICJ Renders First Environmental Compensation Decision: A Summary of the Judgment' (2018) at: https://www.iucn.org/news/world-comm-ission-environmental-law/201804/icj-renders-first-environmental-compensation-decision-summary-judgment.

5   THE BENEFITS OF RAMSAR DESIGNATION

Surveys of Ramsar Site managers in Africa, Canada, and the United States identified multiple benefits associated with Ramsar designation, chief among them increased support for protection of the Sites and surrounding areas.[126] For example, Ramsar Site managers in Canada found that Ramsar designation contributed to maintaining a Site's ecological character by communicating to the public and decisionmakers the importance of the Site; influencing environmental assessments, land use plans and development projects; and improving management through a focus on long-term ecological conditions.[127]

The surveys of Ramsar Site managers also found that the designation helped with enhancing public awareness of Sites and wetlands in general.[128] Furthermore, the international recognition can help to raise the morale of Site staff. One US Site manager reported that the Ramsar 'designation is the highest honor a wetland area can receive, and it reinforces the feeling of pride in the site, by the staff and supporters'.[129] Other benefits identified in the surveys include increased opportunities for funding (eg, through grants), ecotourism, and scientific research.[130] Some respondents in Africa also identified poverty alleviation as a benefit, with the Site manager of Songor Lake in Ghana citing a micro-credit funding project that promoted sustainable farming, fishing and kenkey production.[131]

However, it is important to emphasise – as noted by a Mediterranean Wetlands Observatory study – 'that merely placing a site on the Ramsar list does not ensure the conservation of the natural wetland habitats within it'.[132]

---

[126] See P Lynch-Stewart, *Wetlands of International Importance (Ramsar Sites) in Canada: Survey of Ramsar Site Managers 2007* (Canadian Wildlife Service, Environment Canada, Quebec, 2008) at: https://www.ramsar.org/sites/default/files/documents/library/wurc_canada_survey_2007.pdf; Gardner and Connolly, note 13; RC Gardner, KD Connolly, and A Bamba, 'African Wetlands of International Importance: Assessment of Benefits Associated with Designations under the Ramsar Convention' (2009) 21 *Georgetown International Law Review* 257.

[127] Lynch-Stewart, note 126, 30–31.

[128] Ibid, 111; Gardner, Connolly, and Bamba, note 126, 284.

[129] Gardner and Connolly, note 13, 10095.

[130] Ibid, 10095–97; Lynch-Stewart, note 126, 11; Gardner, Connolly, and Bamba, note 126, 287–89.

[131] Gardner, Connolly, and Bamba, note 126, 289–90.

[132] Mediterranean Wetlands Observatory, *Land Cover: Spatial dynamics in Mediterranean coastal wetlands from 1975 to 2005 (Thematic collection, Special Issue #2)* (Tour du Valat, France, 2014) 38.

For example, a study of 172 Ramsar Sites in 74 countries found that from 1970 to 2011, average trends in vertebrate populations showed a 40% increase.[133] Yet when disaggregated by region, the news was more mixed. While trends were positive in temperate Ramsar Sites, located primarily in Europe, tropical Ramsar Sites in the Afrotropics, the Neotropics, and Asia exhibited declines.[134]

A Ramsar designation may not always be welcomed by Indigenous peoples and local communities, however. The African surveys revealed instances where concerns were raised when protected area status restricted former uses of a Site.[135] Chapter 6 examines the evolution of Indigenous peoples and local communities' participation in Ramsar processes in more detail.

6 OTHER SITE-BASED RECOGNITIONS

Over time, Ramsar has recognised other types of site-based designations beyond Ramsar Sites, including Transboundary Ramsar Sites and Wetland City Accreditation. While not expressly provided for in the Convention's text, the former have an implicit connection to international cooperation under Article 5, while the latter relate to Article 3.1's wise use obligation.

6.1 Transboundary Ramsar Sites

Article 5 contemplates international cooperation with respect to wetlands extending over the territories of more than one contracting party, as well as shared water systems and species.[136] Beginning in 2001 with the Domica-Baradla Cave System, which is located in Hungary and Slovakia, the Secretariat formally recognised the concept of a Transboundary Ramsar Site as 'an ecologically coherent wetland [that] extends across national borders' when 'the Ramsar Site authorities on both or all sides of the border have formally agreed to collaborate in its management, and have notified the Secretariat of this intent'.[137] Although establishing a Transboundary Ramsar Site is one way in which a contracting party may implement its Article 5 responsibilities, the label amounts to an acknowledgement of a cooperative management arrange-

---

[133] RC Gardner et al, 'State of the World's Wetlands and their Services to People: A compilation of recent analyses' *Ramsar Briefing Note 7* (Ramsar Convention Secretariat, Gland, Switzerland, 2015) 9.

[134] Ibid, 10.

[135] Gardner, Connolly, and Bamba, note 126, 290.

[136] Convention on Wetlands, note 1, Art 5.

[137] Ramsar Convention, *The List of Wetlands of International Importance* (2023) 1 and 24, at https://www.ramsar.org/sites/default/files/documents/library/sitelist.pdf.

ment. The transboundary designation does not have a distinct legal status and thus 'does not create additional international legal obligations beyond those already imposed by the Ramsar Convention'.[138]

At the time of COP14, there were 22 Transboundary Ramsar Sites, encompassing 65 individual Ramsar Sites across 26 contracting parties.[139] Thus far, they are confined to Europe, which has the bulk, and Africa. This recognition could be more widely used (assuming that the relevant contracting parties are amenable to cooperative management), as Rosenblum and Schmeier have identified more than 300 wetlands that are transboundary in nature.[140]

6.2   Wetland City Accreditation

A more recent development is the creation of a Wetland City Accreditation programme. In 2015, COP12 adopted a resolution submitted by the Republic of Korea and Tunisia to recognise cities that demonstrate a positive relationship with wetlands.[141] Resolution XII.10 outlined eligibility criteria for this voluntary scheme, which included 'one or more Ramsar Sites or other significant wetlands fully or partly situated' in the city's territory, adoption of wetland conservation measures, and a public awareness programme about the value of wetlands.[142] Unlike Ramsar Site designation, the accreditation scheme is subject to a review process where an Independent Advisory Committee reviews nominations and makes selections.[143] Accreditation lasts for two COP cycles (typically six years) and can be renewed for another two COP cycles, after review by the Independent Advisory Committee.[144]

Wetland City Accreditation is proving popular. At COP13, 18 cities from seven contracting parties were recognised under the programme.[145] The cere-

---

[138] RC Gardner, 'Ramsar Convention: Transboundary Ramsar Sites' in CM Finlayson et al (eds), *The Wetland Book I: Structure and Function, Management, and Methods* (Springer Nature, Dordrecht, 2018) 467, 468.

[139] Ramsar Convention, note 83, [7].

[140] ZH Rosenblum and S Schmeier, 'Global Wetland Governance: Introducing the Transboundary Wetlands Database' (2022) 14:3077 *Water* 1, 10.

[141] Ramsar Convention, *Resolution XII.10: Wetland City Accreditation of the Ramsar Convention* (2015).

[142] Ibid, 4.

[143] Ibid, 5.

[144] Ibid.

[145] International Institute for Sustainable Development, 'Summary of the Thirteenth Meeting of the Conference of the Parties to the Ramsar Convention on Wetlands: 22–29 October 2018' (2018) 17:48 *Earth Negotiations Bulletin* 5, at: https://enb.iisd.org/ramsar/cop13.

mony 'was well received and appreciated by Contracting Parties' and was well attended by mayors and representatives from the accredited cities.[146] Similarly, at COP14, 25 more cities from 12 contracting parties were honoured[147] and the event was a ray of positivity in a COP dominated by the Russia-Ukraine conflict. At this rate, there will soon be more accredited Wetland Cities than Ramsar Sites on the Montreux Record, suggesting that contracting parties are more interested in positive recognition, rather than highlighting management challenges. More importantly, the Wetland City Accreditation programme illustrates that the Convention is open to exploring new site-based approaches to deliver the goals of wetland conservation and wise use.

7 CONCLUSION

The drafters of the Ramsar Convention were indeed visionary, putting a mechanism in place to establish internationally designated sites and instilling in contracting parties the obligation to maintain the ecological character of these sites. The network of Ramsar Sites – Wetlands of International Importance – began largely as a Europe-based effort but has grown to be truly global. Although the World Heritage Sites network now eclipses the Ramsar Sites network in terms of total area of coverage, the Ramsar Convention remains responsible for the most extensive list of sites focusing on wetland conservation, which has contributed significantly to progress on meeting some global biodiversity targets.

The Convention is, however, beginning to show its age, with gaps in implementation becoming apparent. Approximately half of all Ramsar Sites lack a management plan, which does not bode well for maintaining their ecological character. Unlike later international regimes that require a periodic review for a designation to remain valid, the Ramsar Convention has no such formal process. Instead, it relies on its contracting parties to monitor and report. But the contracting parties often fail to provide the Ramsar Secretariat with updated RISs, which should be done at least every six years. Many contracting parties also appear hesitant (or do not have proper procedures in place) to report changes in ecological character at Ramsar Sites, underscored by the number of NGOs that independently contact the Secretariat to alert it of threats to Ramsar Sites. The Montreux Record, which is designed to bring international attention to Ramsar Sites with adverse changes in ecological character, has fallen into disuse. While Ramsar COPs no longer take collective action to

---

[146] Ramsar Convention, SC57, *Doc. 26: Wetland City Accreditation: Guidance for the 2019–2021 triennium* (2019) 2.

[147] International Institute for Sustainable Development, note 34, 3.

offer recommendations on individual Ramsar Sites (with the notable exception of COP14's Resolution on Ukrainian Ramsar Sites), RAMs are a useful tool to assist contracting parties in responding to adverse change in Ramsar Sites. Nevertheless, RAMs could be better utilised across all Ramsar regions.

Although the development of other site-based designations, such as the Wetland City Accreditation programme, can promote wetland conservation, the core obligation of protecting and strengthening Ramsar Sites should not be overlooked. The designation of Ramsar Sites was a cornerstone of the Convention for its first 50 years. Effective management of these wetlands must be a focus for its next half-century.

4. The 'wise use' of wetlands
Edward J. Goodwin

1 INTRODUCTION

Since the adoption of the Ramsar Convention in 1971,[1] the wise use of wetlands has seemingly developed into the lodestar towards which the endeavours of conservationists and other wetland stakeholders have been set. This represents an alternative strategic emphasis for Ramsar, since – as is asserted in this Chapter – its drafters had originally been primarily motivated by the delivery of another strategy for the conservation of wetlands. Nevertheless, under Article 3.1, contracting parties to the Convention committed themselves to promote, 'as far as possible, the wise use of wetlands in their territories' through their 'formulation and implementation of planning'.[2] 'Wise use' is now referenced in countless resolutions, recommendations, and guidelines emanating from the Conference of the Parties (COP);[3] yet despite its modern reach within Ramsar, establishing the precise contours of this type of utilisation can seem challenging. However, a clear and workable framing of 'wise use' is possible when it is conceptualised as taking the form of an obligation of conduct associated with a specific objective, as opposed to an obligation of result.[4]

This Chapter takes a fresh approach to establishing this framing by utilising a historical contextualisation of the treaty's negotiations, while also benefiting from a newly released tranche of draft documents used in the negotiations of

---

[1] Convention on Wetlands of International Importance especially as Waterfowl Habitat adopted 2 February 1971, entered into force 21 December 1975, 996 UNTS 245 (amended 1982 & 1987).
[2] Ibid, Art 3.1: 'The Contracting Parties shall formulate and implement their planning so as to promote the conservation of the wetlands included in the List, and as far as possible the wise use of wetlands in their territory.'
[3] Ibid, Art 6.
[4] For an illuminating account of the distinction, drawing upon a civil law tradition, see B Mayer, 'Obligations of Conduct in the International Law of Climate Change: A Defence' (2018) 27 *Review of European, Comparative and International Environmental Law* 130.

Ramsar.[5] This places 'wise use' within the post-war movement that sought to achieve global acceptance of the idea that conservationism should be mutually supportive of states' development. However, the negotiations also played out within a decades-old conservation tradition of utilising protected enclaves to secure natural areas, originally in a state minimally disturbed by humankind. This fact is drawn upon to propose an understanding of wise use's 'bedfellow' in Article 3.1 – namely, the 'conservation' of listed Wetlands of International Importance.[6] Thus, while Ramsar signals the beginning of a more modern approach to multilateral legal regulation of the environment,[7] it was also formulated during a significant period of change for conservation science. While caught in the confluence of these currents, and reflecting the two strands of conservation practice concerning development and preservation within enclaves, Ramsar generated a core set of obligations in Article 3.1 that today, on first reading, can seem hard to reconcile. This Chapter offers a new reconciliation.

The Chapter begins by charting the insertion of 'wise use' into the draft convention and considers the development of its content and meaning up until today. 'Conservation' of listed wetlands is addressed later as a counterpoint to wise use.

2   WISE USE THEN AND NOW

2.1   Rising Waters

Southern Africa, November 1860: '[A]t Kariba, a basaltic dike, called Nakabele, with a wide opening in it, dangerous only for canoes, stretches like an artificial dam across the stream.'[8]

The replacement of this natural barricade against the waters of the Zambezi River with an artificial dam was completed almost a century after David Livingstone recorded this sight. After a number of years spent in construction, the Kariba Dam was finally sealed at the start of December 1958.[9] As a source

---

[5]   Provided as supplementary material in DA Stroud et al, 'Development of the Text of the Ramsar Convention 1965–1971' (2022) 73 *Marine and Freshwater Research* 1107.

[6]   For text of Article 3.1, see note 2.

[7]   MJ Bowman, PGG Davies, and C Redgwell, *Lyster's International Wildlife Law* (Cambridge University Press, Cambridge, 2nd edition, 2010) 13 (hereinafter, 'Lyster').

[8]   D Livingstone and C Livingstone, *Narrative of an Expedition to the Zambezi and its Tributaries* (Harper & Brothers, New York, 1866) 352.

[9]   'Less Water for Wildlife', 1 December 1958, *The Times*, 11.

of thousands of megawatts of electricity to what is now Zambia and Zimbabwe, the project was indicative of mid-twentieth century efforts to deliver development within post-war, and eventually post-colonial, African states.

However, there were consequences. There was the forced resettlement of over 50 000 Tonga people living near the banks of the Zambezi – in Thayer Scudder's words, refugees from the development.[10] And they were not the only refugees. Even in the final days as the last boulders were sealing the dam, *The Times* reported concerns about the imminent eight-month reduction in downstream water levels driving hippopotami to raid gardens and crops.[11] However, an international scandal was about to develop upstream. In mid-February 1959, reports began to surface of animals being marooned on newly formed islands.[12] The Southern Rhodesian government, under opposition pressure, implemented a programme of using tranquiliser darts to move larger animals.[13] But in Northern Rhodesia, the government was less responsive to the plight.[14] In March, the London-based Fauna Preservation Society (FPS) launched what became known as 'Operation Noah' to raise £10 000 to fund rescue units to assist with relocating captured animals.[15] Money and equipment were soon being donated from the United States, the United Kingdom, and South Africa.[16]

As William Adams states, the Kariba Dam was important since 'it symbolised the impacts of development to the environment'.[17] These impacts had been keenly felt some 50 years previously, as the rapid expansion of colonial development in Africa prompted conservationists to scramble to establish game reserves.[18] These early responses seemed to assume that 'nature conservation and economic development were inherently inimical to one another';[19] but in the United States, a different perspective had taken hold. There, the question was framed as a contest between preservation and conservation, with John Muir advocating the management of natural areas under the former,

---

[10] T Scudder, 'Pipe Dreams: Can the Zambezi River supply the Region's Water Needs?' 31 July 1993, 17(2) *Cultural Survival Quarterly* at: https://web.archive.org/web/20070927222405/http://209.200.101.189/publications/csq/csq-article.cfm?id=971.
[11] 'Less Water for Wildlife', note 9.
[12] 'Rescue of Marooned animals', 19 February 1959, *The Times*, 9.
[13] Ibid; WM Adams, *Against Extinction: The Story of Conservation* (Earthscan, Abingdon, 2004) 174.
[14] Ibid.
[15] Ibid; see also 'Saving Animals from Flood', 23 March 1959, *The Times*, 6.
[16] See R Reynolds, 'Rescuing Animals at Kariba', 4 June 1959, *The Times*, 4.
[17] Adams, note 13, 175.
[18] Ibid, 159.
[19] Lyster, note 7, 16.

which involved little to no industrial profit from such land.[20] However, Gifford Pinchot – the first Chief of the US Forest Service – favoured a utilitarian interpretation of 'conservation', which he defined as 'the wise use of the earth and its resources for the lasting good of men'.[21]

Indeed, Pinchot reworked Bentham's utilitarian maxim in stating that 'conservation' should be 'the greatest good to the greatest number for the longest time'.[22] Pinchot's view was favoured by President Theodore Roosevelt in the early twentieth century, as reflected in such decisions as the approval of the Hetch Hetchy Valley dam,[23] and later by President Franklin D Roosevelt as one of the bases for post-war peace.[24] Pinchot devoted his later years pushing for a global conference to discuss conservation as a pillar of peace and development; but it was not until shortly after his death that the Economic and Social Council to the United Nations (UN) accepted a US offer to hold a scientific conference looking at techniques for resource conservation and development.[25] Held in 1949 at Lake Success, the United Nations Scientific Conference on the Conservation and Utilization of Resources largely mirrored the approach of Pinchot and the United States to conservation. In his opening address, Julius Krug stated that the cause of the conference was 'the improvement of man's standard of living, particularly in the under-developed areas of the world, through the protection and wise use of man's common heritage ... Conservation and wise development of our resources would help ensure world peace'.[26]

While Aldo Leopold and John Krutilla worked to improve the calculations that underlay Pinchot-style utilitarian calculations so as to build in values held by preservationists,[27] the sense that 'conservation' of natural resources meant

---

[20] HS Banzhaf, 'The Environmental Turn in Natural Resource Economics: John Krutilla and "Conservation Reconsidered"' (2019) 41(1) *Journal of the History of Economic Thought* 27; R Hudson Westover, 'Conservation versus Preservation?', 22 March 2016, US Dept of Agriculture/Forest Service at: https://www.fs.usda.gov/features/conservation-versus-preservation.

[21] G Pinchot, *Breaking New Ground* (Harcourt Brace, New York, 1947) 505.

[22] Ibid, 327.

[23] See the address of Cornelia Bryce Pinchot in United Nations, *Proceedings of the United Nations Scientific Conference on the Conservation and Utilization of Resources Volume I: Plenary Meetings* (1950), 318–19, at: https://babel.hathitrust.org/cgi/pt?id=mdp.39015027769119&view=1up&seq=75&q1=wise.

[24] J McCormick, *Reclaiming Paradise: The Global Environment Movement* (Indiana University Press, Bloomington and Indianapolis, 1991) 25.

[25] Ibid, 27.

[26] *Proceedings*, note 23, 5.

[27] See Banzhaf, note 20.

wise management to support peace, poverty reduction, and present and future development started to gain a global toehold as negotiators began drawing up new international regulations for the conservation of nature. For example, the 1968 African Convention on the Conservation of Nature and Natural Resources ('Algiers Convention')[28] was a product of a 1953 international conference held in Bukavu in the Belgian Congo.[29] The conference called for a new convention to reflect the modern needs of Africans.[30] The eventual Algiers Convention replaced a previous treaty negotiated by the colonial powers in 1933 that had focused on strict game protection within reserves.[31] The new Convention better reflected the post-war thinking on conservation and was legally innovative in seeking to integrate conservation of species, habitats, water, and soil with development.[32] It did, however, seem to disassociate conservation and utilisation in a way that Pinchot's conception had avoided. For example, Article VII(1) obliged states to 'ensure conservation, wise use and development of faunal resources and their environment'.[33] At best, it seems that in the 1960s, legal terminology had not yet settled into a clear form that reflected the reconception of 'conservation' as including utilisation.

2.2   Diminishing Waters

It was in these currents that the negotiations for Ramsar were taking place. The problem for wetland stakeholders was not so much rising waters as the removal of water from wetlands through excessive extraction or total drainage to convert them into agricultural land.[34] This was exacerbated by public opinion, which saw wetlands as waste and breeding grounds for life-threatening diseases.[35] Such ill-judged opinions and responses led to the establishment of Project

---

[28]   Adopted 15 September 1968, entered into force 16 June 1969, 1001 UNTS 3.

[29]   Adams, note 13, 167–68.

[30]   Ibid, 170.

[31]   1933 Convention Relative to the Preservation of Fauna and Flora in their Natural State, adopted 8 November 1933, entered into force 14 January 1936, 172 LNTS 241.

[32]   Algiers Convention, note 28, Art XIV(1); Lyster, note 7, 264–65.

[33]   Algiers Convention, note 28.

[34]   GVT Matthews, *The Ramsar Convention on Wetlands: Its History and Development* (Ramsar Bureau, Gland, 1993) 6.

[35]   Ibid; see also MJ Bowman, 'The Ramsar Convention on Wetlands: Has it Made a Difference?' in O Schram Stokke and ØB Thommessen (eds), *Yearbook of International Cooperation on Environment and Development 2002–2003* (Routledge, London, 2002) 61.

MAR[36] by the International Union for the Conservation of Nature (IUCN) in association with the International Waterfowl and Wetlands Research Bureau (IWRB) and the International Council for Bird Protection, and the convening of the MAR Conference in 1962.[37] The MAR Conference issued a number of recommendations, including Recommendation IX, which called for IUCN to compile a list of wetlands of international importance to serve 'as a foundation for an international convention on wetlands'.[38]

Recommendation IX ultimately led to the generation of the MAR list of internationally important wetlands in Europe and North Africa ('MAR List'),[39] and the seeds of an international treaty would be built on this inventory.[40] That said, the MAR Conference contemplated the list as being a foundation and did not rule out other dimensions to a treaty, such as the application of wise use to a wider set of wetlands.[41]

2.3 Agreeing to the Wise Use of Wetlands (1965–71)

A review of the negotiating documents reveals a sense that the parties were less convinced of the need for all wetlands to fall under the Convention or clear about the possible form of the obligation. As described below, general provisions are included intermittently and technical terms deployed without particular care.

The IWRB took the original initiative, elaborating a first draft text in August 1965, alongside a list of eight proposed subjects for inclusion in the convention in October of the same year.[42] Of that subject list, two related to artificial wetlands and four to the MAR List, while two suggested that wetlands in a wider sense needed addressing. Thus, it called for the following:

1. An agreed statement on the designation and utilisation of Wetlands in modern countries and upon the need for their safeguarding and management in order

---

[36] Matthews, note 34, 7. 'MAR' is derived from the word for 'marshes' in various languages: 'marshes' (English), *marécages* (French) and *marismas* (Spanish); see the introduction to this volume.

[37] Ibid.

[38] IWRB/MAR Bureau, *Project MAR: The Conservation and Management of Temperate Marshes, Bogs and other Wetlands Volume 2* (IUCN, Gland, 1965) 7.

[39] See generally ibid.

[40] Matthews, note 34, 12. See also Chapter 3 in this volume.

[41] Stroud et al, note 5.

[42] Ibid, Files S1 (*Draft for an International Convention on the Conservation of Wetlands*, 24 August 1965) and S2 (*Proposed Subjects for an International Agreement or Convention on Wetlands*, October 1965).

to allow their rational use in a fair balance between the interests of nature conservation, hunting and other recreational, educational, scientific and economic needs ...

6. An undertaking to give neither consent nor subsidies for drainage, in-filling, or modifications of Wetland areas, including those not classified in the MAR List, before a detailed inquiry on the recreational, educational, scientific and economic value of the Wetlands in question has been made by consultation with competent ecologists and other appropriately qualified specialists.[43]

From the outset, the IWRB was seemingly keen to promote not just a form of conservation that operated in tandem with development and other human use (which, as was seen earlier, is consistent with developments within the broader direction of conservation thinking), but importantly also one that ruled out the foolish and irrational policies exemplified by the draining and converting of wetlands.

The first draft text duly included an article to cover (almost word for word) the known instances of what was regarded as 'unwise' use as per Point (6) above;[44] however, it did not go so far as to generally call for 'rational use'. Instead, it rather vacuously stated that '[w]etland areas not classified in the MAR LIST deserve also protection'.[45]

Over 20 states gathered at a conference at Noordwijk, the Netherlands, in May 1966 to consider the draft. Misgivings were expressed about international restrictions on land use policies, as well as a general embargo on draining of wetlands, but there was general support for the development of national wetland plans.[46]

The Dutch government was charged with producing further drafts of the Convention, which it duly did, circulating a 21-article version in October 1967.[47] This draft was entirely focused on maintaining an inventory of wetland reserves in an annex to the Convention and a complex system for notifying a central commission about changes in the ecological character to those listed wetlands.[48] Matthews intimates that the annex was, in practice, to be the MAR List.[49] Consequently, a general, 'wise use'-style obligation applicable to all wetlands had been omitted by the time of this second draft.

Following general criticism of this draft and a meeting held at IUCN headquarters in late 1967, the IWRB board published newly agreed points that

---

[43] Ibid, File S1.
[44] Ibid, File S2, Art 5.
[45] Ibid.
[46] Matthews, note 34, 14–15.
[47] Stroud et al, note 5, File S3.
[48] Ibid.
[49] Matthews, note 34, 15.

the Netherlands was asked to reflect in a second draft for the Convention.[50] This document sketches out the now familiar dual focus on listed wetlands utilising the MAR List and wetlands generally. While stating that the aim of the Convention should be the 'conservation of wetlands in the widest sense', it goes on to say that the 'main provisions of the convention should include that: 1(a) The contracting Governments should have a general policy for the conservation and management of their wetlands'.[51] With the MAR List being the focus of a proposed provision to 'preserve and manage' such areas,[52] it might appear that the wider wetlands undertaking was to be framed as closer to a sustainable use form of conservation. However, there are grounds to suspect that terms such as 'conservation' and 'preservation' were not being used particularly carefully at this point or with any specific content in mind, since the overarching priority was to encourage governments to institute national wetland policies.

The second Dutch draft provided in Article 2.1 that each contracting party would organise its 'nature conservation policy in such a way as to ensure the conservation, the expert management and the due protection of the wetlands within its territory'.[53] These policies would need to include regulations for conservation, management, and protection, including regulations that would see the contracting party become aware within good time of 'any proposed changes in the ecological nature of the wetlands within its territory'.[54]

The intention of this second draft, as openly acknowledged in its accompanying notes, was to produce a convention that principally protected the wetlands on the MAR List, while imposing as little interference as possible on national autonomy, preferring instead international consultation.[55] Draft Article 2.1 focused primarily on the latter concern, which:

> would be a logical consequence of the creation of an international consultative body, which after all would be incapable of producing results unless the States were actively and continuously concerned ... to promote the interests that the Convention itself sought to serve. National policy in its turn would have little effect unless it were integrated in a general nature conservancy policy that was part of the wider concept of physical planning. It can hardly be doubted that the last two aspects of

---

[50] Ibid, 16–17.
[51] Stroud et al, note 5, File S4.
[52] Ibid, Proposal 2(a).
[53] Ibid, File S6.
[54] Ibid, Art 2(2).
[55] Ibid, File S7 (*Introduction to the draft of an international convention on wildfowl and wetlands*) 1–2.

national planning are amongst the most important of the Government activities that may favour the preservation of wetlands.[56]

Here, 'conservation' is used in the draft and 'preservation' in the explanatory document, suggesting that these terms were not applied deliberatively. The main intention seems to have been to get states to devise national policies to support international consultation. Even then, the convention was not to be unduly prescriptive:

> The draft goes no further than the acceptance of the principle stated ... nor does it give directives on the measures the parties are to take ... The draft lays down the barest minimum in this respect ... Therefore it would hardly seem possible to institute international interference with national policies except by way of international consultation ... Only where wetlands of international importance are concerned will the international community take action on its own initiative, and then only in an advisory capacity (Articles 7–9).[57]

The draft itself did not define 'Wetlands of International Importance', but it is clear from the draft that Articles 7–9 refer to a list of wetlands, which appears to be the MAR List. The centrality of an inventory of internationally important sites and a link between this and an increased oversight role of the international community is explicitly set out at this point. Indeed, that centrality would once again be reflected in the next draft that appeared, since the Soviet draft of February 1969 omitted provision for any commitments applicable to wetlands generally, focusing only on an inventory of internationally important wetlands.[58]

The second Dutch draft had been intended for discussion at a conference in the Soviet Union in the autumn of 1968, but the invasion of Czechoslovakia that August prompted multiple countries to withdraw from the planned event.[59] The IWRB – now chaired by Geoffrey Matthews – had to consider the two drafts in hand from the Netherlands and Russia, the review of which was duly completed in May 1969.[60] The Dutch draft was largely favoured and a series of broad directions were passed to Mr Panis of the Dutch Ministry of Culture, Recreation and Social Services for incorporation into a new draft.[61] Of note for current purposes were two observations: that the previous Dutch provisions

---

[56] Ibid, 4.
[57] Ibid.
[58] Ibid, File S8.
[59] Matthews, note 34, 18–19.
[60] IWRB, *Record of Points of Discussion and Decision* (20 May 1969), available at Stroud et al, note 5, File S9.
[61] Matthews, note 34, 20.

on the responsibility for states to conserve and manage all wetlands in their territory had to be included in the draft text (in a reversal from the Russian draft); and further, that 'conservation' was to be preferred over 'preservation'.[62] The latter is perhaps unsurprising given the movement within conservation science away from preservation towards a harmonious form of development that maintained the potential value of natural resources for future generations.

In December 1969, another Dutch-derived draft convention was circulated.[63] A general commitment to both listed and unlisted wetlands was included under Article 3, but it is at this point that the problems highlighted in the introduction can be seen. Draft Article 3.1 stated: 'Each contracting party undertakes to organise its nature conservation policy in such a way as to ensure the conservation, positive management and wise utilisation of the wetlands within its territory.'[64]

In this draft, the suggestion seems to be that 'conservation' and 'wise use' are distinct actions – albeit that, at least at this stage, this provision was applicable to both listed and unlisted wetlands. Like the contemporaneous Algiers Convention, this bucked the trends of the time, whereby 'wise use' was part of 'conservation' according to the Pinchot definition and previously described moves to build development and utilisation (in a sustainable form) into the notion of 'conservation'. Matthews' own account of the negotiations casts no further light on this drafting;[65] while Stroud et al do not recognise draft Article 3.1 as being a forerunner for a wider obligation of wise use.[66]

Between this draft and the final draft, drawn up in the light of a meeting of international experts in Helsinki in March 1969 and presented at the Ramsar Conference in February 1971, final draft Article 3.1 had become an undertaking by states to 'design and implement their conservation policy in such a way as to ensure the conservation and rational management of wetlands in their territory, particularly of those included in the List'.[67]

Thus, 'wise utilisation' had now disappeared from the text submitted before the delegates who met at Ramsar. The cover note explaining the final draft fails to illuminate the context for this formulation.[68]

---

[62] Stroud et al, note 5, File S9.
[63] Ibid, File S11.
[64] Ibid.
[65] Indeed, it only speaks of the new Article 3 addressing conservation of wetlands. Matthews, note 34, 23.
[66] Stroud et al, note 5, 1114.
[67] Ibid, File S13.
[68] Ibid, File S14, 4.

As is now well known, the Ramsar Conference adopted the final text of Article 3.1 as follows: 'The Contracting Parties shall formulate and implement their planning so as to promote the conservation of the wetlands included in the List, and as far as possible the wise use of wetlands in their territory.'

How and why this final draft transformed into the adopted form of Article 3.1 – with its apparent differentiation between 'conservation' and 'wise use', and between listed wetlands and wetlands in general – is unrecorded. Matthews described how Sir Hugh Edwards explained the new wording on the fourth day, and that this was adopted without contention by the states, but specifics about Edwards' explanation are not given.[69] The provision was therefore left in a rather awkward position – not least because the Convention also failed to provide definitions of both 'conservation' and 'wise use'. This was left to later decisions of the parties to Ramsar.

2.4   Subsequent Clarification on Wise Use (1971–present)

As is well understood, the Ramsar Convention included undertakings by the contracting parties that related to all wetlands (as defined), but with a handful of additional commitments for those wetlands designated on the List of Wetlands of International Importance.[70] The latter are covered towards the end of this Chapter, while the former amount to establishing nature reserves;[71] encouraging research regarding wetlands and related flora and fauna;[72] promoting the training of personnel competent in the fields of wetland research and management;[73] and cooperating with other contracting parties with respect to transboundary wetlands.[74]

Of course, foremost among the undertakings relating to all wetlands is that of 'wise use'. What has been seen so far is that in 1971, the precise contours of 'wise use' under Ramsar were poorly framed by the negotiating parties. What is more, the early to mid-twentieth century notions of enduring utilisation of natural resources were only explicitly recognised by the contracting parties as a dimension of 'wise use' some 16 years after the adoption of the provision.

In 1987, at the third COP held in Regina, 'wise use' was defined as the sustainable utilisation of wetlands for the benefit of humans, but compatible with

---

[69]   Matthews, note 34, 29.
[70]   See Convention on Wetlands, note 1, Art 2.
[71]   Ibid, Art 4.1.
[72]   Ibid, Art 4.3.
[73]   Ibid, Art 4.5.
[74]   Ibid, Art 5.1.

maintaining the natural properties of the wetland ecosystem.[75] It cannot be known for sure whether this was always the intention of the negotiating parties, but it is (as described in Section 2.1 of this Chapter) certainly consistent with the early popularisation of environmental conservation efforts needing to work in harmony with development. Sustainable utilisation and development would become far more embedded and dominant within conservation practice following the World Conservation Strategy (1980), the World Charter for Nature (1982), and the Brundtland Report (1987).[76]

Matthews explains that 'wise use' was always meant to encapsulate sustainable exploitation.[77] This is the only hint from someone involved at the time of the adoption of the provision that this was the intended objective of Article 3.1 – although he was then writing with the benefit of hindsight after the 1987 definition. He further describes how, up until the 1950s, the protectionist approach had prevailed, which implied excluding human activity; but that with a growing realisation that most natural areas had already been modified by humankind, conservation became popular, which meant maintenance by well-informed management.[78] This, as we have seen, is broadly correct, though it ignores the catalysing effect of post-war African development.

However, the Ramsar parties subsequently adjusted the definition of 'wise use' because of later multilateral developments. The 1992 Convention on Biological Diversity brought to greater prominence ecosystem approaches and sustainable use;[79] while the Millennium Ecosystem Assessment did the same for ecosystem services.[80] The Ramsar Convention itself lent some prominence to the ecology of wetlands, since Article 3.2 obliges states to monitor for changes in the ecological character of listed wetlands. Following a request to the Convention's Scientific and Technical Review Panel (STRP), updated definitions of 'wise use' and 'ecological character' were introduced in 2005.[81]

---

[75] Ramsar Convention, *Recommendation 3.3: Wise Use of Wetlands* (1987).
[76] Lyster, note 7, 15–17, 55–59.
[77] Matthews, note 34, 53.
[78] Ibid.
[79] Convention on Biological Diversity, adopted 5 June 1992, entered into force 29 December 1993, 31 ILM 851. On the ecosystem approach, see Decision V/6 of the COP; and on sustainable use, see Article 2. See Chapter 8 of this volume.
[80] Millennium Ecosystem Assessment, *Ecosystems and Human Well-being: Synthesis* (Island Press, Washington, DC, 2005).
[81] Ramsar Convention, *Resolution IX.1: Additional scientific and technical guidance for implementing the Ramsar wise use concept* (2005). See also Chapter 3 of this volume.

This development provides us with greater clarity[82] and, as David Pritchard has convincingly argued, tied the maintenance of ecological character into the objectives for the management of all wetlands.[83]

'Wise use' of wetlands now means 'the maintenance of their ecological character, achieved through the implementation of ecosystem approaches, within the context of sustainable development'.[84] In turn, 'ecological character' means 'the combination of the ecosystem components, processes and benefits/services that characterise the wetland at a given point in time'.[85]

This resulted in 'wise use' being aligned with the Brundtland Commission's definition of 'sustainable development'.[86] What might, therefore, have been inferred in 1971 as the meaning of 'wise use' in Article 3.1 has come to be explicitly recognised as linked to sustainable development. However, the parties' actions also reveal how later developments around ecosystem thinking have been easy to incorporate into updated definitions. In this sense, the drafting of Article 3.1 and the reference to wise use seem to demand (and enable) reflection of best practices as they develop over time in wetland conservation management thinking. Nevertheless, this has never struck the author as a particularly satisfactory or practically helpful conceptualisation of the wise use commitment.

3 PROPOSING AN OVERALL LEGAL CONCEPTION OF 'WISE USE'

Philippe Sands and others have suggested that obligations under multilateral environmental agreements can be categorised as substantive, procedural, or institutional.[87] The first covers issues such as limiting greenhouse gas emis-

---

[82] For full details, see Ramsar Convention Secretariat, *Handbook 1: Wise Use of Wetlands: Concepts and Approaches for the Wise Use of Wetlands* (Ramsar Handbooks, 4th edition, Ramsar Convention Secretariat, Gland, 2010) (hereinafter, 'Wise Use Handbook').

[83] D Pritchard, 'The "Ecological Character" of Wetlands: A Foundational Concept in the Ramsar Convention yet Still Cause for Debate 50 Years Later' (2021) *Marine and Freshwater Research* 1127, 1128–29.

[84] Ramsar Convention, note 81, Annex A, [22].

[85] Ibid, [15].

[86] Defined as 'development that meets the needs of the present without compromising the ability of future generations to meet their own needs'. UN World Commission on Environment and Development, *Our Common Future* (1987), Chapter 2.

[87] P Sands et al, *Principles of International Environmental Law* (Cambridge University Press, Cambridge, 4th edition, 2018) 145.

sions; the second the completion of procedures such as environmental impact assessments; and the last activities such as submitting a periodic report to an international secretariat.[88] However, as Benoit Mayer argues, there is another categorisation that can be applied to international undertakings that distinguishes between obligations of conduct and obligations of result.[89] Mayer's approach is far more enlightening when it comes to conceptualising the commitment in Article 3.1, with 'wise use' being best understood as an obligation of conduct.

Such obligations are well known in civil law jurisdictions – for example, where doctors are not expected to achieve a particular result but are expected to treat patients to the best of their abilities.[90] That said, similar forms of duty can be seen in common law jurisdictions where, for example, an individual might be required to exercise investment powers with due skill and care, rather than needing to achieve a particular level of annual profit. Mayer has also described the prevalence of obligations of conduct in international human rights law and argued that such obligations can be seen in customary rules around preventing transboundary harm and under the climate change regime (with the exclusion of the obligations of result set by the Kyoto Protocol).[91] To these can be added the International Tribunal for the Law of the Sea's recognition that the responsibilities of flag states to ensure that vessels flying their flag do not engage in illegal, unreported, and unregulated fishing is an obligation of conduct centred on due diligence, rather than an obligation of result.[92]

Clearly, the conduct sought to be regulated under Ramsar concerns decision-making about utilisation of wetlands and the standard of conduct expected is one of wisdom. For reasons that are explained below, neither the Convention nor the COP has explicitly defined in full what it means to be 'wise'; but the ordinary dictionary meaning of the term does seem to align with the initiatives that are highlighted below.[93] This ordinary meaning can be

---

[88] Ibid.
[89] Mayer, note 4.
[90] Ibid, 130.
[91] Ibid, 132–35.
[92] *Request for an Advisory Opinion Submitted by the Sub-Regional Fisheries Commission: Advisory Opinion, 2 April 2015*, ITLOS Reports 2015, [125], [129], at: https://www.itlos.org/fileadmin/itlos/documents/cases/case_no.21/advisory_opinion_published/2015_21-advop-E.pdf.
[93] 'Having or exercising sound judgement or discernment; capable of judging truly concerning what is right or fitting, and disposed to act accordingly; having the ability to perceive and adopt the best means for accomplishing an end; characterized by good sense and prudence. Opposed to foolish.' Oxford English Dictionary Online (Oxford University Press, Oxford, 2022).

distilled into three principal components. The first is that to conduct oneself with wisdom, one must know what the end objective should be. Second, the decisionmaker must know of the relevant factors that bear upon, and the consequences that will flow from, deciding whether to utilise a wetland. Finally, the decisionmaker must be capable and willing to make their determination over use based on the end objective given the factors and consequences brought to their attention. On this latter component, the Ramsar regime has had little input; but that is not the case for the first two.

The definitions of 'wise use' detailed in Section 2 above have served to establish the end objective for decisionmakers. They establish that the wisest decision is one that is best oriented towards the goal of sustainable development of wetlands and the maintenance of their ecological character in the given circumstances. However, as this is not an obligation of result, a failure to maintain the ecological character of the wetland will not represent a breach of Article 3.1, provided that the original decision to utilise the wetland was taken wisely. As Mayer explains, this is an entirely appropriate tactic in situations where outcomes are not entirely within the control of the state, since luck plays a role.[94] A decisionmaker might have gathered all available facts and been disposed to be led by the evidence, but it still might not have been possible to foresee all consequences of the decision. Given that the provision was designed to encompass all wetlands within a state's territory, an obligation of result would have exposed states to unpredictable and uncontrollable financial liabilities. Thus, the parties have accepted an obligation of conduct, reinforced by the express recognition that their commitment in Article 3.1 is to use wetlands wisely only 'as far as possible'.[95]

The contracting parties to Ramsar have dedicated far more effort to ensuring that decisionmakers know of the relevant factors that bear upon, and the consequences that will flow from, deciding whether to utilise a wetland. In doing so, they have sought to define and promote the gathering of legitimate and effective factors to be woven into the decision-making process. This is a well-known technique in international environmental law, where it is feared that illegitimate interests and pressures can lead to political bargaining, rather than objective reasoning based on evidence.[96]

---

[94] Mayer, note 4, 136–37.

[95] Dave Pritchard also links the scope of the undertaking to the need to soften the commitment with this qualifier, though he does not connect it to the fact that the post-1971 challenges confronting governments and societies are unpredictable. Pritchard, note 83, 1129.

[96] See T Gehring and E Ruffing, 'When Arguments Prevail over Power: The CITES Procedure for the Listing of Endangered Species' (2008) 8(2) *Global Environmental Politics* 123.

In the Convention text, many of the previously mentioned commitments tie into supporting wise decision-making. Thus, the quality of the information available for decision-making is to be enhanced through the Article 4.4 commitment of states to 'encourage research and the exchange of data and publications regarding wetlands and their flora and fauna'.

Furthermore, states are to promote the number of available wetland experts with the capacity to play a role in wise decision-making via the Article 4.5 commitment to 'promote the training of personnel competent in the fields of wetland research, management and wardening'. Where wetlands straddle territories, the sensible way to decide how to use and manage such a resource is in consultation with the states concerned and through coordination of national policy and regulation, as required under Article 5. Finally, the COP itself is charged with generating and sharing knowledge around wetland threats, management, and science[97] – something that has been significantly boosted by the work of the STRP, and which is backed up by an obligation of contracting parties to 'ensure that those responsible at all levels for wetlands management shall be informed of, and take into consideration, recommendations of such [COPs]'.[98]

The efforts of the parties operating under the above competences of the COP have produced far more detailed initiatives and guidelines shaping the potential quality of wetland management and decision-making. Twenty handbooks have been issued that bring together relevant decisions of the parties and findings of the STRP; and that define for practitioners, and generate understanding among them of, internationally agreed best practices in wetland management.[99] These capacity-building guides cover crucial components of wise decision-making about use, including river basin management; water allocation and management; impact assessments; production of inventories of wetlands; changes in wetland character; and data information needs.[100] Further, Ramsar's communication, capacity-building, education, participation, and awareness programme (CEPA) promotes the integration of the voices and concerns of more people within the wise use process so that they become advocates for wise use and involved in policy formulation, planning, and management.[101] The regime

---

[97] Convention on Wetlands, note 1, Art 6.2(d), (e).
[98] Ibid, Art 6.3.
[99] Wise Use Handbook, note 82, 4.
[100] For a full catalogue of these guidelines, visit https://www.ramsar.org/resources/the-handbooks.
[101] Ramsar Convention Secretariat, *Wetland CEPA: The Convention's Programme on Communication, Education, Participation and Awareness (CEPA) 2009–2015* (Ramsar Convention Secretariat, Gland, 4th edition, 2010) 9.

has also responded to discrete events that could lead to unwise decisions on how to manage wetland ecosystems, as exemplified by the COP's response to the spread of avian influenza.[102] As the preambular provisions to COP10's Resolution X.21 describe, the role of waterbirds and wetlands in the spread of H5N1 avian flu was poorly understood, leading to known instances or proposals for destroying waterbirds, their nests and wetland habitats. The Resolution observed that these are 'misguided and ineffective responses to the spread ... which ...do not amount to wise use'.[103]

Built into the current strategic plan for Ramsar is further evidence of attempts to define the content and enhance the quality of wise decision-making around wetland use.[104] Among the priority areas identified for 2016–24, the following would serve to feed into the expected quality of decision-making:

- enhancing the generation of science-based advice and guidance through the STRP and CEPA process;
- ensuring that the importance of wetlands for climate change mitigation and adaptation is understood;
- building greater recognition and integration into planning and decision-making of the drivers for wetland loss and degradation;
- achieving greater integration of wetland services, benefits, values, functions, goods, and products in national development plans;
- improving communication to people of wetland values;
- enhancing cooperation at all levels of planning and decision-making;
- strengthening and supporting full and effective participation of stakeholders in wise use; and
- analysing wetland functions and services from a basin perspective.[105]

Notably, the fourth priority area for the 2016–24 period aligns with the crucial fact that if decision-making around wetland use is to be wisely conducted, it needs to be integrated within and supported by national planning. After all, the undertaking in Article 3.1 is to formulate and implement planning in a way that leads to the wise use of wetlands. Consequently, the guidelines, knowledge,

---

[102] As described in R Cromie et al, 'Responding to Emerging Challenges: Multilateral Environmental Agreements and Highly Pathogenic Avian Influenza H5N1' (2011) 14(3–4) *Journal of International Wildlife Law & Policy* 206.

[103] Ramsar Convention, *Resolution X.21: Guidance on responding to the continued spread of highly pathogenic avian influenza* (2008), [4].

[104] For a guide to the plan, see Ramsar Convention Secretariat, *The Fourth Ramsar Strategic Plan 2016–2024* (Ramsar Handbooks, 5th edition, Ramsar Convention Secretariat, Gland, 2016).

[105] Ibid, 5–6.

and best practice that go into soundly judging what is a right and fitting course of action for the use of wetlands need integrating into government policy. Thus, back in 1990, the contracting parties recommended that if wise use were to be implemented, states should have comprehensive national wetland policies.[106] Production of these national wetland policies has been easy to monitor, although the results are not encouraging. At COP11 in 2012, only 51% of parties reported having a national wetland policy, while they reported differing rates of integration of wetlands into other sectoral policies: wetlands were integrated in poverty reduction policies in only 36% of contracting parties, 57% of national strategies for sustainable development, 41% of national agricultural policies, 64% of water resource management policies, 54% of national forestry programmes, and 83% of national biodiversity strategies and action plans.[107] By 2018, the figures had crept up to 54% of contracting parties having either a full policy in place (73 states) or a partial policy in place (18 states).[108] With just under half of states not having a policy, this is hardly encouraging for wise decision-making around wetland use.

While this all serves to try to dictate the conditions under which decisions should be taken, the Ramsar regime and wetland stakeholders seemingly have a continuing challenge to ensure that foolish decisions around wetland utilisation are not being taken. However, it is hoped that by conceptualising the wise use undertaking as an obligation of conduct with sustainable development as its guiding objective, what can appear to advisers and other stakeholders as an inchoate notion is in fact something very familiar to many lawyers. Further, the array of Ramsar initiatives and guidelines can then be recognised as dovetailing with the wise use commitment, giving it greater form and an overarching logic and purpose.

4   CONSERVATION OF LISTED WETLANDS

As noted earlier, Ramsar provides for obligations that are applicable to all wetlands, with additional commitments applying to the more 'exclusive' group of listed wetlands. Once a contracting party designates a site and it is listed as a Wetland of International Importance, the parties have assumed a duty to

---

[106] Ramsar Convention, *Recommendation 4.10: Guidelines for the implementation of the wise use concept* (1990) Annex.

[107] Ramsar Convention, *Doc 7: Report of the Secretary General on the implementation of the convention at the global level COP11* (2012) 14–15.

[108] RC Gardner and CM Finlayson, *Global Wetland Outlook: State of the world's wetlands and their services to people* (Ramsar Convention Secretariat, Gland, 2018) 62.

conserve the site. This Chapter now turns to the meaning of 'conservation' and its relationship with wise use.

4.1 'Conservation' under Article 3.1

Once listed, Article 3.1 obliges the state to 'formulate and implement their planning so as to promote the conservation of wetlands included in the List'. Given that Article 3.1 also calls for all wetlands in general to be used wisely, there have been attempts to clarify whether the standards of conservation and wise use amount to the same level of protection.[109] Adding importance to the question, under international law, the principle of effectiveness infers that where distinct terms have been used, an interpretation ought to be found that results in an operative outcome for the distinction.[110] Farrier and Tucker have suggested that it might be reasonable to expect that the required management of listed wetlands needed to be more cautious than that of unlisted wetlands.[111] The problem, however, as they go on to note, is that since Ramsar's entry into force, the notions and definitions of these terms have been somewhat equated.

'Wise use' was eventually recognised explicitly as being linked to the sustainable development of wetlands in a way that maintained their ecological character.[112] As Bowman then observes, the modern notions of 'conservation' are almost identical to the Ramsar interpretation of 'wise use'.[113] A clear example of this is that the World Conservation Strategy defines 'conservation' as yielding the greatest sustainable benefit to present and future generations.[114] Thus, conservation management might involve elements of periodic preserva-

---

[109] MJ Bowman, 'The Ramsar Convention Comes of Age' (1995) 42 *Netherlands International Law Review* 1, 10–14; D Farrier and L Tucker, 'Wise Use of Wetlands Under the Ramsar Convention: A Challenge for Meaningful Implementation of International Law' (2000) 12(1) *Journal of Environmental Law* 21, 23–24.

[110] Known as the principle of effectiveness and encapsulated in the Latin phrase '*ut res magis valeat quam pereat*' ('So that the matter may flourish rather than perish'). AX Fellmeth and M Horwitz, *Guide to Latin in International Law* (Oxford University Press, Oxford, 2009).

[111] Farrier and Tucker, note 109, 23–24.

[112] See Pritchard, note 83.

[113] Bowman, note 109, 15.

[114] IUCN, *World Conservation Strategy* (1980) [4], at: https://portals.iucn.org/library/efiles/documents/wcs-004.pdf.

tion, maintenance, sustainable use, restoration, and general enhancement of the natural environment.[115]

Consistent with this, the parties to Ramsar have rejected any pure preservationist (in the John Muir sense) approach to managing listed wetlands, preferring instead the apparent extension of wise use standards to listed wetlands. Thus, the 17th *Ramsar Handbook* on designating Ramsar sites emphasises the continuing need for all wetlands to remain a valuable resource, while designation is a first step towards 'achieving the long-term wise (sustainable) use of the site'.[116] It is therefore tempting to conclude that, in light of the modern focus on sustainable development of natural resources, given the fact that maintenance of ecological character has a significant bearing on listed wetlands under Article 3.2 and following the 2005 definition of 'wise use', there is now no difference between the undertakings for listed and non-listed wetlands. Farrier and Tucker went so far as to assert that these developments undermined 'the icon status of listed wetlands by subjecting them to the same management imperatives as unlisted wetlands'.[117]

4.2 An Agreed Revision?

This apparent elision of 'conservation' and 'wise use' could be explained by Article 31(1) of the 1969 Vienna Convention on the Law of Treaties.[118] This article emphasises the good-faith interpretation of terms where they are to carry their ordinary meaning given their context. Subsequent interpretations agreed between the parties must also be respected according to Article 31(3), even if they amount to a revision of the text.[119] With the contracting parties having approved the various guidelines and adopted the resolutions that have subsequently explained the content of both 'conservation' and 'wise use', it can be argued that the parties now want to reflect a common modern understanding.

---

[115] Ibid; see P van Heijnsbergen, *International Legal Protection of Wild Fauna and Flora* (IOS Press, Amsterdam, 1997) 43–52.

[116] Ramsar Convention Secretariat, *Handbook 17: Designating Ramsar Sites: Strategic Framework and Guidelines for the Future Development of the List of Wetlands of International Importance* (Ramsar Handbooks, 4th edition, Ramsar Convention Secretariat, Gland, 2010) [23].

[117] Farrier and Tucker, note 109, 27.

[118] Vienna Convention on the Law of Treaties, adopted 23 May 1969, entered into force 27 January 1980, 8 ILM 679. The provisions could operate retroactively as they have been said to reflect customary law. R Gardiner, *Treaty Interpretation* (Oxford University Press, Oxford, 2008) 12–19.

[119] On amendment, see Gardiner, note 118, 216–25.

The natural development of this is that 'conservation' is the same obligation of conduct requiring wise decision-making around the management and utilisation of listed sites. Further, the end goal that helps in duly exercising sound judgement is maintaining the ecological character of the listed wetland, including those characteristics that signified the international importance of the wetland in the first place.

This reading of events might not seem to fully deal with the 'as far as possible' qualifier that only attaches to the wise use undertaking in Article 3.1. A counterargument to such objections, however, would be that the qualifier is redundant given that the obligation of conduct is about acting wisely. A wise decisionmaker would hardly be acting foolishly if it brought in national concerns arising from poverty alleviation or other economic imperatives of the day when faced with a choice, or if it simply could not acquire and have at its disposal all the scientific information and data available in the timeframe allowed for them to make a decision. That which is practicably possible seems 'part and parcel' of what goes into making a wise choice.

It could therefore be argued that the contracting parties have agreed that the same obligation of conduct is applicable to listed sites as that for wise use – just that the end goals will look slightly different because of the internationally important characteristics exhibited in the site. However, there is an alternative proposition that can be advanced which makes further use of Mayer's perceptive classification of international obligations, and which sees 'conservation' and 'wise use' as distinct.

4.3   Conservation as an Obligation of Result

In contrast to wise use, an argument that Article 3.1 contains an obligation of result for conservation of listed wetlands can be based on the historical context of the Ramsar negotiations, the history of the negotiations themselves, and the other articles in the adopted Convention text.

As with wise use, understanding what was happening in the mid-1950s up to the first meetings for the Ramsar Convention is enlightening. The idea of enclave strategies for the conservation of natural areas had taken root early in the United States;[120] and by the 1960s, further examples could be found in South Africa, India, the United Kingdom, Kenya, and Tanzania. However, in post-war Africa, the same push for development that was catalysing the first global flourishing of sustainable development was impacting these reserves. For example, in 1954 in Kenya, a proposed dam that would have supplied water to Mombasa was moved after a public outcry that its original location

---

[120] Yellowstone was dedicated in 1872.

would have affected the pools of water frequented by hippopotami in Tsavo West.[121] The FPS claimed that this 'confirmed the principle that a national park is not an area to be at one moment set aside for complete protection, and at the next endangered by claims for the use of its natural resources'.[122]

This leaning towards protectionism within enclaves had also been evident in the turn-of-the-century United States, where compromise was reached between the approaches of Muir and Pinchot. Here preservation was the priority within the national parks; while on forestry service land, which often lay next to larger parks, natural resources were to be conserved according to sustainable wise use.[123] Furthermore, up until 1969 (so for much of the time while Ramsar was being negotiated), the IUCN General Assembly equated national parks to areas not materially altered by human exploitation.[124]

A further cause for concern for conservationists at the time was governments violating the boundaries of protected areas. In early 1956, the government of Tanganyika published a white paper describing a deal that was being made with Masai herdsmen that would see them vacate areas in the Serengeti National Park in return for redrawing the boundaries of the park to open up a central swathe of land that they could use.[125] This prompted an outraged response from the FPS in London[126] and international concern from the United States and IUCN.[127] Pressure on the Colonial Secretary in London resulted in a commission being established to review the scheme; and ultimately, over the remaining years of the 1950s, a compromise plan led to new, ecologically sound boundaries for the Serengeti National Park.[128]

---

[121] Adams, note 13, 93.

[122] Ibid, 94.

[123] Hudson Westover, note 20.

[124] IUCN soon after pushed forward the idea that enclaves could be set up with other objectives that were less preservationist. A Phillips, 'The history of the international system of protected area management categories' (2004) 14(3) *Parks* 4, 6. The Great Barrier Reef Marine Park remains a well-known illustration of this, where zoning is employed within protected areas enabling different levels of resource utilisation. GBRMP, *Great Barrier Reef Marine Park Zoning Plan 2003* (2004), at: https://elibrary.gbrmpa.gov.au/jspui/bitstream/11017/382/1/GBRMP-zoning-plan-2003.pdf.

[125] See F Fraser Darling, 'Beast and Man Have Rival Claims in Serengeti', 17 April 1957, *The Times*, 11–12.

[126] See letter to the Editor of *The Times* from CL Boyle, Secretary to the Fauna Preservation Society (the new name of the SPFE) of 5 March 1956.

[127] Adams, note 13, 68.

[128] Ibid, 68–69.

Against this background in 1960, Luc Hoffmann, Vice-President of the IUCN, gained the support of that organisation to operate a programme on conservation of wetlands. This was the previously mentioned Project MAR.[129] As was recounted earlier, the Project MAR Conference of 1962 resolved to have the IWRB draw up a list of internationally important wetlands which, when finally published in 1965, became known as the MAR List.[130] As previously described, the MAR List then became the focus of the negotiations for the Ramsar Convention. Recalling Section 2.3 of this Chapter, the IWRB's first draft of August 1965 dedicated four of the eight proposed subjects to the MAR List. Crucially for the argument in this section of the Chapter, these subjects included the establishment of inviolate reserves within the listed sites and an undertaking from governments not to subsidise the draining or infilling of these wetlands except in cases of imperative necessity – and even then to seek to minimise losses and compensate where losses were unavoidable.[131] While wetlands in general enjoyed intermittent attention during the negotiation period, this was not the case for conserving internationally important wetlands on the MAR List, which never dropped out of sight. As Stroud et al state, 'the primary role of the proposed wetland conservation convention was to ensure, through government policy actions, the sustained conservation of these few critical sites'.[132]

Subsequent drafts wrestled with how far the inviolable quality of the listed sites could be guaranteed through international oversight and shaping of state action. Early drafts proposed the referral to the COP of sites undergoing ecological change followed by a formal COP recommendation on how to preserve the wetland in whole or in part, but these were watered down for reasons of political expediency and to secure the adoption of a convention.[133]

Nevertheless, the final text and subsequent initiatives have delivered much of the mooted oversight and expectations of compensation where boundaries are changed or ecological characteristics are in danger. Article 2.1 requires the boundaries of a listed wetland to be precisely described and illustrated on a map. These boundaries can be enlarged but can only be restricted (or the site entirely deleted) because of urgent national interests, with the Ramsar Bureau being informed of such moves at the earliest opportunity.[134] If the latter

---

[129] Matthews, note 34, 7.
[130] Ibid, 32.
[131] Stroud et al, note 5.
[132] Ibid, 1113.
[133] Ibid, 7–8.
[134] Convention on Wetlands, note 1, Art 2.5.

happens, a state 'should as far as possible compensate for any loss of wetland resources, and in particular it should create additional nature reserves'.[135]

Article 3.2 further provides that:

> Each Contracting Party shall arrange to be informed at the earliest possible time if the ecological character of any [listed] wetland in its territory ... has changed, is changing or is likely to change ... as a result of human interference.

While information on these changes is to be passed to the Ramsar Bureau, the Convention had dropped the earlier drafting that sought to put in place a detailed system for the COP to review the situation and make recommendations.[136] That said, the general powers of the COP expressed in Article 6.2 contain a competence to consider the information provided under Article 3.2, as well as a competence to make general or specific recommendations to parties regarding the conservation of wetlands. Resolution XI.9 refers to these aspects of Articles 2 and 3 as additional commitments and require responses which are part of an overall framework linked to the avoidance and mitigation of, and compensation specifically for, listed wetland loss and degradation.[137]

The COP has strengthened Article 3.2 via a number of procedural developments. Recommendation 4.8 called for states 'to take swift and effective action' to reverse negative changes to ecological character.[138] The same Recommendation required the Ramsar Bureau to keep a record of all such notifications. This became known as the Montreux Record and states are now obliged in their reports to COPs to cover the conservation status of listed wetlands on the record.[139] This process can assist states by prioritising the state for receiving capacity-building support through either awards of finance under the Ramsar Small Grants Fund[140] or a visit by a team of experts under the Ramsar Advisory Mission process.[141]

---

[135] Ibid, Art 4.2. See also Chapter 5 of this volume.

[136] Stroud et al, note 5.

[137] Ramsar Convention, *Resolution XI.9: An integrated framework and guidelines for avoiding, mitigating and compensating for wetland losses* (2012) [41].

[138] Ramsar Convention, *Recommendation 4.8: Change in ecological character of Ramsar sites* (1990).

[139] Lyster, note 7, 442–46.

[140] Ramsar Convention, *Ramsar Small Grants Fund for Wetland Conservation and Wise Use (SGF): Operational Guidelines 2019* (2019) at: https://www.ramsar.org/document/ramsar-small-grants-fund-for-wetland-conservation-and-wise-use-sgf-operational-guidelines-0.

[141] See Ramsar Convention, *Resolution XIII.11: Ramsar advisory missions* (2018).

Given the historical context, the lack of a qualifier expecting states to conserve only 'as far as possible', the various provisions of the Convention dedicated to listed wetlands, and subsequent regime implementation developments such as the Montreux Record, there is good evidence to support the argument that the commitment to conserve listed wetlands is strongly intended to deliver a result, rather than to simply suggest a goal that informs conduct. Consistent with this, Bowman observes that while there is no obligation to avoid any change in ecological character whatsoever, the opposite extreme cannot be countenanced either, whereby states allow wetlands to fall into complete degradation.[142] Instead, he argues that in managing listed wetlands, ecological values may not fall below a point whereby the site no longer possesses the international importance that qualified it for listing in the first place.[143] The obligation of result is therefore to maintain sufficient ecological qualities for an internationally important wetland.

5 CONCLUSION

Previous academic conclusions around the nature and content of the commitments contained in Article 3.1 – including those offered by the current author – have generally settled at equating both 'wise use' and 'conservation' to 'sustainable development'. Bowman's 1995 analysis was perhaps the most rigorous, but this Chapter has looked to transplant some of the most perceptive insights from that into an expanded and more coherent framework. This has been enabled by an original and much deeper look into the historical setting of the negotiations and by the recent release of some of the preparatory materials for the negotiations of the Convention. Mayer's recent framework for analysing the climate change commitments under the Paris Agreement has also proved indispensable.

The improved understanding offered herein is that 'wise use' had long been wrapped up with the 'conservation in harmony with development' movement, but this only gained traction among the international community

---

[142] Bowman, note 109, 20.

[143] If correct, Ramsar would have parallels with the 1972 World Heritage Convention, which was being negotiated at the same time and with the involvement of IUCN. Under the World Heritage Convention, states are obliged to transmit sites of outstanding universal value to future generations. This amounts to an obligation of result because it has the practical effect that, while some value may be lost over time, states endowed with a site listed on the World Heritage List cannot allow the site to deteriorate to the extent that the site loses all the outstanding value that merited its inscription in the first place. See EJ Goodwin, 'The Consequences of Deleting World Heritage Sites' (2010) 21(2) *King's Law Journal* 283.

in the post-war years. There was ultimate recognition by the negotiating states that wetlands in general needed to be managed in accordance with this; but at the time, there seemed to be uncertainty over how exactly to express this. Expanded definitions emerged later in COP decisions and as the scientific and international community came to refine its own understanding of 'sustainable development'. However, the clear concern was around the quality of decision-making for wetland management and utilisation. Thus, the instincts of the negotiating parties were to express this as a need for wisdom and much that has followed has been aimed at enabling stakeholders to make decisions according to sound judgement based on good data and understanding. An end goal was still needed to inform that wisdom – namely, to sustain the ecological values of wetlands; but given the range of these values across all of a nation's wetland reserves and the potential scale of the challenge, adopting the goal as a result could not be included as a core obligation. Instead, the process needed to be the focus of the obligation; hence, this Chapter concludes that wise use is most naturally understood as an obligation of conduct guided by an end goal of seeking sustainable development and maintaining ecological values.

In contrast, the research has also revealed how and why the inviolability of the MAR wetlands' boundaries and ecological conditions was of such central concern for the negotiating parties. That concern translated into the conservation of listed wetlands obligation under Article 3.1 and subsequent supporting implementation structures. Thus, management was to be calibrated against a minimum expectation for maintaining internationally important ecological characteristics present in a listed site. It is thus suggestive of an obligation of result. The added attraction of this reading is that it satisfies international law's principle of effectiveness.

5. Compensatory mechanisms and the loss of wetlands[1]

Jerneja Penca

1 INTRODUCTION

In international environmental law, the Ramsar Convention[2] is regarded as a remarkable multilateral treaty with a pragmatic approach to the management of wetlands. The Convention has been highly adaptive to external developments to accomplish the ambiguous goal of 'wise use' of wetlands[3] and has also contributed to broader developments in biodiversity conservation, beyond wetlands. Its management approach has reflected the changing knowledge over the past 50 years about wetlands and their contribution to people's livelihoods and wellbeing,[4] as well as changing public perceptions of wetlands from being a nuisance to civilisation's progress[5] to 'kidneys of the landscapes' and 'nature's supermarkets'.[6] Part and parcel of the interactive relationship of the Convention with the external environment is the arrangement for establishing compensatory mechanisms for the loss of wetlands. A specific treaty provision requiring that contracting parties compensate for the loss of wetland

---

[1] The production of this Chapter was supported by the Slovenian Research Agency (Project Code: J5-2562).

[2] Convention on Wetlands of International Importance especially as Waterfowl Habitat, adopted 2 February 1971, entered into force 21 December 1975, 996 UNTS 245 (amended 1982 & 1987).

[3] See Chapter 4 of this volume.

[4] E Maltby and MC Acreman, 'Ecosystem services of wetlands: pathfinder for a new paradigm' (2011) 56(8) *Hydrological Sciences Journal* 1341–59.

[5] J Ruffolo, *The U.S. Supreme Court Limits Federal Regulation of Wetlands: Implications of the SWANCC Decision* (Paper 305, California Agencies, 2002) 1, at: https://digitalcommons.law.ggu.edu/cgi/viewcontent.cgi?article=1301&context=caldocs_agencies.

[6] WJ Mitsch and JG Gosselink, *Wetlands* (5th edition, Wiley, Hoboken, 2015) 3–4.

resources represented a considerable novelty in the context of international environmental law at the time of the adoption of the Convention. Through the decades, the operationalisation of that provision and related processes within the Convention has had far-reaching, paradigmatic implications for biodiversity governance beyond wetlands at the national, regional, and international level. Compensatory mechanisms have become a progressively popular tool in view of ever-increasing pressure on biodiversity and conservation areas and an improved understanding of their value as wildlife habitats and indispensable providers of an array of ecosystem services. Thus, compensatory mechanisms exemplify a down-to-earth rationale for the conservation of nature that acknowledges the dominance of economic, technological, and commercial trajectories, and seeks to lessen the detrimental impact of these on nature, rather than renounce the law's ability to shape socio-ecological outcomes.

This Chapter analyses and evaluates the provisions of the Ramsar Convention related to the idea of compensation for loss of wetlands from an evolutionary perspective. Section 2 begins by reviewing the meaning and significance of Article 4.2 of the Convention, as well as subsequent developments within the framework of the Convention's Conferences of the Parties (COPs), situating these within the context of environmental law and governance. Next, Section 3 reflects critically on compensatory mechanisms by outlining the advantages and shortcomings of this provision. Section 4 presents the practice of compensatory mechanisms within the broader biodiversity (rather than just wetland) context, at both the national and transnational levels. Finally, in Section 5, the Chapter considers the paradigmatic changes and challenges within international biodiversity governance, which the mechanism of compensation for biodiversity loss has helped to consolidate.

While the Chapter takes wetlands and the Ramsar Convention as a starting point, the analysis moves beyond these, reflecting the close interplay that the Convention has with outside structures and processes. The practice of wetland compensatory mechanisms has extended beyond the Ramsar Convention and even beyond the intergovernmental arena. As such, this Chapter depicts the Ramsar Convention as constituent in the shifts related to compensatory mechanisms, as well as biodiversity governance. The latter has, together with the Ramsar Convention, come to be much more diverse in terms of environmental challenges – as well as legal actors, principles, and compliance strategies – than initially conceived by international environmental law.[7] Although the

---

[7] V Heyvaert and T Etty, 'Introducing Transnational Environmental Law' (2012) 1 *Transnational Environmental Law* 1–11; V Heyvaert, *Transnational Environmental Regulation and Governance: Purpose, Strategies, and Principles* (Cambridge University Press, Cambridge, 2018) 1–11.

Ramsar Convention has rarely been exposed as an object of analysis in transnational environmental law, its role is unquestioned in this polycentric and complex landscape.

2 OBLIGATION TO COMPENSATE FOR LOSS: EVOLUTION OF THE TREATY PROVISION

The Ramsar Convention is a short, compact treaty that uses words rationally. Placed rather prominently in the text, Article 4.2 spells out the consequences of the deletion or restriction of a Ramsar Site:

> Where a Contracting Party in its urgent national interest, deletes or restricts the boundaries of a wetland included in the List, it should as far as possible compensate for any loss of wetland resources, and in particular it should create additional nature reserves for waterfowl and for the protection, either in the same area or elsewhere, of an adequate portion of the original habitat.

Article 4.2 is a follow-up of the right bestowed upon contracting parties in Article 2.5, which gives them the right to delete or restrict a Ramsar Site due to 'urgent national interests'.[8] Article 4.2 requires that in such cases, the parties compensate for the ecological value lost because of the deletion or reduction of a Site. While mitigation measures occur *in situ*, compensation measures occur *ex situ* or off-site. Compensation can be implemented by introducing protective measures in the form of establishing a protected area elsewhere, without this area necessarily being linked to the Ramsar Site – that is, to the wetlands listed under Article 2 of the Convention. The size of the replacement area should be of 'an adequate portion' – presumably meaning comparable to the forgone Ramsar Site in size. A potential alternative to the creation of a new protected area elsewhere is to extend the boundaries of the wetland in the direction that does not interfere with the reason for restricting that specific Ramsar Site.

A curious step towards expanding the remit of Article 4.2 occurred in a subsequent resolution of the parties that the obligation for compensation also

---

[8] Convention on Wetlands, note 2, Art 2.5. The conditions for 'urgent national interest' were formulated subsequently. Ramsar Convention, *Resolution VIII.20: General guidance for interpreting 'urgent national interests' under Article 2.5 of the Convention and considering compensation under Article 4.2* (2002).

applies to sites that are nominated by parties for inclusion in the Ramsar List but found to not qualify under any of the criteria established:

> When, following consultation between the Convention Bureau and the Contracting Party concerned, it is agreed that a site failed at the time of designation to qualify under any of the criteria, and that there is no possibility of extension, enhancement, or restoration of its functions or values, it shall instruct the Convention Bureau to remove the site from the List and shall apply the provisions for compensation, as provided in Article 4.2 of the Convention.[9]

The Convention does not prescribe that the replacement sites are to be included in the List, stating only that they are nature reserves. In this manner, the requirement for compensation conflates the two 'tiers' of wetlands created under the Convention: those relating to Article 2.1, which are Ramsar Sites of international importance; and those that are guided by Article 4.1 as 'conventional' nature reserves.[10]

Nevertheless, Article 4.2 does introduce a significant safeguard for wetlands under threat of being lost and an effective hierarchy of intervention measures in those instances. The Convention text anticipated that the interest of conserving nature may be overruled by other interests. The idea for compensation was considered in the early drafts of a treaty and was among the nine key subjects identified for such an agreement.[11] The treaty text was forward-looking in setting up a mechanism to defend the purpose of the treaty, even by accepting that the resulting site might have a lesser ecological value than the original site.

Article 4.2 is an articulation of the so-called 'mitigation hierarchy'. This represents a refinement and application of the principle of prevention – a fundamental principle of environmental law. Prevention dictates tackling the problem as close to the source as possible, thus avoiding harm.[12] A pre-

---

[9] Ramsar Convention, *Resolution 5.3: Procedure for initial designation of sites for the List of Wetlands of International Importance* (1993). The criteria for inclusion were determined in Recommendation 4.2. Ramsar Convention, *Recommendation 4.2: Criteria for identifying wetlands of international importance* (1990). On the requirement to compensate in those cases, see also D Pritchard, *Change in ecological character of wetland sites – a review of Ramsar guidance and mechanisms* (2014) 90 [C.78], at: https://www.ramsar.org/sites/default/files/documents/library/ecological_character_report_long_18112914_e.pdf.

[10] MJ Bowman, 'The Ramsar Convention comes of age' (1995) 42 *Netherlands International Law Review* 1, 21.

[11] GVT Matthews, *The Ramsar Convention on Wetlands: its History and Development* (Ramsar Convention Bureau, Gland, 1993) 13–14, 18.

[12] LA Duvic-Paoli, *The Prevention Principle in International Environmental Law* (Cambridge University Press, Cambridge, 2018) 179–232.

ventative approach, contrary to a curative approach, dictates anticipatory and proactive response to environmental challenges.[13] The requirement of the mitigation hierarchy channels that duty into the field of nature conservation on a pragmatic basis. The underlying logic is that negative impacts on a wetland should, as a priority, be avoided. If such negative impacts cannot be avoided or prevented, measures to minimise or reduce the negative impacts should be put in place on site, such as the introduction of buffer zones, the timing of works, and restrictions on use. Finally, if damage nevertheless remains despite the mitigation measures, actions should be taken to compensate for and offset these residual impacts. While the preventative approach does not explicitly espouse the 'compensate' part, this omission is due to the lack of precision and attention provided by the preventive approach,[14] rather than a departure by the mitigation hierarchy from the essence of the principle. The Ramsar Convention sets out quite effectively what should happen if the preventative approach fails. From that perspective, the articles of the Ramsar Convention providing for the conservation of wetlands on the List (Article 3.1) and mandating adequate compensation if that conservation is threatened (Article 4.2) become more closely related than is implied in the treaty text.

The idea of compensatory mechanisms was pioneered in domestic approaches to managing natural resources. In the US context, for example, mitigation banking evolved through agency regulations as one way to comply with the 1972 US Federal Water Pollution Control Act.[15] It might have been the discussions conducted at the national level that led to the idea of providing mitigation for the loss also at the level of the Ramsar Convention, signed in 1971.[16] Interestingly enough, the concept was limited to the management of wetlands. We cannot speak of its wide dissemination until much later. For instance, the Convention on Biological Diversity (CBD), concluded in 1992, did not explicitly formulate *compensation* as a strategy of choice. However, it could be read indirectly in the obligation of parties to 'rehabilitate and restore degraded ecosystems and promote the recovery of threatened species, inter

---

[13] Ibid, 15–26.
[14] Ibid, 2.
[15] J Salzman and JB Ruhl, '"No Net Loss": Instrument Choice in Wetlands Protection' in J Freeman and CD Kolstad (eds), *Moving to Markets in Environmental Regulation: Lessons from Twenty Years of Experience* (Oxford University Press, New York, 2006) 354–56.
[16] J Penca, 'Biodiversity Offsetting in Transnational Governance' (2014) 24(1) *Review of European, Comparative and International Environmental Law* 93, 94.

alia, through the development and implementation of plans or other management strategies'.[17]

The remit of the requirement to compensate expanded considerably during subsequent treaty developments. Various Ramsar COPs have widened the scope for compensation, clarified implementational aspects, and strengthened the preventative approach applied in the treaty provision. First, in 1984, the COP envisaged national measures to mitigate or exclude any adverse effects of wetland transformation, including compensation measures, if modification of wetlands is planned.[18] Then, in 1999, the COP established an 'avoid-mitigate-compensate sequence' and stated that 'effective wetland protection involves the conservation of wetlands as a first choice within a three-step mitigation sequence, including avoidance, minimisation, and compensation, the latter only as a last resort'.[19] Recalling Article 3.1 of the Convention (requesting promotion, through planning, of conservation of wetlands included in the List), rather than Article 4.2, Resolution VII.24 urged the parties to 'take all practicable measures for compensating any loss of wetland functions, attributes and values, both in quality and surface area', and to establish national rules for compensation of wetland loss, preferably with wetlands of a similar type and in the same local water catchment.[20] The Resolution also announced an inter-institutional effort to issue criteria and guidelines for the compensation of wetland habitats in case of unavoidable losses and submit them for approval at the next COP. A further set of resolutions in 2008 referred to the mitigation hierarchy or a three-stage sequence of avoiding, mitigating (or minimising), and compensating for wetland losses,[21] recognising and reiterating that restoration cannot replace the loss of natural wetlands.[22]

---

[17] Convention on Biological Diversity, adopted 5 June 1992, entered into force 31 December 1993, 1760 UNTS 69, Art 8.f.

[18] Ramsar Convention, *Recommendation 2.3: Action points for priority attention* (1984).

[19] Ramsar Convention, *Resolution VII.24: Compensation for lost wetland habitats and other functions* (1999).

[20] Ibid, [12].

[21] Ramsar Convention, *Resolution X.12: Principles for partnerships between the Ramsar Convention and the business sector* (2008); Ramsar Convention, *Resolution X.19: Wetlands and river basin management: consolidated scientific and technical guidance* (2008); Ramsar Convention, *Resolution X.25: Wetlands and 'biofuels'* (2008); Ramsar Convention, *Resolution X.26: Wetlands and extractive industries* (2008).

[22] Ramsar Convention, *Resolution 4.1: Interpretation of Article 10 bis Paragraph 6 of the Convention* (1990); Ramsar Convention, *Resolution VII.17: Restoration as an element for national planning and wise use* (1999) [10]; Ramsar

By the early 2000s, a comprehensive approach to wetlands – including the follow-up provisions for damage to wetlands – had been consolidated globally. Multiple countries throughout the Ramsar regions introduced legal and policy arrangements to implement the mitigation hierarchy, applicable either to wetlands specifically or to biodiversity management more generally.[23] Environmental impact assessment (EIA) represents the key instrument facilitating consideration of the impacts of development and planning adequate mitigation responses.[24] EIAs are implemented in practically all countries of the world and are likely to be the most widespread environmental management tool.[25]

While fostering dissemination of the mitigation hierarchy across national legislation, the Convention also took steps to align various national approaches. The work on the criteria and guidelines for the compensation of wetland habitats in the case of unavoidable losses (anticipated in 1999 for adoption in 2002) was finally completed in 2012 through the adoption of the Integrated Framework and guidelines for avoiding, mitigating, and compensating for wetland losses.[26] The Framework is a 29-page document, setting out principles and steps to support parties in designing appropriate responses to wetland loss and degradation, as well as guiding them through mitigation and compensation for wetland losses. The Framework has become a reference point for the management of wetland and, more broadly, biodiversity losses due to its comprehensive scope, timing, and intention to being communicated inter-institutionally – it was communicated to the Secretariat of the CBD as a contribution to the CBD's voluntary guidelines on EIAs and strategic environmental assessments.[27] The document is explicit in reiterating the imperative of avoiding wetland losses (or degradation) as the primary step in any wetland management approach. However, in the face of the trend of progressive

---

Convention, *Resolution VIII.16: Principles and guidelines for wetland restoration* (2002) [10].

[23] RC Gardner et al, *Avoiding, mitigating, and compensating for loss and degradation of wetlands in national laws and policies, Ramsar Scientific and Technical Briefing Note no. 3* (Ramsar Convention Secretariat, Gland, 2012).

[24] Ramsar Convention, *Resolution VIII.20: General guidance for interpreting 'urgent national interests' under Article 2.5 of the Convention and considering compensation under Article 4.2* (2002).

[25] T Yang and R Percival, 'The Emergence of Global Environmental Law' (2009) 36 *Ecology Law Quarterly* 615, 627; N Affolder, 'Contagious Environmental Lawmaking' (2019) 31(2) *Journal of Environmental Law* 187.

[26] Ramsar Convention, *Resolution XI.9: An Integrated Framework and guidelines for avoiding, mitigating and compensating for wetland losses* (2012).

[27] Ibid, 19.

loss of wetlands in both scope and condition (which counters the purpose of the Ramsar Convention), the concept of no net loss represents a legitimate approach to wetland management. No net loss allows impacts on the scope or ecological character of wetlands, but incorporates compensation as a key element. The Framework argues that no net loss is built into the text of the Convention and is encouraged by it, and thus may be part of any party's implementation of the Convention-wide wise use obligation, beyond Article 4.2.[28]

3　DILEMMAS OVER COMPENSATORY MECHANISMS: CONTROVERSIES OF A POLICY TOOL

The documents adopted within the framework of the Ramsar Convention do not fully reflect the controversy over the concept of compensatory mechanisms. Neither do Ramsar COP discussions regarding the wording of decisions.[29] Compensatory mechanisms, however, are replete with conceptual and implementational challenges, which are better captured by the academic literature.[30] The concerns encompass ethical, social, technical, and governance aspects of biodiversity trading.

Specifically, ethical and social concerns relate to the questions of how virtuous it is to accept loss of nature and valuate it for further repayment; how to define, measure, and exchange nature across time and space, given its strong relationship with culture and society; whom such transactions harm and benefit; how citizens' (and particularly affected communities') values and socio-economic benefits are taken into account; how multiple benefits of a specific habitat are measured and monitored; and how to ensure transparency, justice, and effectiveness of activities. Technical issues relate to determining the unit for measuring biodiversity; selecting baselines or reference points for measuring progress; ensuring the equivalence of the replacement and the longevity of the positive impact over time; and managing multiple uncertainties related to the offsets.[31] In the multitude of complexities, technical and governance challenges might be easier addressed than social and particularly ethical

---

[28] Ibid, 28–30.

[29] 'Mexico opposed reference to "no net loss," with Argentina expressing concern on offsets and compensation.' See, eg, 'Summary of the Eleventh Conference of the Parties to the Ramsar Convention on Wetlands 6–13 July 2012' (16 July 2012) 17(39) *Earth Negotiations Bulletin* 7–8.

[30] See, eg, M Maron et al, 'Taming a Wicked Problem: Resolving Controversies in Biodiversity Offsetting' (2016) 66(6) *BioScience* 489; Penca, note 16.

[31] JW Bull et al, 'Biodiversity offsets in theory and practice' (2013) 47 *Oryx* 369.

concerns.[32] Procedural and substantive approaches can be sought, such as the development of credible standards, robust methodologies, and compliance and enforcement mechanisms. Ethical divisions over whether it is right to treat nature as an exchangeable value or not, however, may be significantly more divisive and in fact reflect fundamental value judgements or ideologies. Far from a mere technical matter, offsetting is a highly political issue and a contentious and problematic policy measure.[33]

These dilemmas perhaps become particularly salient in the context of the Ramsar Convention. If Ramsar Sites are representative, rare, or unique wetland types or have international importance for conserving biodiversity,[34] should there not be an absolute ban on activities that harm their integrity? The idea of commensurability between wetlands and the feasibility of changing one wetland for another is challenging given that wetlands play an important role in their local scales. It seems particularly questionable whether the deletion of a Ramsar Site can be effectively replaced by the expansion or creation of a wetland elsewhere.

The Ramsar Convention process has provided some surprisingly clear-cut answers to these dilemmas. On the one hand, it has consistently stressed that restoration or creation of wetlands cannot replace the loss or degradation of natural wetlands.[35] On a practical level, the restriction of a Ramsar Site occurred on only two occasions and in both cases, the restriction was followed up by provisions of compensation.[36] The possibility of deleting a Ramsar

---

[32] Maron et al, note 30.

[33] J Penca, 'Marketing the Market: The Ideology of Market Mechanisms for Biodiversity Conservation' (2013) 2(2) *Transnational Environmental Law* 235; E Apostolopoulou and WM Adams, 'Biodiversity offsetting and conservation: reframing nature to save it' (2017) 51(1) Oryx 23.

[34] Ramsar Convention, *Recommendation 1.4: [Criteria for identifying wetlands of international importance]* (1980). See also C De Klemm and I Créteaux, *The Legal Development of the Ramsar Convention* (Ramsar Convention Bureau, Gland, 1995), at: https://www.ramsar.org/sites/default/files/documents/library/the_legal_development_of_the_ramsar_convention_0.pdf.

[35] Ramsar Convention, note 24. See also note 22.

[36] In 2010, the Åkersvika wetland delta in Norway was reduced in size due to a road enlargement, and new areas were added to compensate for areas excluded, in line with Article 4.2 and a Ramsar Advisory Mission (No 64). In 1987, Belgium announced it would reduce the size of the Lower Scheldt river site, and while not using the term of 'urgent national interest' and not formally invoking Article 4.2, it proposed a compensation bigger in size than the area reduced. In 1997, Australia considered a reduction of listed Sites Port Phillip Bay and Bellarine Peninsula, but later withdrew that proposal. See Ramsar Convention, 25th Meeting of the

Site due to urgent national interests has never played out. The only Sites ever deleted from the Ramsar List were not deleted due to urgent national interests but instead 'had been designated prior to the adoption of the Criteria and were then found not to fulfil any of them'. To compensate for that, '[t]hree new sites were designated in compensation'.[37]

Increasingly, the sceptical views and concerns over mitigation measures became overridden by practical considerations and implementation measures not only in wetland governance, but also in transnational governance of biodiversity more broadly. A turn in legitimating nature compensation can be traced to the endorsement of biodiversity offsets by the International Union for Conservation of Nature (IUCN). In a report published in 2004, the IUCN moved from recognising the concerns and risks involved in the implementation of biodiversity offsets to setting out credible and transparent standards, methodologies, and guidelines for developing pilot projects.[38] The adoption of the report was complemented by a broad process, which opened up international environmental negotiations and treaty processes to the private sector.[39] From the perspective of the Ramsar Convention, the mentioned report helped to consolidate the acceptability of compensatory mechanisms as a feasible conservation strategy and allowed their expansion. The Ramsar Convention can thus be considered as being ahead of its time for entrenching compensatory mechanisms in its design. The treaty anticipated that unavoidable pressures on biodiversity would continue and that practical adjustments were likely to have better prospects for successful conservation than insisting on a non-restriction approach to wetlands.

Standing Committee, *DOC SC25-8: Analysis and recommendations of IUCN's Environmental Law Centre (Decision SC24-10) on Revisions to Ramsar sites boundaries, interpretation of Articles 2.5 and 4.2 (Resolution VII.23, paragraphs 9, 10, 11 & 13)* (2000).

[37] Ramsar Convention Secretariat, *An Introduction to the Convention on Wetlands* (previously *The Ramsar Convention Manual*) (Ramsar Handbooks, 5th edition, Ramsar Convention Secretariat, Gland, 2016) 42.

[38] K ten Kate, J Bishop, and R Bayon, *Biodiversity offsets: Views, experience, and the business case* (IUCN and Insight Investment, Gland, Cambridge, and London, 2004), at: https://www.iucn.org/sites/default/files/import/downloads/bdoffsets.pdf.

[39] AJ Bled, 'Business to the Rescue: Private Sector Actors and Global Environmental Regimes' Legitimacy' (2009) 9(2) *International Environmental Agreements: Politics, Law and Economics* 153–71; KI MacDonald, 'The Devil is in the (Bio)Diversity: Private Sector "Engagement" and the Restructuring of Biodiversity Conservation' (2010) 42(3) *Antipode* 513.

4 THE PRACTICE OF COMPENSATORY MECHANISMS: GROWING IN SCOPE BUT WITH DOUBTFUL IMPACT

Compensatory mechanisms have gained prominence globally and found their way into practical arrangements to offset the residual impacts of wetland loss – that is, to compensate losses after measures have sought to prevent and minimise the negative impacts. Compensatory mechanisms for wetlands preceded the development of compensatory mechanisms for ecosystems beyond wetlands but have recently been developing closely with them.

The basic governance level at which compensatory mechanisms are organised is that of countries. As per the guidance by the Ramsar COP,[40] most parties have integrated the rules for compensation of wetland loss into their national policies concerning land and water planning and now have provisions for offsetting the damage done to wetlands and other ecosystems in national laws and strategies.[41] The latest available aggregated data (dated 2016) lists 69 countries with a known national policy of biodiversity offsets, partly overlapping with the EU-wide requirement to 27 member states for compensation to sites protected under the Natura 2000 network and a further five countries with sub-national rules.[42] The requirements are well distributed across the globe; middle or low-income countries are just as well represented as richer countries and there are no blank spots among the regions.[43] Wetlands are among the most traded types of ecosystems and the most studied habitats among the peer review literature on offsets.[44]

But national, subnational, or regional legislation does not portray the full picture. Biodiversity offsets are often undertaken on a voluntary basis by businesses as a follow-on from their social and environmental voluntary commitments, particularly in heavy-impact sectors such as extractive industries, water, and urban development.[45] (Interestingly, no similar practice has developed for agricultural business, which is a significant source of pollution for

---

[40] Ramsar Convention, note 26.
[41] Gardner et al, note 23, 3–8.
[42] Maron et al, note 30.
[43] A Villarroya, AC Barros, and J Kisesecker, 'Policy development for environmental licensing and biodiversity offsets in Latin America' (2014) 9(9) *PLoS One*; S Gelcich et al, 'Achieving Biodiversity Benefits with Offsets: Research Gaps, Challenges, and Needs' (2017) 46(2) *Ambio* 185–87.
[44] Gelcich et al, note 43, 186.
[45] N Doswald et al, *Biodiversity offsets: voluntary and compliance regimes. A review of existing schemes, initiatives and guidance for financial institutions* (UNEP-WCMC and UNEP FI, Cambridge and Geneva, 2012) 9, at: https://www

wetlands.) Businesses may opt for voluntary commitments for various reasons: for reputational purposes; to pre-empt regulation; to reduce operational risk exposure; to take advantage of new business development opportunities; or in response to the requirements of investors or lenders, such as development banks.[46] Compensation is a must under the Equator Principles – a voluntary set of standards for financial institutions for determining, assessing, and managing social and environmental risk in project financing with capital costs exceeding $10 million.[47] About 169 countries in total are subject to these requirements by financial institutions.[48] However, despite widespread financial incentives, most offset projects arise due to regulatory requirements.[49]

In terms of methods for restoring the ecological character of lost wetlands, response options are wetland restoration (promoting a return to original, pre-disturbance conditions and improving wetland functions)[50] and wetland creation (creation of wetlands on land that has never been wetland).[51] The Ramsar Convention stresses the significance of formulating goals, objectives, and performance standards (also dubbed 'success criteria') for individual projects, while allowing the specific compensation measures to be determined by national legislation related to EIAs and land and water planning.[52] While ecosystem functions and properties are the primary goals of compensatory endeavours, their implementation can be merged with various socio-cultural or socio-economic benefits. Indeed, compensatory projects have contributed to local cultural values (spiritual enrichment, recreation, ecotourism, aesthetics, formal education, environmental awareness and appreciation, and cultural heritage).[53] Instances of wetland compensation in South Africa have been com-

---

.unepfi.org/fileadmin/documents/Biodiversity_Offsets-Voluntary_and_Complian ce_Regimes.pdf.

[46] Ibid, 14–16.
[47] Ibid, 10.
[48] Maron et al, note 30, 2.
[49] JW Bull and N Strange, 'The global extent of biodiversity offset implementation under no net loss policies' (2018) 1 *Nature Sustainability* 790.
[50] Ramsar Convention, *Resolution VIII.16*, note 22.
[51] Ramsar Convention, note 26, [84].
[52] Ibid, 27 [82].
[53] R Fish et al, 'Making space for cultural ecosystem services: insights from a study of the UK nature improvement initiative' (2016) 21 *Ecosystem Services* 329; B Fischer, RK Turner, and P Morling, 'Defining and classifying ecosystem services for decision making' (2009) 68 *Ecological Economics* 643; B Clarke et al, 'Integrating Cultural Ecosystem Services valuation into coastal wetlands restoration: A case study from South Australia' (2021) 116 *Environmental Science & Policy* 220. See also Chapters 6 and 10 in this volume.

bined with the goal of poverty alleviation.[54] The development of compensatory mechanisms as well as ancillary services can also offer opportunities for skills development – such as monitoring, legal, insurance, registry, and technical support services – which all contribute to economic development and local employment; and skills development for species identification, conservation management, and sociocultural knowledge.[55]

In terms of legal forms of implementation, the three main options are one-off offsets, in-lieu fees, or habitat banking.[56] One-off offsets are carried out by the developer or by a subcontractor (eg, a conservation NGO), with the developer assuming financial and legal liability and with verification typically undertaken by a government agency or an accredited third party.[57] In-lieu fees require the developer to pay a fee to the offset provider, which takes on the financial and legal responsibility for the offset.[58] Habitat banking relates to a repository where credits from actions with beneficial biodiversity outcomes can be purchased by the developer or permittee to offset the debit from the environmental damage they would create.[59] As under the in-lieu fee arrangement, financial and legal liability are transferred from the developer to the provider, which can be a public or private entity.

Habitat banks have attracted considerable attention in scholarship and policy due to their innovative character in the context of cost-effective instruments or market mechanisms for biodiversity protection.[60] Wetland offsets pioneered the idea for more permanent tradable conservation units. On the one hand, the instrument holds potential for reducing loss of habitats and spurring private investment in habitat restoration, particularly when compensation for loss becomes widely required as part of development projects or extractive industries, thus generating sufficient demand for credits. In some

---

[54] IIED, *South Africa-Working for Wetlands (WfWet)* (2012), at: https://watershedmarkets.org/casestudies/South_Africa_Working_for_Wetlands.html.

[55] A Bovarnick, C Knight, and J Stephenson, *Habitat Banking in Latin America and Caribbean: A Feasibility Assessment* (United Nations Development Programme, 2010), at: https://www.cbd.int/financial/offsets/g-offsethabitatbanklac-undp.pdf.

[56] OECD, *Biodiversity Offsets: Effective Design and Implementation* (OECD Publishing, Paris, 2016).

[57] Ibid, 50–53.

[58] Ibid, 53.

[59] Ibid, 50–53.

[60] eftec et al, *The use of market-based instruments for biodiversity protection – The case of habitat banking – Technical Report* (eftec, IIEP, 2010), at: https://ec.europa.eu/environment/enveco/pdf/eftec_habitat_technical_report.pdf; Bovarnick, Knight, and Stephenson, note 55.

places, strong demand exists on the side of developers and organisations for restoration projects.[61] Habitat banking provides readily accessible restoration opportunities to meet such demand.

On the other hand, as the most complex policy tool of the compensatory mechanisms, habitat banking amplifies the concerns and risks around compensatory mechanisms. In habitat banking, it is particularly difficult to ensure that the ecosystem which is replacing the forgone one is equivalent in quality. Establishing an adequate metric (a 'unit' of habitat destroyed or species affected) is a socio-ecological challenge, lacking standardisation and tool design.[62] A particular concern with habitat banking is the temporal loss of wetlands because credits can be released before ecological benefits start to take effect.[63]

Despite the concerns, habitat banks do not exist widely. The United States, Canada, and Australia have piloted wetland mitigation banking models specifically.[64] More general habitat banks seem to be most widespread in the United States – particularly in California and Florida[65] – and to a more limited extent in the European countries (Germany, France, Italy, the United Kingdom, and

---

[61] J King, T Bromfield and I Milborrow, *Accelerating Finance for Nature: Barriers and recommendations for scaling private sector investment* (PwC, 2023), at: https://www.pwc.com/gx/en/nature-and-biodiversity/nature-fin-accelerator-mode.pdf.

[62] SJ Chiavacci and EJ Pindilli, 'Trends in biodiversity and habitat quantification tools used for market-based conservation in the United States' (2020) 34(1) *Conservation Biology* 125.

[63] H Levrel, P Scemama, and A-C Vaissiere, 'Should we be wary of mitigation banking? Evidence regarding the risks associated with this wetland offset arrangement in Florida' (2017) 135 *Ecological Economics* 136.

[64] S Burgin, '"Mitigation banks" for wetland conservation: a major success or an unmitigated disaster?' (2010) 18 *Wetlands Ecology and Management* 4; KW Cox and A Grose (eds), *Wetland Mitigation in Canada: A Framework for Application*, Sustaining Wetlands Issues Paper 1 (Secretariat to the North American Wetlands Conservation Council, Ottawa, 2000), at: https://nawcc.wetlandnetwork.ca/Wetland%20Mitigation%202000-1.pdf.

[65] J Poudel, D Zhang, and B Simon, 'Habitat conservation banking trends in the United States' (2019) 28(6) *Biodiversity and Conservation* 1629; AC Vaissière and H Levrel, 'Biodiversity offset markets: What are they really? An empirical approach to wetland mitigation banking' (2015) 110 *Ecological Economics* 81.

Spain)[66] and in Latin American and Caribbean countries.[67] Habitat banks have also been explored at the transnational level, meaning that implementation of the offsets could take place in a country other than that in which the damage occurs.[68] A lead here is the Business and Biodiversity Offset Programme (BBOP) – a collaborative programme of over 40 companies, financial institutions, governments, and civil society organisations whose efforts have been recognised by a Ramsar Resolution.[69] For the most part, however, wetland and biodiversity banks are being explored as a potential, rather than a fully fledged policy tool.

Reflecting on the practice of biodiversity offsetting, a key concern relates to its impact. There is a disconnect between the use of biodiversity offsetting around the world and evaluations of the impact of this policy tool. The empirical research that exists has pointed to some of the real risks. On the governance end, countries may be advancing quite detailed offset policies while lacking strong requirements regarding impact avoidance.[70] Concerns have been raised over the process and lack of inclusiveness, even in well-resourced countries.[71] Furthermore, implementation and enforcement of such transactions are often subject to weak governance frameworks, insufficient monitoring, and poorly defined liabilities,[72] when regulatory responses throughout the process play a crucial role in mitigating the risks.[73] On a practical level, there is a real risk of temporal loss of wetlands and spatial mismatch due to the (growing) distance between impact sites and compensation sites.[74] Some of the biodiversity offsets were found to have been harmful. In such instances, they missed the

---

[66] eftec et al, note 60; S Maestre-Andrés et al, 'Habitat banking at a standstill: The case of Spain' (2020) 109 *Environmental Science & Policy* 54–63; MMJ Gorissen, C Martijn van der Heide, and JHJ Schaminée, 'Habitat Banking and Its Challenges in a Densely Populated Country: The Case of the Netherlands' (2020) 12 *Sustainability* 3756.

[67] Bovarnick, Knight, and Stephenson, note 55.

[68] Penca, note 16; Penca, note 33.

[69] Ramsar Convention, *Resolution X.12: Principles for partnerships between the Ramsar Convention and the business sector* (2008).

[70] Villarroya, Barros, and Kisesecker, note 43.

[71] Maestre-Andrés et al, note 66.

[72] F Quétier and S Lavorel, 'Assessing ecological equivalence in biodiversity offset schemes: Key issues and solution' (2011) 144(12) *Biological Conservation* 2991.

[73] Levrel, Scemama and Vaissiere, note 63.

[74] Ibid.

target of counterbalancing the ecological loss.[75] They were found to displace people and negatively affect livelihoods.[76] While evidence is not conclusive, some of it points to inappropriate practices in attempting to achieve the impressive-sounding environmental targets.[77]

5 COMPENSATION, NO NET LOSS, AND RESTORATION: PRAGMATIC APPROACHES TO THE STRUCTURAL FAILURE OF GLOBAL BIODIVERSITY GOVERNANCE

Compensatory mechanisms for wetlands are among the pivotal approaches of a contemporary biodiversity regime. Yet they are also fundamentally questioned as the appropriate strategy for furthering sustainability. A number of critics point to how compensatory mechanisms, alongside other mitigation measures, provide green credentials to those that perpetuate biodiversity loss and legitimise the practices that miss the important targets.[78] These critics show how, by building on the same principles that are responsible for imperilling socio-ecological systems, compensatory mechanisms do little to challenge unsustainable trajectories. While trying to mitigate the impact of unsustainable activities as a short-term tactic, from a longer-term perspective they facilitate environmental destruction.

The idea of compensation for loss was introduced by the Ramsar Convention somewhat experimentally and probably without anticipation of its subsequent development and application to other terrestrial habitats. It was deployed as a pragmatic approach to the ongoing conflict between economically profitable

---

[75] DB Lindenmayer et al, 'The anatomy of a failed offset' (2017) 210 *Biological Conservation* 286.

[76] LJ Sonter et al, 'Biodiversity offsets may miss opportunities to mitigate impacts on ecosystem services' (2018) 16 *Frontiers in Ecology and Environment* 143.

[77] Levrel, Scemama and Vaissiere, note 63; Maestre-Andrés et al, note 66; Gorissen, Martijn van der Heide and Schaminée, note 66.

[78] Apostolopoulou and Adams, note 33; P Le Billion, 'Crisis conservation and green extraction: biodiversity offsets as spaces of double exception' (2021) 28(1) *Journal of Political Ecology* 854–88; B Neimark and B Wilson, 'Re-mining the collections: from bioprospecting to biodiversity offsetting in Madagascar' (2015) 66 *Geoforum* 1–10; B Büscher, S Sullivan, K Neves, J Igoe, and D Brockington, 'Towards a synthesized critique of neoliberal biodiversity conservation' (2012) 23 *Capitalism, Nature, Socialism* 4–30.

activities and conservation.[79] The popularity of compensatory mechanisms in other regimes exemplifies how wetland management and research have been influential in the paradigm shift for biodiversity governance.[80] The existence and spread of compensatory mechanisms are also reflective of wetland governance, much like biodiversity governance more broadly, being embedded in discourses, institutions, and other structures of power that support, rather than alter, the existing unsustainable course of action.[81] These structures have prevented more transformative actions and deeper systemic changes in the direction of sustainability.

A less critical perspective may represent the offsets as a logical consequence of the growing research on the indispensable role of biodiversity in sustaining life. Efforts have been made to categorise, quantify, and valuate nature's benefits for people, dubbing them 'ecosystem services'.[82] Valuation approaches have been applied at the global level.[83] Research has pointed to unprecedented biodiversity losses, compromising the health of ecosystems and their future productivity, as well as the health and wellbeing of humans.[84] Important

---

[79] M Virah-Sawmy, J Ebeling, and R Taplin, 'Mining and biodiversity offsets: a transparent and science based approach to measure "no-net-loss"' (2014) 143 *Journal of Environmental Management* 61–70; W M Adams, 'Sleeping with the enemy? Biodiversity conservation, corporations and the green economy' (2017) 24 *Journal of Political Ecology* 243–57. https://doi.org/10.2458/v24i1.20804.

[80] E Maltby and MC Acreman, 'Ecosystem services of wetlands: pathfinder for a new paradigm' (2011) 56(8) *Hydrological Sciences Journal* 1341.

[81] MTJ Kok et al, 'Enabling Transformative Biodiversity Governance in the Post-2020 Era' in IJ Visseren-Hamakers and MTJ Kok (eds), *Transforming Biodiversity Governance* (Cambridge University Press, 1st edition, 2022) 341–60; A Agrawal et al, 'From Environmental Governance to Governance for Sustainability' (2022) 5(6) *One Earth* 615–21.

[82] R Costanza et al, 'The Value of the World's Ecosystem Services and Natural Capital' (1997) 387(6630) *Nature* 253; C Daily (ed), *Nature's Services: Societal Dependence on Natural Ecosystems* (Island Press, Washington, 1997); EB Barbier, M Acreman, and D Knowler, *Economic valuation of wetlands: A guide for policy makers and planners* (Ramsar Convention Bureau, Gland, 1997); National Academy of Sciences USA, *Valuing Ecosystem Services: Toward Better Environmental Decision-Making* (National Academies Press, Washington, 2004).

[83] United Nations Environment Programme, *Millennium Ecosystem Assessment: Ecosystems and Human Well-being: Synthesis* (UNEP, 2005); The Economics of Ecosystems and Biodiversity (TEEB), *Mainstreaming the Economics of Nature: A Synthesis of the Approach, Conclusions and Recommendations of TEEB* (2010), at: http://www.teebweb.org/publications/teeb-study-reports/synthesis.

[84] S Diaz et al, 'Biodiversity loss threatens human well-being' (2004) 4(8) *Plos Biology* 1300–05; B Cardinale et al, 'Biodiversity loss and its impact on humanity' (2012) 486(7401) *Nature* 59.

implications flow from the recognition of close connectivity across terrestrial, inland water, and marine systems, and the consideration of people as an intrinsic part of ecosystems, rather than separate from them.[85]

In response to the advances in scientific knowledge, policy sought to devise increasingly integrative, resilience-focused responses that attempt to mainstream biodiversity conservation in everyday decisions, rather than just creating multiple fenced, siloed protected areas. Alongside strictly protected areas, consideration of habitats, resources, and livelihoods outside designated areas has gained prominence.[86] Imposing strict limitations and prohibiting interference have proven to be politically unfeasible interventions.[87] Instead, the rationale developed to essentially permit negative impacts on habitats, including wetlands, but require action to counterbalance these impacts. The current approach is thus to reconcile conservation with threats from economic development and unsustainable use. To this end, the continuing alerts over the degradation of nature are met with propositions of a positive narrative and hope for change.

The mitigation hierarchy and compensatory mechanisms (or biodiversity offsets) have come to play a key role as tools, while no net loss and restoration targets act as goals, in the architecture and paradigmatic orientation of contemporary biodiversity governance. They offer a normative orientation to reverse the trends – albeit in the form of distant goals which may not be accomplished.

The requirement of *no net loss* has consolidated in the broader biodiversity management regime. The EU 2020 Biodiversity Strategy (2011–20), for example, aimed at no net loss and provided for compensatory mechanisms, including offsets, to play a crucial role in it.[88] Allegedly, more than 80 countries require some form of no net loss – albeit without uniform definition[89] and with offsets far from homogeneous in implementation.[90] Many businesses

---

[85] E Östrom, 'A general framework for analysing sustainability of social–ecological systems' (2009) 325 *Science* 419; S Díaz et al, 'Assessing nature's contributions to people' (2018) 259(6373) *Science* 270.

[86] JN Pretty and MP Pimbert, 'Beyond conservation ideology and the wilderness' (1995) 19 *Natural Resources Forum* 1, 5–14.

[87] Ibid.

[88] European Commission (EC), *Our life insurance, our natural capital: an EU biodiversity strategy to 2020*, COM/2011/0244 final (2011).

[89] M Maron et al, 'The many meanings of no net loss in environmental policy' (2018) 1 *Nature Sustainability* 19; JW Bull et al, 'Seeking convergence on the key concepts in "no net loss" policy' (2016) 53 *Journal of Applied Ecology* 1686.

[90] Bull and Strange, note 49.

have also formulated their corporate strategies relating to this.[91] Some strategic goals have gone beyond the ambition of no net loss and aim to accomplish *net gain* or *net positive gain*. For example, the BBOP aims to apply offsets in a way that achieves no net loss and preferably a net gain of biodiversity on the ground with respect to species composition, habitat structure, ecosystem function, and people's use and cultural values associated with biodiversity.[92]

However, for a *net gain* target to be more advanced and ambitious than *no net loss*, more clarity is needed on the exact requirements.[93] Both versions – even when mandatory – suffer from important shortcomings in their implementation, including leading to a temporary reduction in biodiversity loss with only a *promise* of long-term improvement,[94] rather than actual guarantees of conservation gains in addition to those that would have occurred in the absence of the offset.[95] They also do not create adequate long-term reference scenarios reversing the baseline trajectory.[96]

The *no net loss* approach has been critiqued for entrenching ongoing biodiversity loss unless the principle is merged and aligned with national biodiversity targets.[97] Indeed, targets have become a prominent approach at the highest level. They motivated the Strategic Plan for Biodiversity 2011–20 (Aichi Targets) and the Sustainable Development Goals (Goal 15 – Life on Land is insufficiently precise), and seem to underpin the development of the Kunming-Montreal Global Biodiversity Framework. Target-based approaches

---

[91] HJ Rainey et al, 'A Review of Corporate Goals of No Net Loss and Net Positive Impact on Biodiversity' (2015) 49 *Oryx* 232.

[92] Business and Biodiversity Offsets Programme (BBOP), *To No Net Loss and Beyond: An Overview of the Business and Biodiversity Offsets Programme (BBOP)* (Washington, 2013), at: www.forest-trends.org/biodiversityoffsetprogram/guidelines/Overview_II.pdf.

[93] JW Bull and S Brownlie, 'The transition from No Net Loss to a Net Gain of biodiversity is far from trivial' (2017) 51 *Oryx* 53.

[94] FLP Damiens, A Backstrom, and A Gordon, 'Governing for "no net loss" of biodiversity over the long term: challenges and pathways forward' (2021) 4(1) *One Earth* 60.

[95] TA Gardner et al, 'Biodiversity offsets and the challenge of achieving no net loss' (2013) 27(6) *Conservation Biology* 1254.

[96] M Maron et al, note 89; JS Simmonds et al, 'Moving from Biodiversity Offsets to a Target-based Approach for Ecological Compensation' (2020) 13(2) *Conservation Letters*; SOSE zu Ermgassen et al, 'The Role of "No Net Loss" Policies in Conserving Biodiversity Threatened by the Global Infrastructure Boom' (2019) 1(3) *One Earth* 305–15.

[97] Simmonds et al, note 96.

may be additionally determined at smaller scales – for example, at the level of particular programme, habitat, or area.

Most recently, biodiversity governance vision began promoting the notion of *restoration* of ecosystems. The idea relates to initiating or accelerating the recovery of an ecosystem from a degraded state with a view to regaining its ecological functionality and improving the productivity and capacity of ecosystems to meet the needs of society.[98] As a governance objective, it gained prominence in 2011 with the Bonn Challenge, launched by the German government and IUCN[99] and recognised as facilitating the implementation of many existing international commitments, including fighting climate change, enhancing food security, providing clean water, protecting biodiversity, alleviating poverty, and improving human wellbeing.[100] Indeed, the goal of restoration now underlies the efforts of various conventions and processes, including the CBD's Strategic Plan for Biodiversity 2020 and Aichi Targets 14 and 15; the UN Framework Convention on Climate Change; the UN Convention to Combat Desertification; and the Sustainable Development Goals (particularly Target 15.3 on land degradation neutrality).[101] Political support for ecosystem restoration culminated with the launch of the UN Decade on Ecosystem Restoration (2021–30), with the purpose of scaling up efforts to prevent, halt, and reverse the degradation of ecosystems worldwide and raise awareness of the importance of successful ecosystem restoration.[102] Interestingly, the

---

[98] Intergovernmental Science-Policy Platform on Biodiversity and Ecosystem Services, *The IPBES assessment report on land degradation and restoration* (Secretariat of the Intergovernmental Science-Policy Platform on Biodiversity and Ecosystem Services, Bonn, 2018); UNEP, *New UN Decade on Ecosystem Restoration offers unparalleled opportunity for job creation, food security and addressing climate change opportunity* (2019), at: https://www.unenvironment.org/news-and-stories/press-release/new-un-decade-ecosystem-restorationoffers-unparalleled-opportunity.

[99] IUCN, *The Bonn Challenge*, at: https://www.iucn.org/theme/forests/our-work/forest-landscape-restoration/bonn-challenge.

[100] MARN (Ministerio de Medio Ambiente y Recursos Naturales), *UN decade of ecosystem restoration 2021–2030. Initiative proposed by El Salvador System (SICA). Concept Note.* (2019) (Ministerio de Medio Ambiente y Recursos Naturales, El Salvador with the Support of Countries from the Central American Integration).

[101] UNGA Resolution 73/284 *United Nations Decade on Ecosystem Restoration (2021–2030)* (2019).

[102] Ibid; see also the EU Proposal for Regulation on Nature Restoration, 2022/0195 (COD), 22 June 2022, at: https://environment.ec.europa.eu/publications/nature-restoration-law_en, proposing legally binding targets to restore degraded

Ramsar Convention proposed restoration as an element of national planning for wetland conservation and wise use in 1999 – considerably earlier than the concept rose to global prominence.[103]

The restoration goals are certainly ambitious and potentially far-reaching, but their impact is uncertain. At the operational level, their success is dependent on states adopting adequate policy instruments at national scale and following up with enforcement. But from the perspective of governance for sustainability, their success will be judged by their ability to introduce a reprioritisation of nature-positive goals.

If contemporary biodiversity governance has consolidated around finding compromises to the constant pressures on the environment, the influence of the Ramsar Convention in framing those rules can be assessed as very positive. The treaty process has contributed a framework which is science-based, detailed, dynamically evolving, and unequivocal about impact avoidance being an urgent priority. This can be strengthened through strong attention to the application and enforcement of that framework and an insistence on the rule of law for nature. Without the onus on procedural and substantive oversight, the goals of ending biodiversity loss stand to be subordinated to the pressure of developmental exigencies.[104] This includes a focus on enforcing prevention. In other words, avoidance of harm – rather than mitigation – needs to (increasingly) become the dominant value, practice, and goal.

6 CONCLUSION: CONTRIBUTION TO THE STATUS OF WETLANDS

From an evolutionary perspective, the Ramsar Convention text established the avoid-mitigate-compensate principle and pioneered the approach of compensatory mechanisms in international environmental law. This has had gradual but important ramifications. The Convention required compensation in a strictly defined and limited context, but the subsequent processes foremost outside the Convention further shaped the idea of compensation. The principle has been refined and its use extended – to wetlands beyond Ramsar Sites and to biodiversity more broadly. The idea of compensation for loss and various tools

---

EU ecosystems – in particular, those with the most potential to remove and store carbon and to prevent and reduce the impact of natural disasters.

[103] Ramsar Convention, *Resolution VII.17*, note 22. See also Ramsar Convention, *Resolution VIII.16: Principles and guidelines for wetland restoration* (2002).

[104] Ermgassen et al, note 96; G Tucker et al, 'Conclusions: Lessons from Biodiversity Offsetting Experiences in Europe' in W Wende et al (eds), *Biodiversity Offsets* (Springer, 2018).

to effectuate it became part of the mainstream policy for countries, businesses, and the global biodiversity regime.

The Ramsar Convention introduced the mitigation hierarchy as an extension or operationalisation of the preventive approach. It was always adamant about compensation being a last resort in effective wetland conservation and management, should other options fail. The contribution by the Convention was to present compensatory mechanisms as a compromising tool and work on providing normative guidance and technically sound specification on how to accomplish them. In doing so, the Convention forged its reputation as a pragmatic and highly adaptive international treaty. Among other activities, the development of approaches to avoidance, mitigation, and compensation for wetland losses has helped to expand the Convention's remit and rationale to a more human-centric one, with direct implications for people's livelihoods and wellbeing through climate, culture, food security, clean water, and poverty implications.

Nevertheless, the mechanism of compensation for loss established both through the work of the Ramsar Convention and beyond it has overall not reversed the downward trends of wetland loss. The *Global Wetland Outlook*'s latest edition (2021) reports on the continuing deterioration of wetland extent and condition globally and reveals that natural wetland losses since 1970, where data became available, stand at 35%.[105] The loss of natural wetlands has not been compensated by the creation of human-made wetlands, such as rice paddies and reservoirs.[106] In many places in the world, loss of productive land to developmental projects is still not entirely regulated and avoided. This could be said to be due to the underdevelopment and limited application of compensatory mechanisms, and a lack of enforcement. However, it is predominantly due to the fact that compensatory mechanisms as a conceptual tool do not challenge unsustainable trajectories.

Thus, we celebrate Ramsar Convention's anniversary with mixed feelings. On the one hand, we can be inspired by its remarkable power to influence national and international governance as well as its resilience in the face of ever-increasing pressures on biodiversity. Its pragmatic approach, as illustrated by its compensatory mechanisms, continues to encourage the search for a sustainability balance in a largely growth and wealth-focused world. On the other hand, we regret the weakness – of the Convention, of the policy tools it

---

[105] M Courouble et al, *Global Wetland Outlook: Special Edition 2021* (Ramsar Convention Secretariat, Gland), at: https://www.global-wetland-outlook.ramsar.org/report-1., 21.

[106] Ibid, 21.

devised, and of societies – in arresting the decline of precious biomes, including by enacting more radically different visions for their place in our planet.

6. Indigenous peoples and local communities: Wetlands management and the Ramsar Convention

Simon Marsden

1 INTRODUCTION

The Ramsar Convention was the first global agreement to conserve a particular habitat.[1] 'Wetlands'[2] and 'waterfowl'[3] are defined, and states are required to designate at least one wetland to add to the List of Wetlands of International Importance ('the List'), chosen 'on account of their international significance'.[4] Once a site is inscribed, parties must promote its 'conservation'. A Conference of the Parties (COP) meets every three years,[5] with members to include 'experts on wetlands or waterfowl by reason of *knowledge and experience gained in* scientific, administrative or *other appropriate capacities*'.[6] Despite (traditional and local) knowledge and experience, there is no direct mention in the text of the role of Indigenous peoples or local communities in wetlands management, reflecting the early date of the Convention's adoption. The objective of this Chapter is to uncover the parameters of that role in wetlands management from the text, resolutions, reports, plans, and guidance, together with underlying contextual international law.

In marked contrast to the pioneering work undertaken by Ramsar concerning the designation of protected areas, the advancement of wise use and sustainable development, and elaboration of compensation and mitigation

---

[1] Convention on Wetlands of International Importance especially as Waterfowl Habitat, adopted 2 February 1971, entered into force 21 December 1975, 996 UNTS 245 (amended 1982 & 1987).
[2] Ibid, Art 1.1.
[3] Ibid, Art 1.2.
[4] Ibid, Art 2.2.
[5] Ibid, Art 6.
[6] Ibid, Art 7.1 (emphasis added).

measures in instances in which human development within wetland habitats is inevitable, the Convention has proved to be strikingly less prominent in addressing Indigenous peoples or local communities. Indeed, the only references to these actors in the Ramsar text are indirect (my emphasis in italics throughout). In addition to reference to COP membership, the '*interdependence of Man and his environment*' is recognised;[7] and parties 'shall ensure that *those responsible at all levels for wetlands management shall be informed of, and take into consideration,* recommendations of such conferences *concerning the conservation, management and wise use* of wetlands and their flora and fauna'.[8] The general exclusion of Indigenous peoples and local communities from COP activities[9] suggests top-down consultation with Indigenous peoples and local communities only. Moreover, Ramsar criteria for identifying wetlands[10] are contained in two groups, based on representativeness/uniqueness and biodiversity;[11] in 2005, a ninth criterion was added: wetland-dependent non-avian animal species. None of the criteria includes Indigenous peoples or local communities, notwithstanding any traditional, cultural, or economic connections with wetlands and associated species. Together with limitations on consultation, this further weakens opportunities for Indigenous peoples and local communities to engage in wetlands management.

---

[7] Ibid, Preamble, first Recital.

[8] Ibid, Art 6.3. See Ramsar Convention Secretariat, *Handbook 18: Managing Wetlands: Frameworks for managing wetlands of international importance and other wetland sites* (Ramsar Handbooks, 4th edition, Ramsar Convention Secretariat, Gland, 2010). For an up-to-date commentary focused largely on management, see PA Gell, NC Davidson, and CM Finlayson (eds), *Ramsar Wetlands: Values, Assessment, Management* (Elsevier, Amsterdam, 2021).

[9] Contrast the role of Indigenous peoples on the Arctic Council as 'Permanent Participants'; see Declaration on the Establishment of the Arctic Council, Ottawa, 19 September 1996 (1996) 35 ILM 1387 (Ottawa Declaration), Art 2, which states: 'The category of Permanent Participation is created to provide for active participation and full consultation with the Arctic indigenous representatives within the Arctic Council.' For historical background, see ET Bloom, 'Establishment of the Arctic Council' (1999) 93 *American Journal of International Law* 712; A Yefimenko, 'How Arctic Indigenous Peoples Negotiated a Seat at the Table', 10 May 2021, *Arctic Council* at: https://arctic-council.org/news/a-seat-at-the-table-how-arctic-indigenous-peoples-negotiated-their-permanent-participant-status/.

[10] Ramsar Convention, *Ramsar Information Paper no.5: The Criteria for Identifying Wetlands of International Importance*.

[11] See Ramsar Convention, *Resolution VII.11: Strategic framework and guidelines for the future development of the List of Wetlands of International Importance* (1999).

This Chapter is organised in seven sections. Section 2 distinguishes between Indigenous peoples and local communities with reference to Ramsar and contextual international law. Section 3, in the absence of specific provisions in the Ramsar Convention, examines rights (briefly) and participation in other international regimes. It also analyses roles under the resolutions, reports, plans, and guidance of Ramsar. Section 4 reviews relevant cultural considerations under Ramsar. Section 5, given the centrality of wise use to management, examines the relationship between this concept and traditional knowledge under Ramsar. Section 6 evaluates the relationship between Ramsar and other conservation treaties – particularly measures relevant for Indigenous peoples and, potentially, local communities – focusing on the World Heritage Convention (WHC).[12] Conclusions and recommendations follow in the final section.

2 DEFINITIONS, DIFFERENCES, AND RELATIONSHIPS

By way of wider international context, the United Nations (UN) has wrestled with definitional issues of 'Indigenous peoples' for many years, concluding that a single universal definition is not appropriate. Since 2002, Ramsar has used the term 'Indigenous peoples' in plural, corresponding to UN terminology employed since the UN Declaration on the Rights of Indigenous Peoples (UNDRIP).[13] While UNDRIP does not define 'Indigenous peoples' in detail, the Convention Concerning Indigenous and Tribal Peoples in Independent Countries 1989[14] ('ILO Convention 169') does include a clear comprehensive definition:

1. This Convention applies to:
   (a) tribal peoples in independent countries whose social, cultural and economic conditions distinguish them from other sections of the national community, and whose *status is regulated wholly or partially by their own customs or traditions or by special laws or regulations*;

---

[12] Convention Concerning the Protection of the World Cultural and Natural Heritage, adopted 16 November 1972, entered into force 17 December 1975, 1037 UNTS 151 (WHC).

[13] United Nations General Assembly, *Resolution 61/295: United Nations Declaration on the Rights of Indigenous Peoples*, UN Doc A/RES/61/295 (2007) 4, 32 (UNDRIP).

[14] ILO Convention 169, Concerning Indigenous and Tribal Peoples in Independent Countries, adopted 27 June 1989, entered into force 5 September 1991, 28 ILM 1382.

(b) peoples in independent countries who are regarded as indigenous *on account of their descent from the populations which inhabited the country, or a geographical region to which the country belongs, at the time of conquest or colonisation or the establishment of present state boundaries and who, irrespective of their legal status, retain some or all of their own social, economic, cultural and political institutions.*

2. *Self-identification as indigenous or tribal shall be regarded as a fundamental criterion* for determining the groups to which the provisions of this Convention apply.[15]

The Arctic Council has also examined issues of definition relevant for Indigenous peoples.[16] 'Traditional' or 'Indigenous knowledge' has been defined by the Arctic Council in the context of the Conservation of Arctic Flora and Fauna as follows:

> Indigenous Knowledge is a systematic way of thinking and knowing that is elaborated and applied to phenomena across biological, physical, cultural and linguistic systems. Indigenous Knowledge is owned by the holders of that knowledge, often collectively, and is uniquely expressed and transmitted through indigenous languages. It is a body of knowledge generated through cultural practices, lived experiences including extensive and multi-generational observations, lessons and skills. It has been developed and verified over millennia and is still developing in a living process, including knowledge acquired today and in the future, and it is passed on from generation to generation.[17]

---

[15] Questions of identity are complicated by interaction with colonists. See, for example, E Sidorova, 'Circumpolar Indigeneity in Canada, Russia, and the United States (Alaska): Do Differences Result in Representational Challenges for the Arctic Council?' (2019) 72 *Arctic* 71; M Fitzmaurice, 'Indigenous Peoples and Identity' (2021) 10 *Cambridge International Law Journal* 170.

[16] See further Arctic Council, 'Indigenous Peoples' Secretariat', at: https://arctic-council.org/about/indigenous-peoples-secretariat/; Arctic Council Indigenous Peoples' Secretariat, 'We are the Indigenous Peoples of the Arctic Council', at: https://www.arcticpeoples.com/#intro.

[17] Ottawa Traditional Knowledge Principles 2015 (revised October 2018), at: https://static1.squarespace.com/static/58b6de9e414fb54d6c50134e/t/5dd4097576d4226b2a894337/1574177142813/Ottawa_TK_Principles.pdf. See further Conservation of Arctic Flora and Fauna, *Traditional Knowledge: Progress Report, 2017–2019*, May 2019, at: https://oaarchive.arctic-council.org/handle/11374/2328. Note the Arctic Council Strategic Plan 2021–30 includes Goal 6 – Knowledge and Communications, '[To] generate, collect, analyze and communicate science, and traditional knowledge and local knowledge, as appropriate, and enhance understanding of the Arctic within and beyond the region to inform policy shaping and decision making.' Source: https://oaarchive.arctic-council.org/bitstream/handle/11374/

The United Nations Permanent Forum on Indigenous Issues (UNPFII) has also defined 'Indigenous traditional knowledge', which 'generally means traditional practices and culture and the knowledge of plants and animals and of their methods of propagation. It includes: expressions of cultural values, beliefs, rituals and community laws, knowledge regarding land and ecosystem management'.[18]

A further example is from the International Council for Science, which has defined 'traditional knowledge' as:

> A cumulative body of knowledge, know-how, practices and representations maintained and developed by peoples with extended histories of interaction with the natural environment. These sophisticated sets of understandings, interpretations and means are part and parcel of a cultural complex that encompasses language, naming and classification systems, resource use practices, ritual, spirituality and worldview.[19]

International law and institutions beyond Ramsar therefore provide useful context for definitions of 'Indigenous peoples' and 'Indigenous knowledge' which, given limitations in detail in the Convention text, should be referred to in any definitional questions.

The Ramsar COP did not consider the definitions of 'Indigenous peoples' and 'local communities' until the late 1990s. Resolution VII.8 states vaguely for both:

> [T]he *term 'indigenous people' may vary from country to country.* Furthermore, 'local' is a relative term; *some stakeholders may* live at a distance from the wetland (such as migrating fisherfolk or pastoralists) and still *have traditional claims* to its resources.[20]

2601/MMIS12_2021_REYKJAVIK_Strategic-Plan_2021-2030.pdf?sequence=1&isAllowed=y.

[18] United Nations Permanent Forum on Indigenous Issues, Report of the Secretariat on Indigenous traditional knowledge, UN Doc E/C.19/2007/10 (20 March 2007).

[19] International Council for Science, Science and Traditional Knowledge, 'Report from the ICSU Study Group on Science and Traditional Knowledge,' Paper delivered to 27th General Assembly of ICSU, Rio De Janeiro, Brazil, September 2002, 3.

[20] Ramsar Convention, *Resolution VII.8: Guidelines for establishing and strengthening local communities and indigenous people's participation in the management of wetlands* (1999). My emphasis in italics – an approach continued throughout the text.

This indicates potential overlap between Indigenous peoples and local communities, as the latter may also be Indigenous. Moreover, at COP8 in 2002, Indigenous peoples and local communities were still considered 'stakeholders' in wetlands management.[21] This phrasing is now out of date regarding Indigenous peoples.

A 2010 Ramsar Handbook on participatory skills defines 'community' clearly, but 'Indigenous people' imprecisely:

> The term *community* as used in this Handbook can be understood at two levels. On the one level it represents a more or less homogenous group that is most often defined by geographical location (e.g., a village), but possibly by ethnicity ... On another level, it represents a *collection* of different interest groups such as women and men, young and old, fisherfolk and farmers, wealthy and poor people, and different ethnic groups. Even in relatively unified communities, it is likely that these sub-groups have different interests and perspectives that need to be taken into account in the participatory management process.[22]

Additionally, '*[I]ndigenous people may have been the sole managers of wetlands for many centuries*, so in these contexts *it is more appropriate to speak of acknowledging and strengthening their management role than involvement per se*'.[23] While this formulation of Indigenous 'people' lacks precision, this final sentence nevertheless highlights the significance of their role in the conservation and wise use of wetlands.

---

[21] 'Wetland management, and particularly the planning process, should be as inclusive as possible. Legitimate stakeholders, particularly local communities and indigenous people, should be strongly encouraged to take an active role in planning and in the joint management of sites.' Ramsar Convention, *Resolution VIII.14: New Guidelines for management planning for Ramsar Sites and other wetlands* (2002) [V.29]. For general information on stakeholders as distinct from rights holders, see P Goriup, 'Stakeholder Participation in Management Planning' in CM Finlayson et al, *The Wetland Book* (Springer, Dordrecht, 2017) 1917.

[22] Ramsar Convention Secretariat, *Handbook 7: Participatory skills: Establishing and strengthening local communities' and indigenous people's participation in the management of wetlands* (Ramsar Handbooks, 4th edition, Ramsar Convention Secretariat, Gland, 2010) 6.

[23] Ibid.

3   RIGHTS, PARTICIPATION, AND ROLES

Management of Ramsar Sites and other wetlands is essentially one of management planning (Resolutions 5.7 and VIII.14),[24] via the 'establishment and implementation of a management plan'[25] to 'promote conservation'[26] and maintain their 'ecological character'.[27] Contextual international law is again essential for filling interpretive gaps in Ramsar regarding rights and participation, the latter of which is also relevant for non-Indigenous local communities.

3.1   Contextual International Law

Rights and participation are central to wetlands management for Indigenous peoples and local communities. Significantly, UNDRIP provides that 'Indigenous peoples have the right to the conservation and protection of the environment and the productive capacity of their lands or territories and resources';[28] and that free, prior, and informed consent (FPIC) is required for 'the approval of any project affecting their lands or territories and other resources, particularly in connection with the development, utilization or exploitation of mineral, water or other resources'.[29] These measures go beyond

---

[24] Ramsar Convention, *Resolution 5.7: Management planning for Ramsar sites and other wetlands* (1993); Ramsar Convention, note 21.

[25] Ramsar Convention Secretariat, note 8, 6. Impact assessment is also recognised as part of this process. See Ramsar Convention Secretariat, *Handbook 16: Impact assessment: Guidelines on biodiversity-inclusive environmental impact assessment and strategic environmental assessment* (Ramsar Handbooks, 4th edition, Ramsar Convention Secretariat, Gland, 2010). For discussion of both environmental impact assessment (EIA) and related participation, see R Slootweg, 'Environmental Impact Assessment for Wetlands: Stakeholders and Public Participation' in CM Finlayson et al, *The Wetland Book* (Springer, Dordrecht, 2018) 2059. Additionally, and including reference to FPIC, see F Vanclay, 'Social Impact Assessment for Wetlands' in CM Finlayson et al, *The Wetland Book* (Springer, Dordrecht, 2018) 2077.

[26] Ramsar Convention, note 21, [1]; Convention on Wetlands, note 1, Art 3.1.

[27] Ramsar Convention, note 21, [2]; Convention on Wetlands, note 1, Art 3.2; Ramsar Convention Secretariat, note 8, Strategy 2.4, 5. See Chapter 3 in this volume.

[28] UNDRIP, note 13, Art 29.1.

[29] Ibid, Art 32. 'FPIC describes processes that are free from manipulation or coercion, informed by adequate and timely information, and occur sufficiently prior to a decision so that Indigenous rights and interests can be incorporated or addressed effectively as part of the decision-making process – all as part of meaningfully aiming to secure the consent of affected Indigenous peoples.'

Indigenous peoples as stakeholders, recognising an enhanced status as rights holders. Recognition and respect for traditional knowledge are also crucial to UNDRIP.[30] While UNDRIP may be classed as soft law, national implementation may impose hard law obligations.[31]

ILO Convention 169 contains similar, if weaker, provisions for FPIC.[32] The UNPFII is also of recognised importance.[33] Among other things, the UNPFII 'provides expert advice and recommendations on indigenous issues to the UN Council ... [and] promotes respect for and full application of the provisions of ... UNDRIP'.[34] Collaboration and engagement with multilateral environmental agreements, institutions and processes are high priorities, which is significant for synergies between them, as considered in Section 6 below.

In addition to providing definitions, the Arctic Council has considered the role of Indigenous peoples, specifically regarding wetlands management and stewardship. In the context of the Arctic Council definition of 'Indigenous knowledge', it recently emphasised that 'the participation of and co-management with Arctic indigenous communities is vital, in part as a matter

---

Government of Canada, 'Backgrounder: *United Nations Declaration on the Rights of Indigenous Peoples Act*' (modified 12 October 2021), at: https://www.justice.gc.ca/eng/declaration/about-apropos.html (emphasis in original).

[30] UNDRIP, note 13, Art 31.

[31] In Canada, the UN Declaration on the Rights of Indigenous Peoples Act, 21 June 2021, requires all federal laws to be consistent with UNDRIP.

[32] Cittadino notes that '[a]lthough participation and consultation have been framed as the "cornerstone" of the ILO Convention 169, the requirement of achieving consent has never been interpreted by the ILO Committee as implying an obligation to obtain consent before the initiation of any project'. See F Cittadino, 'The Public Interest to Environmental Protection and Indigenous Peoples' Rights: Procedural Rights to Participation and Substantive Guarantees', in E Lohse and M Poto, *Participatory Rights in the Environmental Decision-Making Process and the Implementation of the Aarhus Convention: A Comparative Perspective* (Duncker and Humblot, Berlin, 2015) 75, 84.

[33] Established by Economic and Social Council Resolution 2000/22, 28 July 2000, at: https://www.un.org/development/desa/indigenouspeoples/unpfii-sessions-2.html.

[34] Cittadino, note 32, 86, highlights furthermore how, in consideration of Indigenous participation in public decision-making affecting the environment, some bodies go beyond merely acknowledging the existence of participatory mechanisms, focusing on effectiveness – notably the Human Rights Committee's General Comment No. 23 on the right of minorities to enjoy their culture. This is illustrated with examples from Inter-American Courts and the Commission.

of rights, due to the importance of *culturally-embedded indigenous knowledge*, and due to the *links between traditional livelihoods and wetlands*'.[35]

Participatory opportunities for local communities and Indigenous peoples have developed significantly in international and domestic law.[36] The reason public participation in wetland conservation is important has been expressed as follows:

> Conservation regimes are likely to be more appropriate and robust when local knowledge of wetlands ecosystems and their specific local uses are known and incorporated into planning and regulatory considerations ... Nearby residents are also more likely to support and abide by protections where they have been involved in establishing them in the first place.[37]

While it deals with neither definitional questions nor the position of Indigenous peoples, the Aarhus Convention is a leading instrument of global application that seeks to entrench and implement participatory rights.[38] It contains procedural rights of information access, public participation, and environmental access to justice, supporting local communities where in force.[39] Yet scholar-

---

[35] Conservation of Arctic Flora and Fauna (CAFF), *Scoping for Resilience and Management of Arctic Wetlands: Resilience and management of Arctic wetlands: Phase 2 Report* (Conservation of Arctic Flora and Fauna International Secretariat, Akureyri, 2021) 23 (my emphasis).

[36] In a transboundary context, see S Marsden, 'Public Participation in Transboundary Environmental Impact Assessment – Closing the Gap between International and Public Law?' in B Jessup and K Rubenstein, *Environmental Discourses in Public and International Law* (Cambridge University Press, Cambridge, 2012) 238.

[37] CAFF, note 35, 44.

[38] Convention on Access to Information, Public Participation in Decision-making and Access to Justice in Environmental Matters, adopted 25 June 1998, entered in force 30 October 2001, 2161 UNTS 447 (Aarhus Convention).

[39] Note also the focus on safeguarding the rights of environmental defenders, who include Indigenous peoples. See Draft Declaration on Environmental Democracy for Sustainable, Inclusive and Resilient Development ECE/MP.PP/WG.1/2021/18 [II.8], which states in relation to the Convention as a tool for improving infrastructure development and spatial planning: 'At the same time, such projects, and spatial planning more generally ... also influence a number of social issues linked to the public's rights, such as *displacement, land ownership, cultural heritage, the rights of indigenous peoples and local communities* ... The assurance of transparency and the rule of law, inclusive, transparent and effective public participation in decision-making, and adequate and effective remedies throughout the planning and development process is paramount in this regard.' See also ibid, [III.26]. Article 3(7) of Aarhus obliges parties to promote the application

ship highlights its inherent weaknesses, contrasting it with the more directed measures of the Escazú Agreement in Latin America.[40] The latter contains provisions specific to Indigenous peoples.[41] The Escazú Agreement specifically aspires to be a Latin American agreement,[42] rather than an extension of Aarhus, with a higher degree of Indigeneity readily apparent in this region (and with one of the official languages of the text being Quechua).[43]

Cittadino notes that the participatory requirements of Aarhus:

> are functional to the realisation of substantive rights, including those rights that purport to preserve the very existence of indigenous peoples. Further, if the requirement of free, prior and informed consent were to be fully realised, this would give indigenous peoples very significant powers over land and natural resources.[44]

Additionally, she notes how Indigenous rights have been restricted by the creation of nature reserves: 'In this context, participation into decision-making has been invoked by indigenous applicants to mitigate the effects of States' unilateral decisions on their rights and their natural environment.'[45] She also emphasises that in their collective stewardship, 'the participatory rights of indigenous peoples may reveal another rationale for public participation,

---

of these principles in other international forums (which could include Ramsar). Aarhus Convention, note 38, Art 3(7).

[40] Regional Agreement on Access to Information, Public Participation and Justice in Environmental Matters in Latin America and the Caribbean, Escazú, adopted 4 March 2018, entry into force 22 April 2021, 3398 UNTS 56654.

[41] E Barrit, 'Theme and Variations: The Aarhus Convention and the Escazú Agreement', 13 August 2021, at: https://gnhre.org/?p=14087.

[42] Specific attention is also given to Indigenous people's needs. Article 7.9 sets out that decisions related to EIA should be progressed appropriately, including 'customary methods', and Article 8.4(d) that materials should be published in non-official languages.

[43] Like Ramsar, the wording of the Aarhus Convention is far-sighted but a product of its time. The more recent Escazú Agreement therefore draws upon commitments concerning Indigenous peoples that have subsequently matured Principle 10 of the Rio Declaration 1992 in the light of contextual international law.

[44] Cittadino focuses in her analysis on how Indigenous peoples and the protection of the environment are linked in the decisions of human rights bodies, highlighting how Indigenous peoples have challenged development with negative environmental impact on their territories. Cittadino, note 32, conclusions, 92.

[45] Ibid, 82.

pointing at the role that the invocation of collective rights may play for the preservation of the environment'.[46]

3.2 The Role of Indigenous Peoples and Local Communities in Ramsar Processes

In the absence of specific treaty provisions, Ramar COP resolutions – such as those pertaining to wise use – outline the role of Indigenous peoples and local communities.[47] For example, Resolution VII.8 urges Parties:

- 'to *include extensive consultation with local communities and indigenous people in the formulation of national wetland policies and legislation*';[48]
- 'to create, as appropriate, *the legal and policy context to facilitate indigenous peoples' and local communities' direct involvement in national and local decision-making for the sustainable use of wetlands*, including the provision of necessary resources';[49]
- 'to ensure that the stakeholders, *including local communities and indigenous people, are represented on National Ramsar Committees*';[50] and
- 'to provide for transparency in decision-making with respect to wetlands and their conservation and *ensure that there is full sharing with the stakeholders of technical and other information related to the selection of Ramsar sites and management of all wetlands, with guarantees of their full participation in the process.*'[51]

The Ramsar approach to Indigenous peoples and local communities evolved significantly in the 1980s and 1990s, from simple recognition to more active involvement. In 1999, the Ramsar COP expanded on the term 'involvement'

---

[46] Ibid, 80.
[47] Generally, see Ramsar Convention Secretariat, *Handbook 3: Laws and institutions: Reviewing laws and institutions to promote the conservation and wise use of wetlands* (Ramsar Handbooks, 4th edition, Ramsar Convention Secretariat, Gland, 2010). The trigger for this is Ramsar Convention, *Resolution VII.7: Guidelines for reviewing laws and institutions to promote the conservation and wise use of wetlands* (1999).
[48] Ramsar Convention, note 20, [14].
[49] Ibid, [15].
[50] Ibid, [16].
[51] Ibid, [17].

and made the fundamental shift to recognising community-based governance as a legitimate option for wetlands management:

> The Parties acknowledged that *local and indigenous people have a particular interest in ensuring that the wetlands within their region are managed wisely* and, in particular, that *indigenous people may have distinct knowledge, experience and aspirations in relation to wetland management.* They also noted that the wise use of wetlands will benefit the quality of life of local and indigenous people who, in addition to their involvement in site management, should derive the benefits that result from conservation and sustainable use of wetlands.[52]

Updated Ramsar guidance on wetland management in 2002, while preceding UNDRIP rights-holder status, nonetheless understood differences between Indigenous peoples and local communities in wetland management, especially where both were the managers of wetlands themselves:

> The involvement and understanding of local communities and indigenous peoples in the management of wetlands is of particular importance where the wetland is under private ownership or in customary tenure, *since then the local communities are themselves the custodians and managers of the site, and in these circumstances it is vital that the management planning process is not seen as one imposed from outside upon those who depend on the wetland for their livelihoods.*[53]

Three schematic streams are now central to involving Indigenous peoples and local communities in wetland management: first, wise use;[54] second, Ramsar's *Programme on communication, capacity building, education, participation, and awareness* (CEPA);[55] and third, the cultural values of wetlands.[56] The concept of involvement may range from consultation to devolution of management authority. Community-based governance in particular is a legitimate governance option for wetlands as an effective form of stewardship and should be a preferred model in the context of Indigenous lands. Furthermore,

---

[52] Ibid, [4.2]; see further 'Participation in Management', in Ramsar Convention Secretariat, *Handbook 1: An Introduction to the Convention on Wetlands* (Ramsar Handbooks, 5th edition, Ramsar Convention Secretariat, Gland, 2016) 41.

[53] Ramsar Convention, note 21, [V.32]. Note that 'local communities' appears to include Indigenous peoples in the highlighted text.

[54] Ramsar Convention, *Recommendation 3.3: Wise use of wetlands* (1987); see Chapter 4 in this volume.

[55] Ramsar Convention, *Resolution X.8: The Convention's Programme on communication, education, participation and awareness (CEPA) 2009–2015* (2008).

[56] Ramsar Convention, *Resolution VIII.19: Guiding principles for taking into account the cultural values of wetlands for the effective management of sites* (2002). See Section 4 below and Chapter 10 in this volume.

Indigenous peoples and local communities must be properly represented in management structures at various levels and states must create opportunities for this.

The Ramsar Strategic Plan 2016–24 (Goal 3, Target 10) emphasises the central role of knowledge in wise use.[57] Traditional or Indigenous knowledge must be identified and applied alongside Western science for effective wetlands management. Target 10 requires that:

> traditional knowledge, innovations and practices of Indigenous peoples and local communities relevant for the wise use of wetlands and their customary use of wetland resources, are documented, respected, subject to national legislation and relevant international obligations are fully integrated and reflected in the implementation of the Convention with a full and effective participation of indigenous peoples and local communities at all relevant levels.

The Strategic Plan also encourages parties 'to *promote, recognize and strengthen active participation of indigenous peoples and local communities, as key stakeholders* for conservation and integrated wetland management'.[58] Additionally, 'the *wise and customary use of wetlands by indigenous peoples and local communities can play an important role in their conservation*'.[59]

At the request of COP12, Oviedo and Kenza Ali produced a report in 2018 on the relationship of Indigenous peoples and local communities with wetlands.[60] An analysis of 150 National Reports and nearly 2300 Ramsar Information Sheets in the report showed significant interest in active involvement, including Indigenous peoples and local communities.[61] The report noted that a majority of national focal points, CEPA focal points and administrative authorities responding to a questionnaire recommended that the Convention develop new or different instruments to support parties to achieve more effective participation. Suggestions for the way forward included:

- adopting rights-based approaches;

---

[57] Ramsar Convention, *Resolution XII.2, The 4th Strategic Plan 2016–2024* (2015) Target 10, 10.

[58] Ramsar Convention, *Resolution XII.2: The Ramsar Strategic Plan 2016–2024* (2015) [19].

[59] Ibid, [20]. Note that footnotes 57 and 58 refer to two distinct documents.

[60] G Oviedo and M Kenza Ali, *The relationship of indigenous peoples and local communities with wetlands* (Ramsar Convention Secretariat, 2018), at: https://www.ramsar.org/sites/default/files/documents/library/indigenous_peoples_local_communities_wetlands_e.pdf.

[61] Ibid, 22.

- strengthening governance to further promote participatory governance and management;
- acknowledging UNDRIP and potential links between its provisions and Ramsar's approach to engagement with Indigenous peoples; and
- updating/creating new tools as policies evolve and lessons from experiences enrich institutional frameworks and strategies.[62]

4 CULTURAL VALUES

The significance of culture is subject to ongoing deliberations within Ramsar; and as with rights and participation, is also central to wetlands management for Indigenous peoples and local communities. Although not established as one of the nine ecological Ramsar designation criteria, cultural values are nevertheless relevant for Ramsar designation;[63] and they are of fundamental importance for Indigenous peoples and local communities, given their traditional knowledge and practices, customary governance, value systems, and cultural expressions.

Parties are advised to use cultural values principles[64] in conserving and enhancing wetlands, 'within their national and legal frameworks and available resources and capacity'.[65] The 2010 *Ramsar Handbook* emphasises that:

> wetlands have special attributes as part of the cultural heritage of humanity – they are related to religious and cosmological beliefs and spiritual values, constitute a source of aesthetic and artistic inspiration, yield invaluable archaeological evidence from the remote past, provide wildlife sanctuaries, and form the basis of important local social, economic, and cultural traditions.[66]

Significantly, parties are furthermore encouraged to apply the principles 'for the designation of new Wetlands ... or when updating ... existing Ramsar sites, *taking into account ... customary law, and the principle of prior informed*

---

[62] Ibid, 43–53.
[63] Ramsar Convention, *Resolution IX.21: Taking into account the cultural values of wetlands* (2005).
[64] Ramsar Convention, note 56. See D Pritchard, 'Human Culture and its Evolving Place within the Ramsar Convention' in PA Gell, NC Davidson, and CM Finlayson (eds), *Ramsar Wetlands: Values, Assessment, Management* (Elsevier, Amsterdam, 2022) 417.
[65] Ramsar Convention, note 56, [20].
[66] Ramsar Convention Secretariat, note 52, 11.

consent'.[67] Additionally, and mindful of obligations in the Convention on Biological Diversity (CBD),[68] they are encouraged:

> to integrate cultural and social impact criteria into environmental assessments ... [to] carry out such efforts *with the active participation of indigenous peoples* ... and to consider *using the cultural values of wetlands as a tool to strengthen this involvement, particularly in wetland planning and management*.[69]

Emphasis is put on 'the strong link between wetland conservation and benefits to people ...; positive correlation between conservation and the sustainable use of wetlands ...; [and] *involvement of indigenous peoples*'.[70] These are clear illustrations of connections between wise use and traditional knowledge – for example, integration of cultural criteria and participation in environmental assessment highlight advocacy for cultural knowledge transfer into policy/ plan making.

Ramsar Resolution IX.21 emphasises the integration of cultural values, recognising wetlands as 'places where ... indigenous peoples have developed strong cultural connections and sustainable use practices [and] are especially important to ... indigenous peoples and that the[y] must have a decisive voice in matters concerning their cultural heritage'.[71]

Relationships between human wellbeing and wetlands are further recognised; although, other than in passing (at paragraph 9), this does not address Indigenous peoples (or traditional knowledge).[72] However, Ramsar Resolution XI.7 goes on to encourage '*active participation of indigenous peoples ... taking fully into account the ethical implications of cultural and historical*

---

[67] Ibid, [18].

[68] Convention on Biological Diversity, adopted 22 May 1992, entered into force 29 December 1993, 1760 UNTS 79 (CBD); see further Chapter 8 in this volume.

[69] Ibid, [18 d) and e)].

[70] Ibid, Guiding principle 2; see also Guiding principle 4 – To learn from traditional approaches; Guiding principle 5 – To maintain traditional sustainable self-management practices; Guiding principle 16 – To safeguard wetland-related oral traditions; Guiding principle 17 – To keep traditional knowledge alive; and Guiding principle 18 – To respect wetland-related religious and spiritual beliefs and mythological aspects in the efforts to conserve wetlands.

[71] Ramsar Convention, note 63.

[72] Ramsar Convention, *Resolution X.3: The Changwon Declaration on human well-being and wetlands* (2008).

issues of indigenous peoples and ... the involvement of such communities in decision-making.[73]

Despite these exhortations, the Ramsar Culture Working Group (WG) highlighted that the cultural dimension had 'lagged behind' economic, scientific, and recreational values.[74] To be sure, the WG and Strategic Plan (2009–15) provided 'some support for greater recognition of cultural (and spiritual) values in decision-making', but these efforts fell 'short of ... strategic direction on the issue which ... is increasingly sought in today's evolving context of ecosystem services and broader partnership working'.[75]

To set the scene for the remainder of this Chapter, the 2018 report by Oviedo and Kenza Ali asks what can be learned about the practice of involving Indigenous peoples and local communities in the management of wetlands; to what extent implementation has happened on the ground; what policy and practice issues must be addressed; and what technical tools of the Convention are needed to strengthen the inclusive, participatory approach.[76] Understanding the linkages between wise use and traditional knowledge is arguably central to providing an answer.

5 WISE USE AND KNOWLEDGE

Wetlands management with reference to wise use is an inherent part of the Ramsar Convention, closely related to conservation.[77] Initially, COP3 defined 'wise use' of wetlands as 'their sustainable utilization for the benefit of humankind in a way compatible with the maintenance of the natural properties of the ecosystem'.[78] The establishment of national wetland policies was a key part

---

[73] Ramsar Convention, *Resolution XI.7: Tourism, recreation and wetlands* (2012), [16].

[74] Ramsar Culture Network, *Culture and wetlands in strategic planning for the Ramsar Convention* (2013), at: https://wli.wwt.org.uk/wp-content/uploads/2020/12/Ramsar-and-culture-strategic-planning-aspects-Jan-2013b.pdf.

[75] Ibid.

[76] Oviedo and Kenza Ali, note 60, 43.

[77] Convention on Wetlands, note 1, Art 3.1. See Ramsar Convention Secretariat, *Handbook 1: Wise use of wetlands: Concepts and approaches for the wise use of wetlands* (Ramsar Handbooks, 4th edition, Ramsar Convention Secretariat, Gland, 2010). See also CM Finlayson et al, 'The Ramsar Convention and Ecosystem-Based Approaches to the Wise Use and Sustainable Development of Wetlands' (2011) 14 *Journal of International Wildlife Law and Policy* 176.

[78] Ramsar Convention, note 54, 1. For an overview of the relationship, see M Everard, 'Traditional Knowledge and Wetlands' in CM Finlayson et al, *The Wetland Book* (Springer, Dordrecht, 2018) 1379.

of the approach, as well as actions to increase knowledge and awareness of wetlands and their values. However, it is not clear from early Ramsar guidance whether the asserted application of traditional knowledge was contemplated.[79] Recently, however, it has become an essential part of the process for certain properties and must be included.[80]

After identifying that '[s]pecial attention needs to be given to the local populations who will be the first to benefit from improved management of wetland sites [and] *[t]he values that indigenous people can bring to all aspects of wise use need special recognition*',[81] the COP adopted an updated version of the 'wise use' concept. This confirmed that the '[w]ise use of wetlands is the maintenance of their ecological character, achieved through the implementation of ecosystem approaches, within the context of sustainable development'.[82] Whether the wise use principle has been effectively implemented has, however, been questioned;[83] and whether the guidance accurately reflects the position of Indigenous rights holders is also open to debate.[84]

The establishment of nature reserves is a key part of the Ramsar approach.[85] The type of reserve established in each state is dependent upon the particular qualities of each site. There has been some proven success with co-management

---

[79] See Ramsar Convention, *Recommendation 4.10: Guidelines for the implementation of the wise use concept* (1990); Ramsar Convention, *Resolution 5.6: Additional guidance for the implementation of the wise use concept* (1996).

[80] F Berkes, C Folke, and M Gadgil, 'Traditional Ecological Knowledge, Biodiversity, Resilience and Sustainability' in C Perrings et al (eds), *Biodiversity Conservation* (Kluwer, Dordrecht, 1995) 281. Note the relevance of Indigenous knowledge to the Universal Declaration of the Rights of Wetlands. See, eg, GT Davies et al, 'Towards a Universal Declaration of the Rights of Wetlands' (2021) 72 *Marine and Freshwater Research* 593, 596, and Chapter 13 in this volume.

[81] Ramsar Convention, *Recommendation 6.3: Involving local and indigenous people in the management of Ramsar wetlands* (1996); and Ramsar Convention, note 20.

[82] Ramsar Convention, *Resolution IX.1: Additional scientific and technical guidance for implementing the Ramsar wise use concept* (2005), in particular Annex A, *Conceptual framework for the wise use of wetlands and the maintenance of their ecological character* 6.

[83] D Farrier and E Tucker, 'Wise Use of Wetlands under the Ramsar Convention: A Challenge for Meaningful Implementation of International Law' (2000) 12 *Journal of Environmental Law* 21.

[84] By contrast, note efforts under the WHC. S Marsden, 'The World Heritage Convention in the Arctic and Indigenous People: Time to Reform?' (2014) 6 *The Yearbook of Polar Law* 226.

[85] Convention on Wetlands, note 1, Art 4.1.

by Indigenous peoples and public reserve managers,[86] although this has not always been successful. The reasons are found in underlying law, policy, and management: 'Poor collaboration between communities and conservation agencies is often rooted in the lack of supportive laws and policies, despite the existence of broad and vague intentions to "enhance community participation."'[87]

Current wise use guidelines recognise the importance of 'local' knowledge.[88] Parties, technical experts, and local and Indigenous peoples are encouraged *'to work together in the planning and management of wetlands to ensure that the best available science and local knowledge are taken into consideration in making decisions'*.[89] Although the guidelines do not recognise rights-holder status (because they were adopted prior to UNDRIP), ILO Convention 169 was applicable in some circumstances.[90] The Ramsar Strategic Plan (2009–15) also recognised *'participation of the local indigenous and non-indigenous population* and *making use of traditional knowledge'* and 'more participative management of wetlands' to achieve the goal of wise use.[91]

In the broader international law context, UNDRIP '[r]ecogniz[es] that respect for indigenous knowledge, cultures and *traditional practices* contributes to sustainable and equitable development and proper management of the environment'.[92] Similarly, Ramsar 'RECOGNIZES that the *wise and customary use of wetlands by indigenous peoples and local communities can play an*

---

[86] See 'World's first Ramsar Site Turns 40' (8 May 2014), *Ramsar*, at: https://www.ramsar.org/news/worlds-first-ramsar-site-turns-40#:~:text=On%208%20May%201974%2C%20Australia,Importance%20under%20the%20Ramsar%20Convention.

[87] G Borrini-Feyerabend, A Kothari, and G Oviedo (eds), *Towards Equity and Enhanced Conservation Guidance on Policy and Practice for Co-managed Protected Areas and Community Conserved Areas*, Best Practice Protected Area Guidelines Series No 11 (World Commission on Protected Areas, IUCN, 2004) 93. See also A Kotari, 'Collaboratively Managed Protected Areas', in M Lockwood, G Worboys, and A Kothari, *Managing Protected Areas: A Global Guide* (Routledge, 2006) 539–40; L Talbot, 'Engaging Indigenous Communities in World Heritage Declarations: Processes and Practice' in P Figgis et al (eds), *Keeping the Outstanding Exceptional: The Future of World Heritage in Australia* (Australian Committee for IUCN, 2012) 134.

[88] Ramsar Convention, note 20, [4].

[89] Ibid, [18].

[90] Ibid, [3].

[91] Ramsar Convention, *Resolution X.1: The Ramsar Strategic Plan 2009–2015* (2008), [5].

[92] UNDRIP, note 13, eleventh Recital.

important role in their conservation, [and] ENCOURAGES relevant parties to provide that information to the Secretariat'.[93] As a means of implementing wise use, the application of traditional knowledge in ascertaining the customary use of wetlands has been extensively analysed.[94] Traditional knowledge principles relevant to Arctic Indigenous peoples have also been developed, as noted above.[95] The extent to which traditional knowledge might influence conservation outcomes and be shaped by conservation efforts depends, in part, on 'mutual trust and respect'.[96] More remains to be done with regard to Ramsar, as the following discussion indicates.

6 RAMSAR AND OTHER BIODIVERSITY-RELATED CONVENTIONS

The Ramsar Convention is part of a cluster of biodiversity-related Conventions which, through 'a network of formal and informal contacts ... facilitates the exchange of information, interoperability between databases and institutional mechanisms, and best practices'.[97] As such, the practices of these sister regimes can provide insight and lessons with respect to collaborating with Indigenous peoples and local communities. Most relevant in this context are the CBD[98] and the WHC.[99]

---

[93] Ramsar Convention, note 58, [20].

[94] See, eg, NJ Turner, M Boelscher Ignace, and R Ignace, 'Traditional Ecological Knowledge and Wisdom of Aboriginal Peoples in British Columbia' (2000) 10 *Ecological Applications* 1275.

[95] See Arctic Council, note 16; Arctic Council Indigenous Peoples' Secretariat, note 16.

[96] See JM McPherson et al, 'Integrating traditional knowledge when it appears to conflict with conservation: lessons from the discovery and protection of sitatunga in Ghana' (2016) 21 *Ecology and Society* 24.

[97] Chapter 8 in this volume. See Ramsar Convention Secretariat, note 52, 17–18; Ramsar Convention Secretariat, *Handbook 5: Partnerships: Key partnerships for implementation of the Ramsar Convention* (Ramsar Handbooks, 4th edition, Ramsar Convention Secretariat, Gland, 2010) 7–8. See additionally NC Davidson, 'Biodiversity-Related Conventions and Initiatives Relevant to Wetlands' in CM Finlayson et al, *The Wetland Book* (Springer, Dordrecht, 2018) 433.

[98] See D Coates, 'Convention on Biological Diversity (CBD) and Wetland Management' in CM Finlayson et al, *The Wetland Book* (Springer, Dordrecht, 2018) 487.

[99] See R McInnes, M Ali, and D Pritchard, *Ramsar and World Heritage Conventions: Converging Towards Success* (Ramsar Convention Secretariat, Gland, 2017). A Memorandum of Understanding was signed between the Ramsar

The CBD includes goals regarding the protection of biodiversity and *'traditional knowledge, innovations and practices'*,[100] with further synergies between Aichi Biodiversity Target 18[101] and Ramsar Target 10. Given traditional knowledge held in relation to plant resources and their commercial exploitation, the Nagoya Protocol to the CBD is pertinent as well,[102] and this may include wetland species.[103] It has been acknowledged that fair and equitable benefit-sharing *'has originated elsewhere, for instance, in the practices of indigenous peoples and local communities on the ground'*.[104] The CBD further advances impact assessment provisions 'for the *conduct of cultural*, environmental and social *impact assessment* regarding developments proposed to take place on, or which are likely to impact on, *sacred sites and on lands and waters traditionally occupied or used by indigenous and local communities,*'[105]

Secretariat and the World Heritage Centre in 1999, and 'a fruitful working relationship continues, with a view to: promoting the nominations of wetland sites under the two Conventions; coordinating the reporting on sites listed under both Conventions; and in many cases collaborating on advisory missions to those sites, as needed'. Ramsar Convention Secretariat, note 52, 15.

[100] CBD, *Decision VI/10: Article 8.J and related provisions* (2002) [C] and [D]. See further Chapter 8 in this volume.

[101] 'Aichi Biodiversity Targets' in the global CBD, *Decision X/2: Strategic Plan for Biodiversity 2011–2020* (2010). Target 13 also refers to maintenance of the genetic diversity of plants and animals 'including culturally valuable species'. The relevance of this is highlighted in Ramsar Resolution XII.2 (Strategic Plan), note 58, which includes Annex 2.

[102] Nagoya Protocol on Access to Genetic Resources and the Fair and Equitable Sharing of Benefits Arising from Their Utilization to the Convention on Biological Diversity, adopted 29 October 2010, entered into force 12 October 2014, 3008 UNTS 3.

[103] See RJ Coombe, 'The Recognition of Indigenous Peoples' and Community Traditional Knowledge in International Law' (2001–2002) 14 *St. Thomas Law Review* 275; G Aguilar, 'Access to Genetic Resources and Protection of Traditional Knowledge in the Territories of Indigenous Peoples' (2001) 4 *Environmental Science and Policy* 241.

[104] L Parks and E Morgera, 'The Need for an Interdisciplinary Approach to Norm Diffusion: The Case of Fair and Equitable Benefit Sharing' (2015) 24 *Review of European Community and International Environmental Law* 354.

[105] CBD, *Resolution VIII.1: Guidelines for incorporating biodiversity-related issues into EIA legislation and/or processes and in SEA, and their relevance to the Ramsar Convention*; see also note 25.

who should also be involved in both the development of guidelines and their assessment process, including decision-making.[106]

The WHC provides that 'ensuring the identification, protection, conservation, presentation and transmission to future generations of the cultural and natural heritage' is a duty.[107] States must nominate potential sites for inclusion on the World Heritage List,[108] protecting and managing them in compliance with the Convention. Properties may be domestic (within a state), transboundary (between states), serial national (more than one location within a state), or serial transnational (more than one location in more than one state). There are potential synergies with Ramsar here in terms of wetland transboundary properties also being developed into sub-categories such as these.[109]

In 1992, the WHC was the first regime to recognise and protect cultural landscapes,[110] which are particularly noteworthy for Indigenous peoples.[111] The World Heritage Committee adopted guidelines for inscription, acknowledging that they represent the 'combined works of nature and of man'.[112] Significantly from the standpoint of protecting Indigenous heritage and knowledge, the parties acknowledged that '[c]ultural landscapes often reflect specific techniques of sustainable land use, considering the characteristics and limits of the natural environment they are established in, and may reflect a specific spiritual relationship to nature', whose protection 'is *therefore helpful in maintaining biological diversity*'.[113]

The role of Indigenous peoples in nomination[114] and evaluation is not directly acknowledged in the WHC. However, the process of information gathering to evaluate the authenticity of cultural heritage does encompass tradi-

---

[106] In relation to participatory rights, see U Etemire, 'The Convention on Biological Diversity Regime and Indigenous Peoples: Issues Concerning Participatory Rights and Impact Assessment' (2013) 4 *City University of Hong Kong Law Review* 1.

[107] WHC, note 12, Art 4.

[108] Ibid, Art 11.1.

[109] See further Chapter 3 in this volume.

[110] A Luengo and M Rössler (eds), *World Heritage Cultural Landscapes* (Elche, Ayuntamiento de Elche, 2012).

[111] Note the UNESCO policy on engaging with Indigenous peoples, at: https://en.unesco.org/indigenous-peoples/policy. See also UNESCO, *World Heritage and Indigenous Peoples*, at: https://whc.unesco.org/en/activities/496/.

[112] WHC, note 12, Art 1.

[113] UNESCO World Heritage Centre, *The Operational Guidelines for the Implementation of the World Heritage Convention* (UNESCO, Paris, 2023), II.A, [47], at: https://whc.unesco.org/en/guidelines/.

[114] Ibid, annex 5.

tional knowledge and the recording of such information.[115] The importance of different 'information sources' is recognised[116] – particularly 'in *communities maintaining tradition and cultural continuity*',[117] which include Indigenous societies. Statements of authenticity (for cultural properties)[118] and integrity (for cultural, natural, and mixed properties)[119] are therefore significant components of the information required for nomination,[120] with subsequent management based on 'the *active participation of the communities and stakeholders* concerned with the property'.[121]

Despite and because of these measures, participation conducive to the involvement of Indigenous peoples was found wanting – notably regarding reference to 'stakeholders' rather than 'rights holders'.[122] It was therefore significant that, following a 2012 report,[123] the Committee encouraged states to involve Indigenous peoples in decision-making, monitoring, and evaluation of the state of conservation of World Heritage Sites; and to respect the rights of Indigenous peoples when nominating, managing and reporting on Sites within Indigenous peoples' territories.[124] This 'Call to Action',[125] which called for clearer consistency with UNDRIP, enhanced support for the Copenhagen Expert Meeting later that year and prompted revisions to the Operational

---

[115] Ibid, [80–84].
[116] Ibid, [84].
[117] Ibid, [83].
[118] Ibid, [80].
[119] Ibid, [88].
[120] Ibid, Part. II.E.
[121] Ibid, [119].
[122] The International Work Group for Indigenous Affairs, *Report of International Expert Workshop on the World Heritage Convention and Indigenous Peoples, 20–21 September 2012 – Copenhagen, Denmark* (2013) 19, in reference to FPIC, at: https://www.iwgia.org/images/publications/0610_International_orkshop_-H_and_IP_-Report_eb.pdf.
[123] Ibid.
[124] UNESCO, Convention Concerning the Protection of the World Cultural and Natural Heritage, World Heritage Committee, 35th Session, Paris, France, 19 June–29 June 2011, Decision 35 COM 12E, [15 e) and f)].
[125] International Work Group for Indigenous Affairs, note 122, 60–63.

Guidelines.[126] Notably – and with direct reference to UNDRIP[127] – the Guidelines recognise 'rightsholders' in preparation of Tentative Lists (for proposed inscriptions) including local communities and Indigenous peoples. For sites affecting the lands, territories, or resources of Indigenous peoples, 'States Parties shall consult and cooperate in good faith with the indigenous peoples concerned through their own representative institutions in order to obtain their free, prior and informed consent before including the sites on their Tentative List'.[128] These are significant changes and Ramsar must follow them to ensure that its procedures are also consistent with UNDRIP.

WHC criteria for the assessment of Outstanding Universal Value were merged from two separate sets for natural and cultural properties into one set of ten unified criteria. This has had the benefit of emphasising connections between natural and cultural properties, which is demonstrated by traditional ownership. Relevant criteria for properties include those which:

> (iii) bear a *unique or at least exceptional testimony to a cultural tradition or to a civilization which is living* or which has disappeared ... (v) [are] an *outstanding example of a traditional human settlement, land-use, or sea-use which is representative of a culture (or cultures), or human interaction with the environment* especially when it has become vulnerable under the impact of irreversible change; and (vi) [are] *directly or tangibly associated with events or living traditions*, with ideas, or with beliefs, with artistic and literary works of outstanding universal significance.[129]

These properties must also meet the test of 'authenticity', which is 'judged primarily within the cultural contexts to which it belongs'.[130] These criteria – particularly (v), land-use, or sea-use – are relevant in the context of certain wetlands and should be considered as additional Ramsar listing criteria, where there is 'human interaction with the environment'.

Participatory management planning is a key part of the WHC, as it is under Ramsar; and in promoting an 'effective management system', the Guidelines

---

[126] As noted by the Committee at the 37th Session in Phnom Penh, 'the results of the International Expert Meeting on World Heritage Convention and Indigenous Peoples (Denmark, 2012) [should] re-examine the recommendations of this meeting following the results of the discussions to be held by the Executive Board on the UNESCO Policy on indigenous peoples for further steps'. See UNESCO, Convention Concerning the Protection of the World Cultural and Natural Heritage, World Heritage Committee, 37th Session, Phnom Penh, Cambodia, 16–27 June 2013, WHC13/37.COM/20, 5 July 2013.

[127] *Operational Guidelines*, note 113, I.I [40]; Decision 39 COM 11.

[128] *Operational Guidelines*, note 113, II.C [64]; Decision 43 COM 11A.

[129] *Operational Guidelines*, note 113, II.D [77]; Decision 6 EXT.COM 5.1.

[130] *Operational Guidelines*, note 113, II.E [81]; Decision 39 COM 11.

refer to 'traditional practices' of local communities and Indigenous peoples.[131] In relation to 'sustainable use' (with synergies with Ramsar's 'wise use'), parties must *'promote and encourage the effective, inclusive and equitable participation of the [local] communities, [and] indigenous peoples'*.[132] As indicated above in relation to Tentative Lists, participation in the nomination process of properties also recognises the rights-holder status of Indigenous peoples, specifying the requirement for FPIC.[133] In each of these instances, the processes of the WHC have made major progress in addressing the concerns of the 2012 report, demonstrating that the Call to Action has been heard. Ramsar needs to listen, learn, and change, building on the experience of the WHC.[134] By incorporating Indigenous peoples' traditional connection within inscription types (eg, a mixed property or cultural landscape) and utilising criteria for definition and listing, the WHC recognises traditional ownership and the stewardship that has protected and conserved sites for generations. As such, there are clear benefits for Ramsar parties to emulate this approach – above all, to ensure compliance with UNDRIP.

7 CONCLUSIONS

While there has been progress in recognising the role of Indigenous peoples and local communities under Ramsar, much remains to be done. Indigenous peoples are yet to be afforded the recognition and rights provided under UNDRIP, Indigenous knowledge remains undervalued in Ramsar processes, and these processes are yet to fully enable the participation of Indigenous

---

[131] *Operational Guidelines*, note 113, II.F [110], [111a)]; Decision 39 COM 11 and Decision 43 COM 11A.

[132] *Operational Guidelines*, note 113, II.F [119]; Decision 43 COM 11A.

[133] *Operational Guidelines*, note 113, III.A [123]; Decision 39 COM 11 and Decision 43 COM 11A.

[134] Similarly, the related UNESCO regime on intangible cultural heritage mandates 'the widest possible participation of communities, groups and, where appropriate, individuals that create, maintain and transmit such heritage, and to involve them actively in its management', including consideration of *'knowledge and practices concerning nature and the universe'*. Convention for the Safeguarding of the Intangible Cultural Heritage, adopted 17 October 2003, entered into force 20 April 2006, 2368 UNTS 3, Arts 15 and 2.2(d). See further P Kuruk, 'Cultural Heritage, Traditional Knowledge and Indigenous Rights: An Analysis of the Convention for the Safeguarding of the Intangible Cultural Heritage' (2004) 1 *Macquarie Journal of International and Comparative Environmental Law* 111. Both UNESCO treaties are relevant to Indigenous peoples in definition, inscription, and participatory processes.

peoples in listing and protecting wetlands. Furthermore, local communities are yet to be fully involved in decision-making processes that impact their livelihoods. Although Ramsar has not kept pace with contextual international law, related international environmental agreements – in particular the WHC – show a path forward, providing opportunities for potential learning and legal borrowing.

The difference between implementation of the recommendations of the WHC International Work Group for Indigenous Affairs (updating the WHC Operational Guidelines), and the failure to implement the recommendations in the 2018 report by Oviedo and Kenza Ali (for Ramsar), is starkly contrasting and disappointing. Updating the current Ramsar arrangements is essential and will require:

- adopting rights-based approaches;
- strengthening governance;
- acknowledging UNDRIP and potential links; and
- updating and creating new tools.

Each of these matches the findings of this Chapter overall, and, via the examples considered in the 2018 report, there are evidently many opportunities to enable positive change – particularly where there are clear synergies with the WHC. It was furthermore acknowledged in the 2018 report that local communities are also key stakeholders and must be both provided with information and encouraged to provide information, to enable full participation in wetland decision-making processes.

Additional recommendations are that the cultural values of wetlands should be strengthened by adding to existing Ramsar Sites criteria, in similar ways to the operation of cultural landscapes/mixed properties under the WHC. Adding an explicit role for Indigenous peoples in nominating wetlands and related management measures would ensure consistency with UNDRIP. Incorporation of Indigenous knowledge in wise use decision-making must also be formalised. This complements, and is as important as, Western science. Indeed, both are mutually supportive forms of information, as well as a significant part of the cultural value of wetlands recognised via impact assessment processes elsewhere (eg, CBD). Furthermore, acceptance of, and respect for, Indigenous knowledge is an essential part of recognising Indigenous peoples as rights holders. Without this, on Indigenous lands, wetlands management is meaningless.

PART II

Wetlands and the international legal ecosystem

7. Birds of a feather? The inter-relationship between the Ramsar Convention and the migratory waterbird regimes

Melissa Lewis[1]

1 INTRODUCTION

As reflected in its full title, the origins of the Convention on Wetlands of International Importance Especially as Waterfowl Habitat[2] ('Ramsar Convention') can largely be traced back to states' desire to conserve waterbirds. These species are vulnerable to the deterioration and loss of wetlands, and their tendency to congregate in large numbers at key sites means that localised threats may have significant impacts at the population level.[3] In addition, many waterbird species migrate between multiple countries, resulting in a need for international cooperation in their conservation.

Over the past half-century, the Ramsar Convention's focus has broadened considerably. In parallel to its evolution, myriad multilateral initiatives for waterbird conservation have emerged – including but not limited to those developed under the Convention on the Conservation of Migratory Species of

---

[1] This Chapter was concluded in December 2022 and addresses relevant developments up to that date. The author is grateful to this text's editors, whose valuable suggestions helped to strengthen the Chapter.

[2] Convention on Wetlands of International Importance especially as Waterfowl Habitat, adopted 2 February 1971, entered into force 21 December 1975, 996 UNTS 245 (amended 1982 & 1987). Various migratory bird initiatives have also developed outside of the CMS framework. Examples of such initiatives are provided in Sections 4 and 5 of this Chapter.

[3] W Hagemeijer, 'Site networks for the conservation of waterbirds' in GC Boere, CA Galbraith, and DA Stroud (eds), *Waterbirds around the world: A global overview of the conservation, management and research of the world's waterbird flyways* (The Stationary Office, Edinburgh, 2006) 697.

Wild Animals[4] (CMS). Questions consequently arise regarding Ramsar's role in waterbird conservation and its relationship with these migratory waterbird regimes.

This Chapter starts by providing a brief overview of the types of measures needed to conserve migratory waterbirds, which should ideally be promoted and coordinated through multilateral arrangements. It then proceeds:

- to outline how these species featured in Ramsar's development and subsequent evolution, and the key contributions that Ramsar makes to their conservation; and
- to provide an overview of the various multilateral initiatives for migratory waterbird conservation and how their approaches compare to those of Ramsar.

The remainder of the Chapter examines ways in which Ramsar and the migratory waterbird regimes have deliberately aligned with and supported one another, and examples of collaborations between them. While a variety of multilateral initiatives are touched upon throughout this Chapter, its section on collaborative examples focuses specifically on the inter-*treaty* synergies between Ramsar, the CMS, and the Agreement on the Conservation of African-Eurasian Migratory Waterbirds (AEWA).[5]

2 PRIORITY MEASURES FOR THE CONSERVATION OF MIGRATORY WATERBIRDS

The long-term conservation of migratory waterbirds relies on a variety of measures. Chief among them are measures to ensure the availability of adequate habitat at all lifecycle stages – that is, both at and between breeding areas and non-breeding destination areas. Because many waterbird species are congregatory, site-based measures play an especially important role in their conservation. For *migratory* species, it is necessary to identify and conserve not only individual sites, but also *networks* of sites along their flyways.[6] However,

---

[4] Convention on the Conservation of Migratory Species of Wild Animals (CMS), adopted 23 June 1979, entered into force 1 November 1983, 1651 UNTS 67.

[5] Agreement on the Conservation of African-Eurasian Migratory Waterbirds (AEWA), adopted 16 June 1995, entered into force 1 November 1999, 2365 UNTS 203.

[6] Hagemeijer, note 3, 698. A 'flyway' can be defined as 'the entire range of a migratory bird species (or groups of related species or distinct populations of

site networks alone are inadequate for species that are dispersed for part(s)/ all of their annual cycles. Measures are therefore also needed to maintain the ecological functions of the broader landscapes on which they rely.[7]

A variety of other anthropogenic threats may cause waterbird population decline through increased mortality or other negative impacts, and therefore require targeted conservation action. Examples include unsustainable harvesting; poisoning; disease; human disturbance; the introduction of non-native species; physical barriers such as wind turbines and powerlines; incidental taking (eg, as bycatch); depletion of food resources (especially through overfishing); and climate change.[8]

To achieve their objectives, international initiatives to conserve migratory waterbirds need – either independently or jointly – to promote, support, and coordinate the implementation of measures to address both 'habitat threats' and 'species threats',[9] as well as various ancillary measures, such as data collection, capacity-building, and awareness-raising.

3 THE RAMSAR CONVENTION'S FOCUS ON AND CONTRIBUTIONS TO MIGRATORY WATERBIRD CONSERVATION

3.1 Ramsar's Initial Avifaunal Emphasis and Subsequent Evolution

In the early 1960s, initial discussions on developing a wetlands convention were largely directed towards conserving a network of wildfowl refuges.[10] This focus subsequently widened. However, avifauna continued to feature promi-

---

a single species) through which it moves on an annual basis from the breeding grounds to non-breeding areas, including intermediate resting and feeding places as well as the area within which birds migrate'. GC Boere and DA Stroud, 'The flyway concept: what it is and what it isn't' in Boere, Galbraith, and Stroud, note 3, 40.

[7] T Dodman and G Boere, 'Module 2: Applying the Flyway Approach to Conservation' in T Dodman and G Boere (eds), *The flyway approach to the conservation and wise use of waterbirds and wetlands: A training kit* (Wings Over Wetlands Project, Wetlands International, and BirdLife International, Ede, 2010) 68.

[8] See further G Boere and T Dodman, 'Module 1: Understanding the Flyway Approach to Conservation' in ibid, 81–88.

[9] Ibid.

[10] JJ Swift (ed), *Proceedings of the First European Meeting on Wildfowl Conservation*, 16–18 October 1963 (Nature Conservancy, London, and International Wildfowl Research Bureau (IWRB), France, 1964).

nently throughout the evolution of Ramsar's legal text[11] and this influence is evident in the adopted text's preamble, its operative provisions,[12] and indeed its very title. This approach was favoured by the Soviet Union in particular, which also advocated the inclusion of additional text on waterfowl (including their hunting).[13] However, delegates ultimately agreed that attempting to advance *all* waterfowl conservation aims in one convention would increase the risk of failure,[14] and that it would be more appropriate for Ramsar later to be supplemented by separate but complementary international agreements.[15]

This initial emphasis on waterfowl had certain advantages. As recognised at the 1962 MAR Conference, the importance of wetlands to migratory birds 'makes their continued existence a matter of international significance appropriate to international cooperation';[16] thus, the focus on wetlands as waterfowl habitat helped to justify the adoption of a multilateral wetlands convention. It also enabled states to agree on Ramsar's exceptionally wide definition of 'wetlands'.[17] As Bowman observes: '[I]t is hard to believe that, without this overarching ornithological perspective, it would ever have been considered appropriate to devise a single instrument for the protection of such a diverse variety of habitats as the Convention embraces.'[18]

That said, wetlands obviously offer humans a multitude of additional services, and although avian conservation remains an aspect of Ramsar, it has – over time – steadily been de-emphasised under the Convention. This shift can be seen in the incremental refinement, expansion, and reordering of Ramsar's

---

[11] See GVT Matthews, *The Ramsar Convention on Wetlands: Its History and Development* (Ramsar Convention Bureau, Gland, 1993), Chapter 3.

[12] Convention on Wetlands, note 2, Arts 2.1, 2.2, 2.6, 4.1, 4.2, 4.4, 7.1.

[13] Z Salverda (ed), *Proceedings of the Second European Meeting on Wildfowl Conservation*, 9–14 May 1966 (Ministry of Cultural Affairs, Recreation and Social Welfare, The Netherlands, State Institute for Nature Conservation Research, and IWRB, France, 1967) 171–172.

[14] Ibid, 174, 181, 184.

[15] Matthews, note 11, 50–51.

[16] IWRB/MAR Bureau, *Project Mar – the conservation and management of temperate marshes, bogs and other wetlands* (IUCN Publications new series No 3, 1963) 29.

[17] Per Article 1.1 of the Convention, 'wetlands are areas of marsh, fen, peatland or water, whether natural or artificial, permanent or temporary, with water that is static or flowing, fresh, brackish or salt, including areas of marine water the depth of which at low tide does not exceed six metres.'

[18] MJ Bowman, 'The Ramsar Convention Comes of Age' (1995) 42 *Netherlands International Law Review* 1, 6.

site designation criteria.[19] It is also evident in the expansion of topics covered by Ramsar's resolutions[20] and in reduced references to species-specific actions in the Convention's successive strategic plans. Bridgewater and Kim go so far as to assert that, by the tenth meeting of Ramsar's Conference of the Parties (COP) in 2008, 'it had effectively become a wetlands and water, not a wetlands and waterfowl convention'.[21] They point, in particular, to the priority areas identified by Ramsar's current Strategic Plan, observing that '[w]ater appears 72 times in this list, waterfowl/waterbirds/migratory species once'.[22]

3.2 Continued Contributions to Waterbird Conservation

3.2.1 Habitat conservation and associated data collection

Despite Ramsar's shift in emphasis, it continues to provide important tools for migratory waterbird conservation. Its most significant contribution to this endeavour has been, and remains, the identification and conservation of sites that are internationally important for such species. Article 2 of the Convention places explicit emphasis on designating sites that have importance as waterfowl habitat.[23] Two of Ramsar's nine site designation criteria currently focus on waterbirds as bio-indicators of a site's international importance: Criterion 5 focuses on the absolute number of waterbirds regularly supported by the site (>20 000), whereas Criterion 6 is concerned with the site's proportionate importance for a single population (1% of individuals in a population of one species or subspecies).[24] The qualitative criteria specified in Criteria 2, 3, and 4 can also be relied on to designate sites that are important to waterbirds but that fail to satisfy Criterion 5 or 6. The Convention, in other words, provides a mechanism for drawing attention to, and promoting the conservation of, sites for these species – including not only the most important 'mega-sites' at which waterbirds congregate, but also smaller sites (eg, those constituting stepping stones between larger areas).[25]

---

[19] See Matthews, note 11, 42–46. For the current set of criteria, see Ramsar Convention, 'The Ramsar Site Criteria', at: https://www.ramsar.org/sites/default/files/documents/library/ramsarsites_criteria_eng.pdf.

[20] P Bridgewater and RE Kim, '50 Years on, w(h)ither the Ramsar Convention? A case of institutional drift' (2021) 30 *Biodiversity and Conservation* 3919, 3923.

[21] Ibid, 3929.

[22] Ibid.

[23] Convention on Wetlands, note 2, Arts 2.1, 2.2, 2.6.

[24] Ramsar Convention, note 19.

[25] For further discussion, see M Lewis, 'Migratory Waterbird Conservation at the Flyway Level: Distilling the Added Value of AEWA in Relation to the Ramsar Convention' (2016) 34 *Pace Environmental Law Review* 1, 19–23.

The data requirements for the application of Ramsar's criteria have stimulated waterbird monitoring globally (especially Criterion 6's reliance on both site population data and the sizes of biogeographic populations). To date, 1138 Ramsar Sites, covering 144 961 660 hectares, have been designated using one or both of the Convention's quantitative waterbird criteria. This amounts to 46% of Ramsar Sites and over 56% of the total hectarage covered by Ramsar designations.[26] The Ramsar COP has repeatedly urged contracting parties to cooperate in identifying and designating networks of Ramsar Sites along migratory bird flyways[27] (see Section 5.1 below). In addition, the wise use commitment[28] and the extensive body of guidance by which it is accompanied are obviously relevant for undesignated sites – including the complexes of wetlands across landscapes that provide habitat for dispersed populations.

3.2.2 The relevance of species-based approaches

Flyway conservation requires a combination of species and ecosystem-based approaches.[29] Ramsar is especially well positioned to support the latter,[30] including by promoting both the integration of wetland conservation into wider landscape management and the intersectoral cooperation on which this depends.[31] Nevertheless, species-based approaches are also relevant under the Convention. Where the presence of waterbirds at a site constitutes an important feature of its ecological character, the maintenance of this feature should be reflected in the site's management objectives and parties should take measures to address species-specific threats.[32] Indeed, Ramsar's various guidance documents recognise that a site's ecological character may be negatively

---

[26] These numbers were drawn, and the associated percentages calculated, from data in Ramsar, 'Ramsar Sites Information Service', at: https://rsis.ramsar.org/ris-search/.

[27] Eg, Ramsar Convention, *Recommendation 2.5: Designation of the Wadden Sea for the List of Wetlands of International Importance* (1984); Ramsar Convention, *Resolution X.22: Promoting International Cooperation for the Conservation of Waterbird Flyways* (2008) [21].

[28] Convention on Wetlands, note 2, Art 3.1.

[29] Eg, Ramsar Convention, *Resolution X.22*, note 27, [6].

[30] See M Finlayson et al, 'The Ramsar Convention and Ecosystem-Based Approaches to the Wise Use and Sustainable Development of Wetlands' (2011) 14(3) *Journal of International Wildlife Law and Policy* 176.

[31] Eg, Ramsar Convention, *Guidelines for Integrating Wetland Conservation and Wise Use into River Basin Management* (1999).

[32] See discussion in Lewis, note 25, 28–32, 62–63.

impacted by, among other things, the unsustainable harvest of wetland fauna[33] and bird collisions with/electrocutions by energy infrastructure.[34] Moreover, Articles 4.4 and 5 require contracting parties to endeavour 'through management to increase waterfowl populations on appropriate wetlands' and 'to coordinate and support present and future policies and regulations' concerning the conservation of wetland fauna. Ramsar's guidance on implementing Article 5 highlights the cooperative management of shared wetland-dependent species.[35]

Despite the above, Ramsar's body of guidance has done little to advise parties on how to manage wetlands in a manner that meets the needs of migratory waterbirds specifically or how to address threats that are not habitat related.[36] Nor has the Convention endeavoured to coordinate measures to address the latter across entire flyways. The provision of such guidance and coordination has instead been left to international instruments that are more narrowly focused on species conservation.

4 EMERGENCE AND ROLE OF THE INTERNATIONAL MIGRATORY WATERBIRD REGIMES

In the time since Ramsar's adoption, other aspects of international conservation law have also evolved considerably. In Ramsar's early years, the Convention was the principal multilateral framework for migratory waterbird conservation, although a handful of bilateral migratory bird treaties existed.[37] Today, however, an exceptionally wide variety of multilateral cooperative efforts endeavour to conserve migratory waterbirds – some span entire flyways, while others are more restricted in scope.[38]

---

[33] Ramsar Convention, *Guidelines for International Cooperation under the Ramsar Convention* (1999) [54].

[34] Ramsar Convention, *Resolution XI.10: Wetlands and Energy Issues* (2012) annex [(B)(5)(v)].

[35] Ramsar Convention, note 33, [15]–[18].

[36] An exception is the threat posed by highly pathogenic avian influenza. See discussion in Section 6.2.2 below.

[37] See further M Bowman, P Davies, and C Redgwell, *Lyster's International Wildlife Law* (Cambridge University Press, Cambridge, 2010) 212–26.

[38] See generally T Jones and T Mundkur, 'A Review of CMS and Non-CMS Existing Administrative and Management Instruments for Migratory Birds Globally' in *A Review of Migratory Bird Flyways and Priorities for Management* (UNEP/CMS Secretariat, Bonn, 2014), at: https://www.cms.int/sites/default/files/publication/CMS_Flyways_Reviews_Web.pdf.

Writing in 1994, de Klemm argued that the CMS's success would in the future 'be judged on its ability to bring about the conclusion of flyway agreements, especially for the conservation and sustainable exploitation of water birds'.[39] Almost three decades later, the CMS family is well populated with flyway-level instruments for migratory bird conservation. Those dedicated to waterbirds specifically include a treaty covering Africa and western-Eurasia (AEWA); an action plan covering the Central Asian Flyway (CAF) – the CAF Action Plan, which does not yet have an institutional framework;[40] and several species-specific memoranda of understanding[41] (MOUs) and international species action plans.[42] Examples that emerged outside the CMS framework include the East Asian-Australasian Flyway Partnership (EAAFP)[43] (of which the CMS Secretariat is nevertheless a partner); and a large suite of initiatives

---

[39] C de Klemm, 'The Problem of Migratory Species in International Law' in HO Bergesen and G Parmann (eds), *Green Globe Yearbook of International Co-operation on Environment and Development* (Oxford University Press, Oxford, 1994) 74–75.

[40] CMS, *Central Asian Flyway Action Plan for the Conservation of Migratory Waterbirds and their Habitats*, (2005) at: https://www.cms.int/sites/default/files/document/CAF_action_plan_e_0.pdf.

[41] Eg, Memorandum of Understanding on the Conservation of High Andean Flamingos and their Habitats (November 2008), at: https://www.cms.int/flamingos/sites/default/files/basic_page_documents/mou_text_e.pdf. The CMS Raptors Memorandum of Understanding (MOU) is also relevant insofar as it both applies to wetland-dependent raptors and promotes the conservation of several sites of shared importance to raptors and other taxonomic groups. The Action Plan annexed to this MOU identifies a provisional list of areas that are known to be important congregatory sites for birds of prey. Several of these are Ramsar Sites – for example, Atanasovsko Lake in Bulgaria, which is important for a wide range of avifauna. Memorandum of Understanding on the Conservation of Migratory Birds of Prey in Africa and Eurasia (1 November 2008), at: https://www.cms.int/raptors/; Ramsar Information Sheet: Atanasovsko Lake, at: https://rsis.ramsar.org/RISapp/files/RISrep/BG292RIS_2001_en.pdf.

[42] Eg, PK Ndang'ang'a and E Sande, *International Single Species Action Plan for the Madagascar Pond-heron, Ardeola idea* (CMS Technical Series No 20; AEWA Technical Series No 39, Bonn, 2008), at: https://www.unep-aewa.org/sites/default/files/publication/ts39_ssap_madag_pond_heron_0.pdf.

[43] Partnership for the Conservation of Migratory Waterbirds and the Sustainable Use of their Habitats in the East Asian – Australasian Flyway (EAAFP), at: https://www.eaaflyway.net/wp-content/uploads/2020/06/EAAFP-Partnership-Doc-Updated-postMOP10-2020_04.pdf.

in the Americas, such as the Western Hemisphere Shorebird Reserve Network (WHSRN)[44] and the North American Waterfowl Management Plan.[45]

Evidently, the contemporary suite of multilateral initiatives for migratory waterbird conservation is complex and, in some regions, fragmented.[46] In an endeavour to, among other objectives, enhance synergies between these initiatives (both for waterbirds and other taxa), the CMS COP adopted a Programme of Work (POW) on Migratory Birds and Flyways for 2014–23.[47] For the Americas specifically, it also adopted an Americas Flyways Framework[48] and an Action Plan for the Americas.[49]

Overlap inevitably exists between the measures called for by species-based instruments and by Ramsar. However, as one would expect, the former place a firmer emphasis on managing sites and broader habitats in a manner that meets the ecological needs of waterbirds, and endeavour to support such management through guidance and other measures.[50] Many species-based instruments also explicitly promote measures to address the broader range of threats to species. AEWA's legal text, in particular, includes requirements regarding hunting, the planning and construction of structures, human disturbance, lead poisoning, bycatch, and overfishing, among other threats.[51] Several of these threats are also explicitly referred to in the CAF Action Plan,[52] the Action Plan for the Americas Flyways[53] and various international single or multi-species action plans.[54] The CMS and AEWA have developed guidelines

---

[44] https://whsrn.org/.
[45] https://nawmp.org/.
[46] Jones and Mundkur, note 38, 19.
[47] CMS, *Resolution 12.11 (Rev.COP13): Flyways* (2020) annex 1.
[48] Ibid, annex 2.
[49] Ibid, annex 3.
[50] Eg, WHSRN, 'Site Support', https://whsrn.org/site-support/; AEWA Conservation Guidelines No 4, *Guidelines on the management of key sites for migratory waterbirds* (AEWA Technical Series No 18, 2002), at: https://www.unep-aewa.org/sites/default/files/publication/cg_4new_0.pdf; see further Lewis, note 25, 49–58.
[51] AEWA, note 5, annex 3.
[52] CMS, note 40, [2], [4].
[53] CMS, note 47, annex 3, [1.1.4]–[1.1.7].
[54] Eg, the multi-species action plan for Benguela seabirds calls for actions targeting such threats as resource competition with fisheries, bycatch, and oil spills. C Hagen and R Wanless, *International Multi-species Action Plan for the Conservation of Benguela Upwelling System Coastal Seabirds* (AEWA Technical Series No 60, Bonn, 2015), at: https://rsis.ramsar.org/RISapp/files/RISrep/BG292RIS_2001_en.pdf.

on how to address various species threats.[55] The CMS has also established multi-stakeholder taskforces aimed at improving measures to address particular threats;[56] while AEWA has drawn from the experience of the North American Waterfowl Management Plan to establish a platform to coordinate the adaptive harvest management of several European species across their flyways.[57]

5 THE RAMSAR CONVENTION AND THE MIGRATORY WATERBIRD REGIMES' ALIGNMENT AND MUTUALLY SUPPORTIVE NATURE

5.1 Ramsar's Role in Encouraging the Development of and Participation in Flyway Initiatives

By promoting and supporting the implementation and coordination of measures to conserve wetland-dependent species, multilateral initiatives for waterbird conservation complement Ramsar and assist states in satisfying several of its broadly framed provisions (including Articles 3.1, 4.1–4.2,[58] 4.4, and 5). Ramsar's COP has repeatedly encouraged contracting parties to develop, support, and participate in such arrangements.[59]

An especially interesting example concerns the East Asian-Australasian flyway – an area with significant gaps in CMS membership. In the early 1990s, the CMS Scientific Council initiated a process to develop a migratory waterbird treaty for this flyway. It was ultimately unsuccessful.[60] However, two successive Asia-Pacific Migratory Waterbird Conservation Strategies

---

[55] Eg, HAM Prinsen et al, *Guidelines on How to Avoid or Mitigate Impact of Electricity Power Grids on Migratory Birds in the African-Eurasian Region* (CMS Technical Series No 29; AEWA Technical Series No 50; CMS Raptors MOU Technical Series No 3, Bonn, 2012), at: https://www.unep-aewa.org/sites/default/files/publication/ts50_electr_guidelines_03122014.pdf.

[56] CMS, 'Task Forces', https://www.cms.int/en/taskforces.

[57] DA Stroud, J Madsen, and AD Fox, 'Key actions towards the sustainable management of European geese' (2017) 46(2) *Ambio* S328.

[58] On the establishment of nature reserves for waterfowl.

[59] Eg, Ramsar Convention, *Resolution X.22*, note 27, [19].

[60] C Shine and C de Klemm, *Wetlands, Water and the Law: Using law to advance wetland conservation and wise use* (IUCN Environmental Policy and Law Paper No 38, 1999) 293–94, at: https://portals.iucn.org/library/sites/library/files/documents/EPLP-038.pdf.

were developed outside the CMS framework.[61] The first was launched at Ramsar COP6 (1996).[62] The COP encouraged support for the strategy's implementation and highlighted Ramsar's ability to facilitate the development of multilateral approaches to waterbird conservation.[63] To this end, it adopted the 'Brisbane Initiative', calling for 'the establishment of a network of Ramsar-listed and other wetlands of international importance for migratory shorebirds along the East Asian-Australasian Flyway, managed to maintain their suitability for migratory shorebirds'.[64]

Ramsar's Secretariat subsequently sat on the committee responsible for overseeing and promoting the strategies' implementation.[65] The Ramsar COP also continued to encourage contracting parties to support and implement the strategies, to nominate additional sites to the networks established thereunder, and to 'consider actively the development of a multilateral agreement or other arrangement, to provide a long-term conservation framework for migratory waterbirds and their habitats' in this region.[66] In 2005, the EAAFP was endorsed as a Ramsar Regional Initiative[67] and the CMS is a partner thereto.[68]

Out of the EAAFP's 18 government partners, 17 are parties to the Ramsar Convention.[69] Ramsar's strong footing in this region, combined with the EAAFP's status as a Ramsar Regional Initiative, provide opportunities for the EAAFP to influence national policy and decision-making, despite not itself being a legally binding instrument. Indeed, a 2016 review of the EAAFP's effectiveness recommended that more should be done to capitalise on its relationship with Ramsar, and stressed that '[a]ctive engagement at the national and site levels under the banner of the Ramsar Convention will help to raise the

---

[61] T Mundkur, 'Successes and challenges of promoting conservation of migratory waterbirds and wetlands in the Asia-Pacific: nine years of a regional strategy' in Boere, Galbraith, and Stroud, note 3, 81.

[62] Ibid.

[63] Ramsar Convention, *Recommendation 6.4: The 'Brisbane Initiative' on the Establishment of a Network of Listed Sites Along the East Asian-Australasian Flyway* (1996) [12], [14].

[64] Ibid, [13].

[65] Mundkur, note 61, 82.

[66] Ramsar Convention, *Recommendation 7.3: Multilateral cooperation on the conservation of migratory waterbirds in the Asia-Pacific Region* (1999).

[67] Ramsar Convention, *Resolution IX.7: Regional initiatives in the framework of the Ramsar Convention* (2005) annex I.

[68] EAAFP, note 43, appendix I.

[69] Ibid; read with Ramsar, 'Country profiles', https://www.ramsar.org/country-profiles. In comparison, only five are parties to the CMS. CMS, 'Parties and Range States', https://www.cms.int/en/parties-range-states.

profile of the EAAFP' and promote the integration of waterbird conservation priorities into decision-making and planning regarding wetland management.[70]

Even in Africa and western Eurasia, where the CMS has gained more traction and migratory waterbird conservation is coordinated through a dedicated multilateral treaty (AEWA), Ramsar has fewer gaps in membership than either of these instruments.[71] The Convention therefore provides a valuable avenue for encouraging accession/involvement by non-party range states. Indeed, the AEWA Secretariat has used meetings arranged under Ramsar to engage non-party range states in Africa.[72]

5.2 Alignment between Ramsar and the Migratory Waterbird Regimes

Although some bilateral migratory bird treaties existed at the time of Ramsar's adoption, the Convention preceded all flyway-scale instruments. This has facilitated deliberate alignment of these instruments' texts and interpretations with those of the Convention.

This alignment is seen, first, in definitions. Ramsar defines 'waterfowl' as 'birds ecologically dependent on wetlands'[73] and this term is considered synonymous with 'waterbird' under the Convention.[74] The definition was mirrored in the Asia-Pacific Migratory Waterbird Conservation Strategy.[75] AEWA[76] also draws on this definition, with the drafters using Ramsar's definition as a starting point for determining which species AEWA should cover.[77] The

---

[70] RJ D'Cruz, *Independent Review of the East Asian-Australasian Flyway Partnership* (2016) 15, 29, at: https://eaaflyway.net/wp-content/uploads/2017/12/Annex_Doc.2.3.1_Report_on_EAAFP_Independent_Review.pdf. This is not to suggest that the Ramsar Convention is the only treaty whose provisions can be leveraged by the EAAFP, as both the CMS and various bilateral migratory bird treaties are also applicable to some states.

[71] AEWA, 'Parties and Range States', https://www.unep-aewa.org/en/parties-range-states.

[72] AEWA, MOP7, *Report on the Implementation of the AEWA African Initiative and AEWA Plan of Action for Africa 2012–2018* (2018) 11.

[73] Convention on Wetlands, note 2, Art 1.2.

[74] See discussion in Lewis, note 25, 14–16.

[75] Wetlands International and IWRB, *Asia-Pacific Migratory Waterbird Conservation Strategy: 1996–2000* [2.3], at: http://www.jawgp.org/anet/str1996.htm#Preface.

[76] AEWA, note 5, Art I(2)(c).

[77] Minutes of the Informal Negotiation Meeting on the draft Agreement text of AEWA, first session (12–14 June 1994), [38] (copy on file with author).

same is true for the CAF Action Plan,[78] which additionally employs Ramsar's broad definition of 'wetlands'.[79] Initially, this term was not explicitly defined in AEWA, although its negotiating history and guidance documents suggested it should be interpreted to have the same meaning under AEWA as it does under Ramsar.[80] Indeed, the European Union interpreted 'wetland' in this manner for the purposes of implementing its AEWA commitment to phase out the use of lead shot for hunting over wetlands.[81] Clarity was provided in 2022, when a footnote was added to AEWA's legal text, specifying that '[f]or the purposes of the implementation of AEWA, the definition of "wetlands" as provided by Article 1.1 of the Convention on Wetlands of International Importance Especially as Waterfowl Habitat shall apply'.[82]

Second, when one considers the habitat-related provisions of various flyway instruments, it is evident that these were at least partially designed to support and complement the implementation of Ramsar, rather than to introduce entirely new regimes for habitat conservation. For instance, AEWA's legal text requires that parties endeavour to make 'wise use' of wetlands and provide special protection to 'wetlands which meet internationally accepted criteria of international importance'.[83] The latter have been defined to include sites qualifying for populations of AEWA species under the site selection criteria of Ramsar and other international instruments.[84] The Ramsar List is therefore one of the international site inventories currently being used to identify AEWA Flyway Network Sites.[85] The CAF Action Plan similarly envisages the establishment of a network of 'waterbird sites of international importance' across

---

[78] CMS, note 40, 3. Notably, however, the species coverage of both AEWA and the CAF Action Plan is limited to species explicitly listed under, and falling within the geographic scope of, these respective instruments. Neither of their species lists take as inclusive an approach as the Ramsar Convention's guidance. In particular, neither covers non-migratory or endemic species, coucals, or wetland-related raptors and owls.

[79] Ibid, 4.

[80] See Lewis, note 25, 17–18.

[81] Commission Regulation (EU) 2021/57 of 25 January 2021 amending Annex XVII to Regulation (EC) No 1907/2006 of the European Parliament and of the Council concerning the Registration, Evaluation, Authorisation and Restriction of Chemicals (REACH) as regards lead in gunshot in or around wetlands, [24].

[82] AEWA, *Resolution 8.2: Adoption of amendments to the AEWA annexes* (2022) [2].

[83] AEWA, note 5, annex 3 [3.2.2]–[3.2.3].

[84] AEWA, *Inventory of nationally and internationally important sites for migratory waterbirds in the AEWA area: Guidance notes* (2020) 2.

[85] Ibid, 3.

the Central Asian Flyway and calls on range states to endeavour to give special protection to sites that meet internationally accepted criteria of international importance, including Ramsar Sites.[86] The development of a flyway network of sites of international importance for migratory bird conservation is also a key objective of the EAAFP, and the partnership makes use of Ramsar listing Criteria 2, 5, and 6 (among other criteria) in identifying appropriate sites for inclusion therein.[87]

5.3 Challenges Regarding Waterbird Population Estimates and Potential for Synergy

Clearly, both the Ramsar Convention and the various flyway instruments rely on information on waterbird population sizes, population trends, and bird numbers at sites – both for their implementation and for assessing their effectiveness. Since 1994, information on population sizes and 1% thresholds has been compiled and published in several editions of *Waterbird Populations Estimates* (*WPE*).[88] However, a challenge has arisen in that, although the *WPE* is Ramsar's 'official' source of population estimates and 1% thresholds, funding constraints have prevented regular updates (the most recent edition was published in 2012).[89] In contrast, both AEWA and the EAAFP produce conservation status reviews at regular intervals. This misaligned schedule has caused confusion regarding which estimates and thresholds to rely upon for the purposes of Criterion 6 of the Ramsar Criteria.[90]

Clarification was recently provided by the Ramsar COP, which agreed that, until the *WPE* is updated, parties may use alternative data sources to determine 1% thresholds in the context of Criterion 6, subject to certain conditions. The COP also agreed that for migratory species, such thresholds should be derived from estimates 'based on the Conservation Status Reviews (CSR) produced under the auspices of flyway instruments',[91] and encouraged parties to work

---

[86] CMS, note 40, [3.3.1], [3.2.4].
[87] EAAFP, note 43, appendix IV.
[88] Wetlands International, *Waterbird Population Estimates Fifth Edition* (2012), at: https://www.wetlands.org/publications/waterbird-populations-estimates-fifth-edition/.
[89] Ramsar Convention, 59th Meeting of the Standing Committee, *Report of the Chair of the Scientific and Technical Review Panel* (2021) annex 1, [4.1].
[90] Ibid, annex 1, [4.4].
[91] Or other peer-reviewed assessments where a CSR-type assessment does not exist for a migratory population, as well as for non-migratory and endemic populations. Ramsar Convention, *Draft resolution on waterbird population estimates to support new and existing Ramsar Site designations under Ramsar Criterion*

cooperatively with flyway agreements and partnerships to facilitate regular updates to the *WPE*.[92] Indeed, the Ramsar Scientific and Technical Review Panel (STRP) has been considering the potential value of partnering with flyway frameworks with a view to producing more regular and predictable editions of the *WPE*,[93] and the most recent AEWA Meeting of the Parties (MOP) agreed that AEWA should offer to contribute to the *WPE* through its CSR.[94]

5.4 Flyway Instruments' Promotion of Ramsar Designation

Various flyway instruments have actively encouraged states to designate Ramsar Sites. In some instances, this encouragement takes the form of general exhortations.[95] In others, it has been directed towards important habitats for individual species (eg, key breeding sites for the Madagascar pond heron).[96] For species with their own International Species Working Group (ISWG), specific sites are sometimes identified as high priorities for Ramsar designation, helping to promote national action. For instance, out of the three South African sites agreed by the AEWA White-Winged Flufftail ISWG as priorities for Ramsar designation, one was designated in 2021 and efforts are underway to designate the other two.[97]

Despite a high percentage of Ramsar Sites having been designated based on the Convention's waterbird criteria, there remain major gaps in site designations across flyways.[98] In continuing to promote such designations, it should be recalled that Ramsar's legal text explicitly requires contracting parties to consider their 'international responsibilities for the conservation, management and wise use of migratory stocks of waterfowl' when designating or changing

---

6 – use of alternative estimates, Ramsar COP14 Doc 18.21 Rev 1 (as adopted at COP14, 2022) [11].

[92] Ibid, [13].

[93] Ramsar Convention, note 89, annex 1, appendix A.

[94] AEWA, *Waterbird Monitoring Synergies with Other Frameworks*, Doc AEWA/MOP 8.28 (2022) 10, at: https://www.unep-aewa.org/sites/default/files/document/aewa_mop8_28_waterbird_monitoring_synergies.pdf; as endorsed by AEWA, *Resolution 8.5: Further development and strengthening of monitoring of migratory waterbirds* (2022) [9].

[95] See, eg, CMS, note 47, annex 3, [1.1.2.3].

[96] Ndang'ang'a and Sande, note 42, 24.

[97] AEWA White-winged Flufftail International Working Group, *Implementation Plan for 2020–2022* (copy on file with author); read with Ramsar Sites Information Service, 'Ingula Nature Reserve', https://rsis.ramsar.org/ris/2446?language=en.

[98] Lewis, note 25, 26–27.

entries to the List.[99] There is no indication that Ramsar's drafters intended this provision to be limited to 'international responsibilities' articulated in the Convention itself. Indeed, at the time of Ramsar's adoption, some migratory bird treaties already existed and it was intended that the Convention would ultimately be supplemented by additional, species-focused instruments.[100] Ramsar thus arguably requires parties that have also ratified a migratory bird treaty to take these commitments into consideration when designating Ramsar Sites.[101]

6 COLLABORATION BETWEEN RAMSAR, THE CMS AND AEWA

6.1 Agreed MOUs and Joint Work Plans

Overlap in the objectives and provisions of Ramsar and the international migratory waterbird regimes obviously presents considerable scope for collaboration and a need for coordination so as to ensure that actions are mutually supportive yet not duplicative. In recent decades, multilateral environmental agreements have placed increasing emphasis on the importance of synergies[102] and Ramsar, the CMS, and AEWA are no exception.

Since 2004, the Ramsar and CMS Secretariats have participated in the Liaison Group of Biodiversity-Related Conventions, which aims to explore opportunities for synergies and enhanced coordination among the global biodiversity-related conventions.[103] However, even before this group's establishment, Ramsar's first Strategic Plan (adopted in 1996) envisaged 'coop-

---

[99] Convention on Wetlands, note 2, Art 2.6.
[100] See Section 3.1 above.
[101] See also Ramsar Convention, *Resolution XI.8: Streamlining procedures for describing Ramsar Sites at the time of designation and subsequent updates* (2012) annex 2, [77]. This guidance urges Ramsar's parties, when considering Ramsar Site designations, to 'consider the opportunities this may also provide for contributing to other established and developing initiatives under related international and regional environment conventions and programmes', and highlights that this applies 'in particular to the Convention on Biological Diversity and the Convention on Migratory Species and its Agreements, such as the African-Eurasian Waterbirds Agreement'.
[102] See, eg, T Honkonen and E Couzens (eds), *International Environmental Law-making and Diplomacy Review 2011* (University of Eastern Finland, Joensuu, 2013).
[103] CBD, 'Liaison Group of Biodiversity-related Conventions', at: https://www.cbd.int/blg/.

erative arrangements with the [CMS], flyway agreements, networks and other mechanisms dealing with migratory species',[104] and an MOU had been entered into between the Ramsar and CMS Secretariats. To operationalise this MOU, the Ramsar, CMS, and AEWA Secretariats developed a tripartite Joint Work Plan (JWP) for the period 2003–05, which identified detailed actions for the respective Secretariats under several areas of cooperation.[105] This JWP was ultimately too ambitious relative to the capacities of the respective Secretariats.[106] More recent JWPs between Ramsar and the CMS have been considerably less detailed[107] and divide activities into the following categories:

- support of national policy initiatives for coordinated implementation of the CMS and Ramsar;
- facilitation of mutual participation in relevant meetings, including those of regional agreements and initiatives;
- management of species populations and wetland ecosystems;
- monitoring and assessment;
- global science and policy (eg, guidance materials); and
- information, outreach, and capacity-building.[108]

The extent to which Ramsar has collaborated with the CMS and its ancillary Agreements under these headings has varied over time, with contemporary synergies tending to be largely *ad hoc* rather than systematic. The past two decades have, however, seen several noteworthy examples of collaboration.

---

[104] Ramsar Convention, *Strategic Plan 1997–2002* (1996) action 7.2.5, at: https://www.ramsar.org/sites/default/files/documents/pdf/key_strat_plan_1997_e.pdf.

[105] Joint Work Plan 2003–05 between The Bureau of the Convention on Wetlands (Ramsar, Iran, 1971) and the Secretariat of the Convention on the Conservation of Migratory Species of Wild Animals (CMS) and between The Bureau of the Convention on Wetlands (Ramsar, Iran, 1971) and the Secretariat of the Agreement on the Conservation of African-Eurasian Migratory Waterbirds (AEWA), at: https://www.cms.int/sites/default/files/document/Inf17_CMS_JWP_CMS_Ramsar_AEWA_0.pdf.

[106] CMS, 38th Meeting of the Standing Committee, *Cooperation Between CMS and Ramsar* (2011) [3], at: https://www.cms.int/sites/default/files/document/doc_05_cms_ramsar_jwp_e_0.pdf.

[107] Ibid, annex 1; CMS, 44th Meeting of the Standing Committee, *Cooperation Between CMS and Ramsar* (2015) annex 1, at: https://www.cms.int/sites/default/files/document/Doc_18_3_CMS_Ramsar_JWP%2015-17_0.pdf.

[108] CMS, note 106; CMS, note 107.

6.2 Specific Examples of Collaboration

6.2.1 Developing CMS flyway frameworks and promoting interflyway collaboration

The Ramsar Secretariat has previously participated actively in the CMS Scientific Council Working Group on Flyways[109] – a think tank on migratory bird flyways and frameworks that, among other activities, was responsible for drafting a series of global reviews (published in 2014), the CMS Flyways POW, and the Americas Flyways Framework.[110]

As conventions with global scope yet differing membership, Ramsar and the CMS can also collaborate to promote the sharing of knowledge and experience between initiatives in different flyways. In the past, the Ramsar COP has explicitly encouraged such exchanges;[111] and in 2011, an international workshop was convened for this purpose by Ramsar and the CMS, among others.[112] Workshop participants agreed to establish a Global Interflyway Network to facilitate future interflyway cooperation.[113] To date, however, this has not occurred.

6.2.2 Collaboration in scientific and technical bodies and task forces

The scientific and technical bodies of Ramsar, the CMS, and AEWA typically send representation to one another's meetings and remain in contact on issues of common concern (eg, the impact of invasive species and extractive industries on wetlands). Communication between these bodies has also been supported by overlap in their members and the nomination of a liaison officer between the AEWA Technical Committee and the Ramsar STRP.[114] This interaction has, first, enabled the former to learn from the latter's more extensive experience – for instance, regarding the management and prioritisation

---

[109] CMS, 'CMS Flyways Working Group Membership' (as of 1 September 2014), at: https://www.cms.int/en/document/cms-flyways-working-group-membership.

[110] CMS, 'Working Group on Flyways', at: https://www.cms.int/en/workinggroup/working-group-flyways.

[111] Ramsar Convention, *Resolution X.22*, note 27, [24].

[112] CY Choi et al (eds), *Waterbird Flyway Initiatives: Outcomes of the 2011 Global Waterbird Flyways Workshop to promote exchange of good practice and lessons learnt* (AEWA Technical Series No 40, Bonn; CMS Technical Series No 25, Bonn; EAAFP Technical Report No 1, Incheon; Ramsar Technical Report No 8, Gland, Switzerland, 2012), at: https://www.wetlands.org/publications/global-interflyway-network-korea-workshop-2011/.

[113] Ibid, 4.

[114] AEWA, MOP7, *Report of the Technical Committee to the 7th Session of the Meeting of the Parties* (2018) 5.

of its work,[115] and its procedure for horizon scanning for emerging issues.[116] Second, it has enabled the Technical Committee and STRP to provide input on the reports and guidance documents developed under one another's governing treaties. For example, the AEWA Technical Committee has, through the Ramsar STRP, provided input on Ramsar's *Global Wetland Outlook*.[117] Third, it has helped to ensure that each treaty is aware of the guidance being developed under the other and can draw its parties' attention to this. Indeed, it is fairly common for AEWA resolutions to refer parties to, and in some instances explicitly urge them to apply, Ramsar's guidance materials rather than providing new or duplicative guidance.[118] Finally, it facilitates the development of complementary responses to, or joint guidance on, particular conservation threats.

Regarding the last of these points, conservation treaties' ability to respond to rapidly emerging threats and to coordinate such responses between one another is sometimes constrained by both the meeting cycles of individual treaty bodies (eg, where a COP or MOP typically only meets once every three years) and misalignments between the meeting schedules of similar treaties. The triennial AEWA MOP and CMS and Ramsar COPs are not timed to align with each other.[119] Nevertheless, AEWA, the CMS, and Ramsar have demonstrated their ability to quickly coordinate responsive actions when the need arises.

The principal example of this is their response to highly pathogenic avian influenza (HPAI). In 2005, a major HPAI outbreak in wild birds and the westward spread of this disease from East Asia raised pressing economic, human health, and conservation concerns.[120] In response, international organisations and treaties worked closely to develop and maintain a common position, which was articulated in a suite of complementary resolutions adopted under AEWA, Ramsar, and the CMS in 2005 and 2008.[121] Such coordination was achieved

---

[115] AEWA, MOP6, *Report of the Technical Committee* (2015) 3–4.

[116] R Cromie et al, 'Responding to Emerging Challenges: Multilateral Environmental Agreements and Highly Pathogenic Avian Influenza H5N1' (2011) 14 *Journal of International Wildlife Law and Policy* 206, 237.

[117] AEWA, note 114, 12.

[118] Eg, AEWA, *Resolution 5.14: Waterbirds, Wetlands and the Impacts of Extractive Industries* (2012) [1]–[3], [5].

[119] In AEWA's early years, an attempt was made to schedule sessions of its MOP to occur shortly after meetings of the Ramsar COP (see AEWA, *Resolution 2.10: Date, Venue and Funding of the Third Session of the Meeting of the Parties* (2002) [1]). However, this has not continued.

[120] Cromie et al, note 116, 208–16.

[121] For a detailed discussion of this collaborative effort, see ibid.

through a Scientific Task Force on Avian Influenza and Wild Birds, which was established by the CMS (in close collaboration with AEWA) in August 2005 as a liaison mechanism between relevant treaties and organisations,[122] and has since provided opportunities to exchange knowledge and adapt guidance and policy advice in light of the developing understanding of HPAI.[123] Ramsar continues to be represented on this Task Force. Its most recent statement was published in January 2022 in response to new outbreaks of HPAI and aimed to inform various stakeholders about HPAI viruses in wild birds and appropriate responses.[124]

Both the Ramsar and AEWA Secretariats are also represented on the CMS Energy Task Force, which brings together various stakeholders with the aim of avoiding and minimising negative impacts of energy developments on migratory species.[125]

6.2.3 Advisory Missions

In many instances, threats to the ecological character of a Ramsar Site will have implications for a state's compliance with not only Ramsar itself, but also the CMS and/or AEWA. Where this is the case, it makes sense for the threat to be assessed and recommendations made in a manner that draws on the provisions and expertise of all relevant treaties. Over the years, the CMS and AEWA Secretariats have been represented in several Ramsar Advisory Missions (RAMs) – for example, the 2008 RAM to consider a proposed soda ash facility at Lake Natron, Tanzania.[126] Interestingly, a 2009 RAM to Mozambique included representation of both the CMS and AEWA, despite

---

[122] The co-conveners of the Task Force are currently the CMS and the United Nations Food and Agriculture Organisation. See CMS, 'Scientific Task Force on Avian Influenza and Wild Birds', at: https://www.cms.int/en/workinggroup/scientific-task-force-avian-influenza-and-wild-birds.

[123] Cromie et al, note 116, 234.

[124] Scientific Task Force on Avian Influenza and Wild Birds, *H5N1 Highly Pathogenic Avian Influenza in Poultry and Wild Birds: Winter of 2021/2022 with Focus on Mass Mortality of Wild Birds in UK and Israel* (24 January 2022), at: https://www.cms.int/sites/default/files/uploads/avian_influenza_0.pdf.

[125] CMS, 'Energy Task Force', at: https://www.cms.int/en/taskforce/energy-task-force.

[126] *Mission Report, Ramsar Advisory Mission No. 59: Tanzania* (2008), at: https://www.unep-aewa.org/sites/default/files/basic_page_documents/ram_rpt_59e.pdf.

Mozambique not being a party to the latter. One of the RAM's recommendations was that Mozambique accede to AEWA.[127]

Since 2008, when AEWA established its own Implementation Review Process (IRP), several of its IRP cases have involved Ramsar Sites. The AEWA and Ramsar Secretariats have liaised closely in this regard – for example, they have recently been collaborating to organise joint IRP/RAM visits regarding urban development at the Sebkhet Sejoumi Ramsar Site (Tunisia) and powerline construction at the Lake Elmenteita Ramsar Site (Kenya).[128] In 2017, the CMS COP agreed to establish a review mechanism to facilitate compliance with the Convention.[129] The general principles according to which this mechanism operates include that '[r]eviews are done in a synergistic and cooperative manner with other relevant processes both within and outside the CMS';[130] and the Convention's Flyways POW recognises the potential for collaboration with Ramsar in the provision of timely advice and technical support to parties facing compliance challenges.[131]

6.2.4 Collaborative projects

There are several examples of collaboration on multi-partner projects. For instance, both the Ramsar and AEWA Secretariats are currently involved in the RESSOURCE project, which aims to improve knowledge on, and sustainable use of, waterbird populations in sub-Saharan Africa.[132] However, the most significant collaboration to date has been the Wings Over Wetlands (WOW) Project – a joint effort between the Ramsar Secretariat, the AEWA Secretariat, Wetlands International, and BirdLife International, which ran from 2006 to 2010. The project developed an interactive online portal (the 'Critical Site

---

[127] *Mission Report, Ramsar Advisory Missions – No. 62: Marromeau Complex Ramsar Site, Mozambique* (2009) [24], at: https://rsis.ramsar.org/RISapp/files/ RAM/ RAM_062_MZ_en.pdf. Such accession has not, however, subsequently occurred.

[128] UNEP/AEWA Secretariat, *Report on the Implementation Review Process to the 8th session of the Meeting of the Parties* (2022), AEWA/MOP 8.20, at: https:// www.unep-aewa.org/sites/default/files/document/aewa_mop8_20_irp_report.pdf.

[129] The COP mandated the CMS Standing Committee to perform the functions of the review body, which may include, *inter alia*, the provision of 'in-country assistance, technical assessment or a verification mission.' CMS, *Resolution 12.9: Establishment of a review mechanism and a national legislation programme* (2017).

[130] Ibid.

[131] CMS, note 47, annex 1, action 17.

[132] AEWA, note 114, 5–6.

Network Tool'),[133] which provides data on species of migratory waterbirds in Africa and western Eurasia, their migration routes, and the sites on which they rely, thereby supporting implementation of both Ramsar and AEWA. It also developed a Flyway Training Kit to facilitate the training of stakeholders in both wetlands management and waterbird conservation, and supported site-level demonstration projects in 12 countries.[134]

In mid-2010, the WOW partner organisations signed a Memorandum of Cooperation which aimed to build on the institutional partnerships developed during the project and provide a basis for continued collaboration in flyway conservation.[135] This initiative ultimately did not gather momentum. Nevertheless, some collaboration has continued. For instance, in 2019, the AEWA and Ramsar Secretariats (along with the Wadden Sea Flyway Initiative) collaborated to organise a training-of-trainers workshop in flyway conservation for francophone countries in west and central Africa.[136] Notably, the CMS COP has also recognised the need for a Global Critical Site Network Tool, modelled on that produced under the WOW Project.[137] The AEWA MOP has also recognised the desirability of developing a joint programme with Ramsar to promote the Ramsar designation of critical sites along flyways.[138]

7 CONCLUSIONS

The Ramsar Convention's conception was firmly rooted in the desire to conserve waterfowl. Several decades later, it has made important contributions to achieving this objective – most notably through its role in promoting the identification and conservation of some of the most critical sites for waterbirds and its stimulation of waterbird monitoring as a basis for these activities. The provision of guidance and other support for managing sites/habitats for waterbirds and addressing the broader variety of threats these species face has largely been left to the suite of complementary multilateral flyway initiatives that have emerged since Ramsar's adoption. This division of responsibilities is consistent with the recognition by Ramsar's drafters that, rather than the Convention endeavouring to provide a framework for states' responses to

---

[133] This tool is freely available at http://criticalsites.wetlands.org/en.
[134] Wetlands International, 'The Wings Over Wetlands (WOW) Project', at: https://www.wetlands.org/the-wings-over-wetlands-wow-project/.
[135] AEWA, MOP5, *Report of the Secretariat* (2012) 3.
[136] AEWA, 15th Meeting of the Standing Committee, *Report of the UNEP/AEWA Secretariat* (2019) 3.
[137] CMS, note 47, [19].
[138] *AEWA Strategic Plan 2019–2027* (2018) target 3.3.

all threats facing waterbirds, it should ultimately be supplemented by other multilateral regimes.

The Ramsar COP has historically encouraged the development of, and support for, flyway initiatives, and efforts have been made to ensure alignment between these initiatives and the Convention so that they are mutually supportive. Capitalising on linkages with Ramsar is a strategy through which the status of flyway initiatives can potentially be elevated nationally, and through which additional states can be engaged with a view to their participation.

The 1990s and 2000s saw some particularly noteworthy Ramsar activities relating to waterbird conservation, including the adoption of the Brisbane Initiative; the endorsement of the EAAFP as a Ramsar Regional Initiative; collaboration with other treaties and organisations to rapidly develop and align their positions on HPAI; the WOW project; and the agreement of an ambitious JWP with the CMS and AEWA Secretariats.

Over the past decade, collaborative efforts have begun dissipating but have by no means disappeared. Moreover, although a long-term, structured programme for collaboration in flyway-level waterbird and wetland conservation has not yet emerged, discussions regarding a possible international partnership in relation to waterbird population assessments could potentially lead to the development of a well-structured synergy on this topic. The future development of structured synergies for other priority areas is also possible – in particular, for the continued promotion of Ramsar designations along waterbird flyways.

8. The Ramsar Convention and the Convention on Biological Diversity: Proposals for enhanced cooperation, synergies, and interoperability for the Kunming-Montreal Global Biodiversity Framework

Teresa Fajardo[1]

1 INTRODUCTION

The coincidence between the 2021–22 biennium of the anniversaries and Conferences of the Parties (COP) of the major biodiversity-related Conventions[2]

---

[1] This Chapter is the result of the collaboration within the framework of the Research Project PID2020-117379GB-I00, Biodiversity, Climate and Global Public Health: Interactions and Challenges for International Law (BIOCLIHEALTH), directed by Professor Mar Campins i Eritja of the University of Barcelona, and financed by the Spanish Research State Agency, Ministry of Science and Innovation.

[2] In chronological order, these treaties are the International Convention for the Regulation of Whaling, adopted 2 December 1946, entered into force 10 November 1948, 161 UNTS 72 (ICRW); International Plant Protection Convention, adopted 6 December 1951, entered into force 3 April 1952, 150 UNTS I-1963 (revised 1997; 2367 UNTS A-1963) (IPPC); Convention on Wetlands of International Importance especially as Waterfowl Habitat, adopted 2 February 1971, entered into force 21 December 1975, 996 UNTS 245 (amended 1982 and 1987) (Ramsar Convention); the Convention Concerning the Protection of the World Cultural and Natural Heritage, adopted 16 November 1972, entered into force 17 December 1975, 1037 UNTS 151 (World Heritage Convention); the Convention on International Trade in Endangered Species of Wild Fauna and Flora, adopted 3 March 1973, entered into force 1 July 1975, 993 UNTS 243 (CITES); the Convention on the Conservation of Migratory Species of Wild Animals, adopted 23 June 1979, entered into force 1

and the Rio Conventions[3] has provided an opportunity to enhance synergies and cooperation between them and to integrate biodiversity into all ongoing United Nations (UN) policy-making processes. The Ramsar Convention, as part of the constellation of biodiversity-related Conventions led by the Convention on Biological Diversity (CBD), was involved in the negotiation of the Global Biodiversity Framework at CBD COP15, which took place in Montreal in December 2022 under the presidency of China, and replaced the Strategic Plan for Biodiversity 2011–20, as well as the Aichi Targets, with the new 2030 Targets and the Kunming-Montreal Global Biodiversity Framework.

Cooperation among the biodiversity-related Conventions to achieve their common goals and shared missions has so far been explored on the proviso that the fulfilment of their specific objectives and mandates is not at stake and that no additional resources are necessary. Based on these premises, cooperation can be achieved through respect for common principles and the establishment of a network of formal and informal contacts that facilitates the exchange of information, interoperability between databases and institutional mechanisms, and best practices. However, when assessing results on the occasion of the 50th anniversary of the Ramsar Convention and in the context of preparation for CBD COP15, the lack of implementation of the different commitments under the various biodiversity-related Conventions has led to an insistence on the need for closer alignment and interoperability of sectoral plans and the search for common objectives to facilitate joint implementation, especially through national biodiversity strategies and action plans (NBSAPs) required to comply with the provisions of these treaties. The most ambitious objective proposed so far in the pre-conferences was to provide a common strategy for all biodiversity-related Conventions to synchronise work plans and headline targets until 2030; thus, the Ramsar Convention's Strategic Plan for 2020–24 has been updated 'to map the new Kunming-Montreal Global Biodiversity Framework targets to align with the Ramsar Strategic Plan goals and targets'.[4]

November 1983, 1651 UNTS 333 (CMS); the Convention on Biological Diversity, adopted 5 June 1992, entered into force 29 December 1993, 1760 UNTS 79 (CBD); and the 2001 International Treaty on Plant Genetic Resources for Food and Agriculture, adopted 3 November 2001, entered into force 29 June 2004, 2004 UNTS 303 (ITPGRFA).

[3] United Nations Framework Convention on Climate Change, adopted 9 May 1992, entered into force 21 March 1994, 1771 UNTS 107; CBD, note 3; and United Nations Convention to Combat Desertification in those Countries Experiencing Serious Drought and/or Desertification, Particularly in Africa, adopted 14 October 1994, entered into force 26 December 1996, 1954 UNTS 3.

[4] See Ramsar Convention, *The 4th Strategic Plan 2016–2024: 2022 update*, at: https://www.ramsar.org/sites/default/files/documents/library/4th_strategic_plan_2

The existing institutional mechanisms on assessment, monitoring, and reporting – among which the Ramsar mechanisms once stood out – can be used as models of best practice by other regimes, since the need to better comply with the biodiversity-related Conventions is urgent. However, so far, there has been no innovative proposal for institutional design to overcome existing cooperation gaps or even to improve existing institutions, such as the Biodiversity Liaison Group. The only significant outcome is a new monitoring framework that will establish indicators from pre-existing data to show, for example, the progression – or, as the case may be, regression – in the extent and condition of protected areas in the results obtained at the national level by NBSAPs implementing multi-year plans.[5]

During the long road to CBD COP15, the Ramsar Convention had much to say. The *Global Wetland Outlooks* have shown that wetlands are the most threatened of all ecosystems.[6] The work pursued by both the Ramsar Convention and the CBD is intrinsically linked at two levels. First, they are linked at the level of nature because, as Wetlands International has highlighted, solutions to biodiversity problems are intrinsically linked to wetlands, as '[w]etlands offer effective nature-based solutions to global problems. But environmental protection of wetland sites is not sufficient since the fluidity of water means that effective action requires coordination across landscapes and national boundaries'.[7] Second, they are linked at the policy level, since improving the implementation and enforcement of these instruments – which appear fragmented despite the intrinsic relationship between their purposes – is essential to enhancing existing institutional cooperation.

Over the years, the Ramsar Convention has developed various avenues of cooperation with the institutions responsible for overseeing the biodiversity-related Conventions – in particular the CBD – as well as vital non-governmental organisations (NGOs) concerned with the protection of wetlands of international and transboundary importance.[8] In the process of

---

022_update_e.pdf.

[5] CBD, *Decision 15/5: Monitoring framework for the Kunming-Montreal Global Biodiversity Framework* (2022).

[6] M Courouble et al, *Global Wetland Outlook: Special Edition 2021* (Ramsar Convention Secretariat, Gland, 2021) 29.

[7] Wetlands International, 'Statement: Wetlands International welcomes the 2030 Action Targets and calls for a focus on wetlands', 30 August 2021, at: https://www.wetlands.org/news/wetlands-international-welcomes-the-2030-action-targets-and-calls-for-a-focus-on-wetlands/.

[8] The contracting parties to the Ramsar Convention have as one of their main commitments the promotion of international cooperation, including 19 Ramsar Regional Initiatives and 22 Transboundary Ramsar Sites.

adopting the new post-2020 Global Biodiversity Framework, representatives of the Ramsar Convention advocated the importance of closer cooperation between the different biodiversity-related Conventions at the pre-conferences, meetings, and workshops[9] prior to CBD COP15. Furthermore, in its statements to the pre-COP15 events,[10] the Ramsar Secretariat requested that its principles and compliance indicators also be taken into account when discussing the shortcomings and gaps of the previous Strategic Plan for Biodiversity 2011–20[11] and the Aichi Targets. As these statements highlighted, the Kunming-Montreal CBD COP15 offered the perfect opportunity to consider a common framework for the biodiversity-related Conventions to align with relevant indicators of the Sustainable Development Goals (SDG) to avoid duplication, build on existing work, and promote collaboration and synergies.[12] The Ramsar Convention's representatives highlighted three aspects that were considered: '[W]here is the Convention regarding its strategic planning processes, the status of collaboration with CBD, and processes of assessment, monitoring and reporting that are relevant to the post-2020 biodiversity framework[?]'[13] In addition, and of particular importance, are the common principles, which – given the improbability of adopting new treaties in the coming years – serve to establish a common regulatory framework, covering the biodiversity-related Conventions and

---

[9] See the Bern Workshops, Consultation Workshop of Biodiversity-Related Conventions on the Post-2020 Global Biodiversity Framework, Bern Switzerland, 2019, 2021, 2022, 'Report of the consultation workshop of biodiversity related Conventions on the Post-2020 Global Biodiversity Framework', Bern I, 10–12 June 2019, at: https://www.cbd.int/conferences/post2020/BRC-WS-2019-01/documents; Bern II, 18 January–2 February, at: https://www.unep.org/events/workshop/bern-ii-consultation-workshop-biodiversity-related-conventions-post-2020-global.

[10] Third meeting of the Open Ended Working Group on the post 2020 Global Biodiversity Framework, Agenda item 4: Statement of the Ramsar Convention on Wetlands to the First Draft of the Post 2020 Framework (24 August 2021), at: https://www.cbd.int/doc/interventions/6124d9b408edd2000181d09d/RamsarStatement.pdf.

[11] Secretariat of the Convention on Biological Diversity and United Nations Environment Programme, *Strategic Plan for Biodiversity 2011–2020 and the Aichi Targets: Living in Harmony with Nature* (2019), at: https://wedocs.unep.org/20.500.11822/27533.

[12] Statement of Martha Rojas Urrego, Secretary-General, Ramsar Convention on Wetlands, Contributions of the Ramsar Convention on Wetlands to the Post 2020 Framework, Consultation Workshop of Biodiversity-Related Conventions on the Post 2020 Global Biodiversity Framework, Bern Switzerland 10–12 June 2019, 1, 3, at: https://www.cbd.int/doc/c/8c90/ef0c/aed13b606b8e648d5e08efc6/ramsar-bernworkshop-en.pdf.

[13] Ibid, 2.

strengthening regulatory linkages with other frameworks, such as the climate change regime.[14]

This Chapter examines these issues from the perspective of the Ramsar Convention's special cooperation with the CBD as one of its main partners, and the prospects and expectations for further cooperation in the Kunming-Montreal Global Biodiversity Framework. Thus, it is necessary to examine the goals and principles inspiring cooperation and facilitating inter-treaty synergies between the Ramsar Convention and the CBD. The institutional structure that will support the coordination of strategies and work plans is then examined, along with the indicators and mechanisms for promotion and monitoring of compliance that will serve to achieve the desired results in combating biodiversity loss.

2   THE VALUE OF INTER-TREATY SYNERGIES AND INTEROPERABILITY

A common problem faced by the biodiversity-related Conventions is that their central obligations are primarily implemented through the adoption of national measures. Moreover, as with most multilateral environmental agreements (MEAs), the obligations, principles, and standards of protection are not addressed only to states parties, but will ultimately have to be fulfilled by citizens, farmers, businesses, and other biodiversity stakeholders. As foreseen in Article 6 of the CBD, each contracting party:

> shall, in accordance with its particular conditions and capabilities, develop national strategies, plans or programs for the conservation and sustainable use of biological diversity or adapt for this purpose existing strategies, plans or programs which shall reflect, inter alia, the measures set out in the Convention relevant to the Contracting Party concerned.

Thus, 'NBSAPs are the principal planning tool for the implementation of the Convention at the national level' and '[a]lmost all Parties (97%) have developed at least one NBSAP since they became a Party'.[15] While respecting different national capacities and circumstances, COPs have established strategic plans that guide states parties in the adoption of their national plans. The clearest examples are provided by the Strategic Plan for Biodiversity 2011–20 and the Aichi Biodiversity Targets, which, despite their non-binding nature,

---

[14] See Chapter 11 of this volume.
[15] CBD, Note by the Executive Secretary, *Update on Progress in Revising/ Updating and Implementing National Biodiversity Strategies and Action Plans, Including National Targets*, CBD/SBI/3/2/Add.1 (25 March 2020) 1.

have had a major effect in promoting adherence to their proposals, but far lesser impact in securing compliance by states and individuals. In 2021, after a decade to deliver on its strategies, the time came to consider what improvements were needed to provide operational continuity but also to address shortcomings.

After the period of stagnation imposed by the COVID-19 pandemic, it was hoped that the COPs would provide the necessary impetus to renew efforts and commitments, given the seriousness of the problems to be encountered, as the latest *Global Biodiversity Outlook* and *Global Wetland Outlook* have shown.[16] Ramsar COP14 was finally held in November 2022. One month later, the second segment of CBD COP15 took place in Montreal, resulting in the adoption of the new Kunming-Montreal Global Biodiversity Framework, which is expected to serve as a roadmap for enhanced cooperation for the next decade. However, despite the large number of proposals that were discussed and adopted at Ramsar COP14 and CBD COP15, their generality does not suggest a radical change from the status quo, which has so far failed to mobilise states sufficiently.

This is why the adoption of the Kunming-Montreal Global Biodiversity Framework has given rise to such profound reflection – among not only the states parties to the CBD, but also the other parties responsible for the biodiversity-related Conventions and the Rio Conventions – questioning the way in which things have been done to date, as the results obtained so far are unsatisfactory in view of the fragmentation of policy instruments, programming, and mechanisms for monitoring and promoting implementation and compliance. This necessitates greater clarity regarding the objectives, expected results, and design of the strategies needed to achieve them, and the mobilisation of all necessary resources, taking into account their limited nature.[17]

At the pre-conferences and the first CBD COP15 session, the Ramsar Convention participated as an observer – as did many other conventions. On these occasions, the Secretary General of the Ramsar Convention pointed out where the Ramsar Convention stands, recalling that the Convention's Strategic Plan for 2016–24[18] is 'fully aligned or compatible with the Aichi Biodiversity

---

[16] Secretariat of the Convention on Biological Diversity, *Global Biodiversity Outlook 5* (Montreal, 2020); Courouble et al, note 6.

[17] CBD, *Preparations for the Post-2020 Biodiversity Framework Synthesis of Views of Parties and Observers on the Scope and Content of the Post-2020 Global Biodiversity Framework*, CBD/POST2020/PREP/1/INF/1 (24 January 2019) 3.

[18] Ramsar Convention Secretariat, *The Fourth Ramsar Strategic Plan 2016–2024* (Ramsar Handbooks, 5th edition, Ramsar Convention Secretariat, Gland, 2016).

Targets and the Sustainable Development Goals' in its pursuit of four goals: '[a]ddressing the [d]rivers of [w]etland [l]oss and [d]egradation', '[e]ffectively [c]onserving and [m]anaging the Ramsar Site Network', '[w]isely [u]sing [a]ll [w]etlands' and '[e]nhancing [i]mplementation'.[19]

The results of Ramsar COP14 were a commitment to synergies and interoperability between the different environmental conventions but with a view to the adoption of national measures and plans by their contracting parties. Thus, it adopted Resolution XIV.16 on Integrating wetland protection, conservation, restoration, sustainable use, and management into national sustainable development strategies that encourage contracting parties to integrate wetland conservation, restoration, sustainable management, and wise use policies and actions into national sustainable development strategies; and to evaluate the role of wetland conservation and restoration in national and global sustainable development strategies in line with the 2030 Agenda for Sustainable Development, NBSAPs under the CBD as well as Nationally Determined Contributions and adaptation plans under the United Nations Framework Convention on Climate Change (UNFCCC) and the Paris Agreement, and the UN Convention to Combat Desertification (UNCCD) and its land degradation targets.[20]

Synergy and interoperability among the different agreements require enhanced cooperation that goes beyond the institutional dimension and reaches the level of implementation and enforcement through national measures. This implies the establishment of shared objectives, principles, and indicators to reach common results. Such measures have been identified in the preparatory reports of the different COPs, and those of Ramsar and the CBD will be the examples to show the process of their development and adoption.

---

[19] Ramsar Convention, *Inputs to the Post 2020 Global Biodiversity Framework* (2021), 1, 7–12, at: https://www.cbd.int/api/v2013/documents/09553304-7649-52 B1-C185-07FC251E88FC/attachments/212097/ramsar.pdf.

[20] Ramsar Convention, *Resolution XIV.16: Integrating wetland protection, conservation, restoration, sustainable use and management into national sustainable development strategies* (2022), 2 [11]. See also Ramsar Convention, *Kunming-Montreal Global Biodiversity Framework: Upscaling wetland conservation, restoration and wise use through National Biodiversity Strategies and Action Plans (NBSAPs)* (Secretariat of the Convention on Wetlands, 2023), at: https://www.ramsar.org/sites/default/files/2023-11/GBF_NBSAP_e.pdf.

2.1 Inclusive Multilateralism and Enhanced Cooperation with Biodiversity Non-State Actors

The concept of inclusive multilateralism – which incorporates international civil society and all its components in international negotiations for the design of global and national public policies in the different COPs and international conferences – is part of the new vision presented in the report of UN Secretary-General Antonio Guterres, *Our Common Agenda*,[21] which was prepared at the request of the General Assembly on the occasion of the 75th anniversary of the UN. *Our Common Agenda* revisits effective multilateralism – understood as the concerted international action of states within the UN – to improve practices so that multilateralism is interconnected and inclusive, in terms of both its methods and main actors.[22]

The Ramsar Convention is an outstanding example of inclusive multilateralism. Fifty years ahead of this new concept, the Ramsar Convention was the result of negotiations between states promoted by the International Union for Conservation of Nature (IUCN), born out of concern about the loss and degradation of wetland habitats for migratory waterbirds.[23] Thus, the Ramsar Convention is one of the first examples of the contribution of international civil society to the preparation and conclusion of an international treaty for environmental protection. IUCN has been involved since Ramsar's inception and has since been joined by other NGOs that have become part of the international negotiation processes in which global policies and regulatory instruments are designed in those sectors – such as biodiversity – that have a transnational dimension or protect common resources of international importance, such as wetlands.[24] Ramsar's memorandum of cooperation with IUCN and other specialised NGOs – such as WWF International, Birdlife International, and Wetlands International – can serve as a model to follow insofar as it provides

---

[21] United Nations, *Our Common Agenda – Report of the Secretary-General* (2021) 62, at: https://www.un.org/en/content/common-agenda-report/assets/pdf/Common_Agenda_Report_English.pdf.

[22] See T Fajardo, 'La Sociedad Civil Internacional y el Multilateralismo Inclusivo en la CoP26 de Glasgow' (2022) XXVI *El Derecho en la encrucijada: los retos y oportunidades que plantea el cambio climático, Anuario de la Facultad de Derecho de la Universidad Autónoma de Madrid* 157, 161.

[23] DA Stroud et al, 'Development of the Text of the Ramsar Convention 1965–1971' (2022) 73 *Marine and Freshwater Research* 1107, 1109.

[24] Memorandum of Cooperation between The Bureau of the Convention on Wetlands (Ramsar, Iran, 1971) and IUCN – The World Conservation Union, 13 September 2003, at: https://www.ramsar.org/news/iucn-and-ramsar-bureau-memorandum-cooperation-september-2003-0.

a framework for exportable cooperation given its effectiveness in promoting information exchange and awareness of the conservation status of wetlands of international importance.[25] In the run-up to CBD COP15, the proposals of IUCN were of particular interest; based on its experience on the ground, they emphasised that 'each national target must "add up" to the relevant global target to make national level contributions transparent and measurable', and also that 'regular "biodiversity stocktakes" will be needed to increase ambition, resources and implementation'.[26]

The growing importance of NGOs was recognised in the first segment of CBD COP15 that took place in October 2021 in Kunming.[27] In the preparatory phase – in which multiple workshops and forums were held[28] in line with the principles of the new inclusive multilateralism – a call was made to international and Chinese well-known experts and scholars, international organisations, national ministries and local governments, research institutions, private sectors, civil society, and media to explore the integration and top design of ecological civilisation and biodiversity conservation for mainstreaming, share best global and domestic practices, and develop a multi-stakeholder collaborative mechanism for mainstreaming biodiversity conservation to cultivate ecological civilisation and biodiversity conservation, building a community with shared future for human and nature.[29]

Of particular importance is the proposed 'Sharm El-Sheikh to Kunming and Montreal Action Agenda for Nature and People'. Intended as both a laboratory and an instrument of social mobilisation on the road to COP15 – but also thinking beyond – the CBD Secretariat's new platform has a mission to foster a 'whole-of-society approach to showcase commitments and actions from non-state actors to put biodiversity on a path to recovery by 2030'.[30] This

---

[25] 'Partners for Wetlands' Memorandum of Cooperation (23 April 2018), at: https://www.ramsar.org/sites/default/files/documents/library/moc_iops_23042018 _e.pdf.

[26] Statement from IUCN Director General Bruno Oberle at OEWG 3.1 (23 August 2021) (recorded in advance and submitted to Secretariat of the CBD).

[27] *Report of the Conference of the Parties to the Convention on Biological Diversity on its Fifteenth Meeting (Part I)* (15 October 2021), CBD/COP/15/4.

[28] See webinars to support discussions at the resumed sessions of SBSTTA-24, SBI-3, and WG2020-3, at: https://www.cbd.int/article/pre-geneva-2022-webinars.

[29] See *Sub-forum 2 Mainstreaming Ecological Civilization and Biodiversity Conservation* at: https://s3.amazonaws.com/cbddocumentspublic-imagebucket-15 w2zyxk3prl8/417ba1f7f8ebfc163e6b845fd3326bd9.

[30] Elizabeth Maruma Mrema, CBD Executive Secretary, said: 'The Sharm El-Sheikh to Kunming Action Agenda for Nature and People is critical to building a whole-of-society approach towards the implementation of the post-2020

proposed action agenda requires the initiatives and actions of non-state and sub-national actors to meet certain criteria, some of which coincide with those that states parties are already expected to meet – including, for instance, 'ways and means to support the mainstreaming of biodiversity considerations across productive sectors addressed at COP13 and COP14'.[31] This platform[32] was finally institutionalised at COP15 and will serve 'to study the options for its animation and the follow-up of commitments, and, *in fine*, to make non-state action a pillar of the action for biodiversity'.[33]

2.2 Shared Principles for the Kunming-Montreal Global Biodiversity Framework

One of the most notable achievements of the cooperation between the Ramsar Convention and the CBD has been the cross-fertilisation of their principles within the framework of biodiversity management and wetland protection, building bridges between these overlapping regimes in terms of the interest protected,[34] and especially in the processes of monitoring and promoting compliance through principles accepted as shared – such as the principle of sustainable development, the ecosystem approach (CBD), and wise use (Ramsar Convention). Both approaches are applied in the management of sites and protected areas covered by both conventions. Thus, De Lucia considers that 'the doctrine of wise use, central to the Ramsar Convention, has been redefined to accommodate the new language of conservation as developed particularly in the CBD, and was consequently deemed to be fully "congruent"

---

global biodiversity framework.' See CBD, 'Press Release: CBD Secretariat's new Action Agenda platform fosters "whole-of-society approach" to showcase commitments and actions from non-state actors to put biodiversity on a path to recovery by 2030', at: https://www.cbd.int/doc/press/2021/pr-2021-08-18-actionagenda-en.pdf.

[31] See CBD, 'Criteria', 6 October 2022, at: https://www.cbd.int/action-agenda/criteria.shtml.

[32] Source: https://www.cbd.int/portals/action-agenda/.

[33] J Landry, 'Dernière escale avant la COP 15 à Kunming: quelles priorités de négociations pour la biodiversité à Genève ?', 10 March 2022, *Billet de Blog*, IDDRI (translated from French), at: https://www.iddri.org/fr/publications-et-evenements/billet-de-blog/derniere-escale-avant-la-cop-15-kunming-quelles-priorites.

[34] T Fajardo, 'Environmental law principles and General principles of international law' in L Kramer and E Orlando (eds), *Principles of Environmental Law*, *Elgar Encyclopaedia of Environmental Law* (Vol VIII) (Edward Elgar Publishing, Cheltenham, 2018) 38, 46.

with the ecosystem approach'.[35] Furthermore, the Ramsar Convention has 'embraced'[36] an ecosystem approach to wetland conservation – as is shown in the *Ramsar Handbook for the Wise Use of Wetlands*, which incorporates an updated definition of 'wise use', taking into account the Convention's mission statement, the terminology of the Millennium Ecosystem Assessment, the concepts of the ecosystem approach and sustainable use applied by the CBD, and the definition of 'sustainable development' adopted by the 1987 Brundtland Commission, which states: 'Wise use of wetlands is the maintenance of their ecological character, achieved through the implementation of ecosystem approaches, within the context of sustainable development.'[37]

Both in the bilateral relationship between the Ramsar Convention and the CBD and in the much more complex and diffuse multilateral framework of the relationship between all the biodiversity-related Conventions, the principles and approaches are instrumental to the achievement of the objectives and purposes of the different Conventions and, 'as such, are the legal expression of the mandates and powers of delegation that are vested in the [governing bodies]'.[38] There is certainly a convergence of the principles and standards of the different biodiversity-related Conventions. Thus, there has been, as Caddell[39]

---

[35] V De Lucia, 'A critical interrogation of the relation between the ecosystem approach and ecosystem services' (2018) 27 *Review of European, Comparative and International Environmental Law* 104, 109.

[36] See L Jing, *Preservation of Ecosystems of International Watercourses and the Integration of Relevant Rules* (Brill/Nijhoff, Leiden, 2014) 9.

[37] A footnote in this definition clarifies that the phrase 'in the context of sustainable development' aims to acknowledge that 'whilst some wetland development is inevitable and that many developments have important benefits to society, developments can be facilitated in sustainable ways by approaches elaborated under the Convention, and it is not appropriate to imply that "development" is an objective for every wetland'. Ramsar Convention Secretariat, *Handbook 1: Wise Use of Wetlands: Concepts and Approaches for the Wise Use of Wetlands* (Ramsar Handbooks, 4th edition, Ramsar Convention Secretariat, Gland, 2010) 16.

[38] T Fajardo, 'Principles and Approaches in the Convention on Biological Diversity and Other Biodiversity-Related Conventions in the Post-2020 Scenario' in M Campins Eritja and T Fajardo del Castillo (eds), *Biological Diversity and International Law: Challenges for the Post-2020 Scenario* (Springer Verlag, Berlin, 2021) 15, 28.

[39] As Caddell pointed out, '[t]here have been varying degrees of endorsement for a "CBD-ification" of strategic management across the various Biodiversity Related Conventions … The Ramsar Convention has ultimately been prepared to amend fundamental working principles to accommodate CBD concerns. Interactions with the CBD have prompted a reformulation of key Ramsar commitments towards the "wise use" of wetlands, away from the "benefit of

and Velázquez Gomar call it, a 'CBD-ification' process,[40] 'whereby older conservation-focused agreements increasingly have embraced the CDB's main mission and principles'.[41]

Similarly, as Gardner suggests, wise use is '[a]n overarching principle'[42] and a unifying factor for cooperation and environmental governance. An equivalent role is played by the ecosystem approach, which – as pointed out by Suietnov – has been 'most consistently developed today under the CBD, but numerous references to it can be found in international environmental agreements adopted much earlier'; these 'created the necessary base for the further formation and development of the ecosystem approach as a holistic concept in international environmental law under the CBD'.[43] This approach has grown since it was developed under the CBD umbrella as:

> soft law operative guidance linked to the conservation and protection of ecosystems, natural habitats, and species. It is now one of the principles embodied in many of the biodiversity-related conventions, and works in combination with the other principles and approaches linked to conservation objectives, such as the precautionary approach, the principle of sustainable use and the principle of integration.[44]

These principles are also at the core of the design of the future cooperation structure resulting from COP15. Of particular relevance is the principle of

---

mankind", to explicitly incorporate the ecosystem approach, which the CBD Secretariat has quietly considered to be one of its greatest collaborative achievements.' R Caddell, 'The Integration of Multilateral Environmental Agreements: Lessons from the Biodiversity-Related Conventions' (2011) 22(1) *Yearbook of International Environmental Law* 37, 69.

[40] See JO Velázquez Gomar, 'Environmental policy integration among multilateral environmental agreements: The case of biodiversity' (2016) 16(4) *International Environmental Agreements: Politics, Law and Economics* 525, 529.

[41] See M Petersson and P Stoett, 'Lessons learnt in global biodiversity governance' (2022) 22 *International Environmental Agreements: Politics, Law and Economics* 333, 339.

[42] See R Gardner, 'Opportunities and challenges for synergies across biodiversity-related conventions in the context of human health and zoonotic diseases: the role of scientific advisory bodies' in M Campins and T Fajardo, *Biological Diversity and International Law: Challenges for the Post-2020 Scenario* (Springer, Berlin, 2021) 57–75.

[43] See Y Suietnov, 'Formation and Development of the Ecosystem Approach in International Environmental Law before the Convention on Biological Diversity' (2021) 1 *Journal of Environmental Law and Policy* 47, 80. See also Chapter 9 in this volume.

[44] Fajardo, note 38, 29–30.

biodiversity integration, which has since inspired an important commitment in the UN system through the adoption of the Common Approach to Integrating Biodiversity and Nature-based Solutions for Sustainable Development into United Nations Policy and Program Planning and Delivery.[45] This Common Approach contributes:

> to the realization of the 2050 Vision for Biodiversity, under which biodiversity is valued, conserved, restored and wisely used, maintaining ecosystem services, sustaining a healthy planet and delivering benefits essential for all people ... It also proposes a set of outcomes that can be achieved by the United Nations system through increased collaboration, as well as an accountability framework for coherent and collective outputs on biodiversity.[46]

This Common Approach was also considered to embrace interinstitutional cooperation since:

> The United Nations system supports parties to multilateral environmental agreements, such as the biodiversity-related conventions and agreements, as well as other relevant multilateral frameworks, and the United Nations goals and targets, which provide a critical component of international cooperation and governance.[47]

In a key document prepared by the Chief Executives Board for Coordination of the United Nations,[48] the Ramsar Convention was considered associated for 'potential collective action' in several of these ways, together with other organs, programmes and international organisations of the United Nations family,[49] to:

---

[45] Chief Executives Board for Coordination, *Common Approach to Integrating Biodiversity and Nature-based Solutions for Sustainable Development into United Nations Policy and Programme Planning and Delivery*, CEB/2021/1/Add.1 (16 August 2021).

[46] Ibid, 2.

[47] See ibid, Annex III, 21.

[48] UN System Chief Executives Board for Coordination (UNSCEB), *50+ ways to integrate biodiversity and nature-based solutions – a UN system commitment to collective action for people and planet*, 5 May 2021, at: https://unsceb.org/sites/default/files/2022-01/Biodiversity_Common_Approach_50%2B_ways_to_integrate_biodiversity_and_nature-based_solutions.pdf.

[49] Together with UNEP, the United Nations Development Programme, the Office of the High Commissioner of Human Rights, the Food and Agriculture Organization, IFAD, CBD, CITES, CMS, UNCCD, the United Nations Economic Commission for Europe, United Nations Educational, Scientific and Cultural Organization designated sites, the International Labour Organization, the United Nations Children's Fund, UNFCCC, the United Nations Economic and Social

2. Fulfil the right to a safe, clean, healthy and sustainable environment ...
13. Prevent risks from climate change, including extreme weather events ...
14. Halt habitat degradation and promote ecosystem restoration ...
16. Prevent soil degradation and combat coastal erosion ...
20. Promote education and the generation, sharing and use of knowledge ...
32. Transform unsustainable agricultural and fisheries practices ...
47. Prevent introduction and spread of invasive alien species ...[50]

In view of the adoption of the Kunming-Montreal Global Biodiversity Framework, the recognition of the principle of biodiversity mainstreaming is especially urgent because '[o]nly a minority of national biodiversity strategies and action plans demonstrate that biodiversity is being mainstreamed significantly into cross-sectoral plans and policies, poverty eradication policies, and/or sustainable development plans'.[51] Coincidentally, on the day that the final CBD COP15 pre-conference began in Geneva, the European Parliament endorsed a new EU action programme that warns that the principle of environmental integration into other policies will not fulfil its objectives if key factors such as anticipating social costs are not taken into account, and if no 'far-reaching reform of its agricultural, trade, transport, energy and infrastructure investment policies' is foreseen.[52]

Other principles that appeared in the final outcomes on implementation and compliance mechanisms are the principles of non-regression[53] and progres-

---

Commission for Western Asia, the United Nations Economic Commission for Africa, and the United Nations World Food Programme; see the Common Approach to Biodiversity website at: https://unsceb.org/un-common-approach-biodiversity.

[50] See UNSCEB, note 48.

[51] See CBD Subsidiary Body on Implementation, *Recommendation 2/1: Progress in the implementation of the Convention and the Strategic Plan for Biodiversity 2011–2020 and Towards the Achievement of the Aichi Biodiversity Targets* (2018) 1, at: https://www.cbd.int/doc/recommendations/sbi-02/sbi-02-rec-01-en.pdf.

[52] European Parliament legislative resolution of 10 March 2022 on the proposal for a decision of the European Parliament and of the Council on a General Union Environment Action Programme to 2030 (COM(2020)0652 – C9-0329/2020 – 2020/0300(COD)).

[53] As highlighted by Viñuales: 'States can thus choose their level of ambition subject to two requirements, namely the regular updating – at least every five years (Article 4(9)) – and an obligation of non-regression (Article 4(3)). The latter is new and signals what perhaps will become a major new principle of international environmental law in the years to come.' J Viñuales, 'The Paris Climate Agreement: An Initial Examination (Part II of III)', 8 February 2016, *EJIL Talk!* at: https://

sion[54] that were first introduced in the Paris Agreement on climate change. These principles will serve to assess compliance with the different elements of the new Kunming-Montreal Global Biodiversity Framework. This will be accomplished through the monitoring mechanism adopted in Decision 15/5 on the Monitoring Framework for the Kunming-Montreal Global Biodiversity Framework, which will use:

> the period from 2011–2020, where data is available, as the reference period, unless otherwise indicated, for reporting and monitoring progress in the implementation of the Kunming-Montreal Global Biodiversity Framework, while noting that baselines, conditions and periods used to express desirable states or levels of ambition in goals and targets should, where relevant, take into account historical trends, current status, future scenarios of biodiversity and available information on the natural state.[55]

The different projects and electronic platforms of the new Kunming-Montreal Global Biodiversity Framework will be developed on the basis of, or complement where appropriate, pre-existing data repositories, such as the digital data reporting tool developed by the UN Environment Programme (UNEP) InforMEA portal;[56] and will provide the necessary applications to enable monitoring of and compliance with the principles of non-regression and progression.

3   BUILDING BETTER GOVERNANCE TO ENABLE INSTITUTIONAL COOPERATION

There has also been a 'CBD-ification' process[57] of international biodiversity governance. Just as there is a constellation of biodiversity-related Conventions that revolve around the CBD, there is also a multitude of bodies at different levels of competence that orbit the CBD institutions. Thus, CBD COPs and pre-conferences have become the venue for the executive leaders of the

---

www.ejiltalk.org/the-paris-climate-agreement-an-initial-examination-part-ii-of-iii/.

[54] See S Maljean-Dubois et al, 'Pour un meilleur suivi du cadre mondial sur la biodiversité pour l'après-2020: options juridiques et possibles arrangements institutionnels' (2022) 3 *Études IDDRI* 10–11.

[55] CBD, note 5.

[56] UNEP InforMEA is the keyword search service covering both environmental treaty texts and the more than 10 000 instruments adopted by COPs and governing bodies, as well as internationally agreed goals and targets, including the SDGs and more than 100 000 pieces of national legislation and jurisprudence (with the help of ECOLEX). Source: https://www.informea.org/en.

[57] See Caddell, note 39, 33; Velázquez Gomar, note 40, 529.

biodiversity-related Conventions to express their positions on new formulas for cooperation and future commitments for coordination of work programmes – as occurred, for instance, during the negotiation of the new Kunming-Montreal Global Biodiversity Framework. This cooperation works both ways, with Ramsar representatives attending CBD COPs and vice versa.[58] In 2015, Ramsar COP12 adopted a Strategic Plan to align with the Aichi Targets[59] and in 2022 COP14 updated it to align it with the new Kunming-Montreal Targets.[60] Thus, the COPs to both Ramsar and the CBD have endorsed or referred to each other's resolutions and decisions (eg, peatland restoration).[61]

Furthermore, enhancing the institutional structures of cooperation between the biodiversity-related Conventions[62] was one of the most important challenges of CBD COP15, after two decades without advancements in initial

---

[58] See R Gardner and A Grobicki, 'Synergies between the Convention on Wetlands of International Importance, especially as Waterfowl Habitat and other multilateral environmental agreements: possibilities and pitfalls' in UN Environment (2016) *Understanding synergies and mainstreaming among the biodiversity related conventions: A special contributory volume by key biodiversity convention secretariats and scientific bodies*, 54, 56, at: https://wedocs.unep.org/bitstream/handle/20.500.11822/17017/understanding-synergies-mainstreaming-biodiv.pdf?sequence=1&isAllowed=y.

[59] See Ramsar Convention Secretariat, note 18.

[60] See note 4. The COP14 also extended the term of the 4th Strategic Plan from 2024 to the occurrence of COP15 to ensure continuity between successive Strategic Plans. Ibid.

[61] In 2018, at Ramsar COP13, the contracting parties adopted Resolution XIII.13, on restoration of degraded peatlands to mitigate and adapt to climate change and enhance biodiversity and disaster risk reduction; and Resolution XIII.14, promoting conservation, restoration and sustainable management of coastal blue-carbon ecosystems. As has been pointed out, '[b]oth resolutions were proposed in synergy with commitments not only under the Ramsar Strategic Plan but also under different multilateral environmental agreements, such as CBD, UNFCCC and the Paris Agreement, and the United Nations Convention to Combat Desertification'. Note by the CBD Executive Secretary, Thematic Workshop on Ecosystem Restoration for the Post-2020 Global Biodiversity Framework, *Considerations on Ecosystem Restoration for the Post-2020 Global Biodiversity Framework, including on a possible successor to Aichi Biodiversity Target 15*, CBD/POST2020/WS/2019/11/3 (30 October 2019), 13.

[62] See R Caddell, 'Inter-Treaty Cooperation, Biodiversity Conservation and the Trade in Endangered Species' (2013) 22 *Review of European, Comparative & International Environmental Law* 264.

proposals such as those of CBD Decision X/20,[63] and Decision XIII/24[64] and Decision XIV/30[65] on cooperation with other conventions, international organisations, and initiatives. A combination of reports prepared by UNEP[66] and the various bodies of the CBD and other biodiversity-related Conventions saw a document with possible options presented in the final phase of the adoption of the Kunming-Montreal Global Biodiversity Framework.[67] Decision 15/13 on cooperation with other Conventions and international organisations will be the backbone of what would otherwise be countless reports of a fragmented reflection process that has as its starting point the SDGs.[68] The different tools and institutions that can be used to enhance cooperation among the different biodiversity-related Conventions are discussed below.

3.1 The Adoption of a Model Memorandum of Cooperation

Over the years, multiple memoranda of cooperation have been adopted between the different biodiversity-related Conventions, often driven by the CBD.[69] In

---

[63] CBD, *Decision X/20: Cooperation with other conventions and international organizations and initiatives* (2010). Of particular interest is that it '*Urges* Parties to establish close collaboration at the national level between the focal points for the Convention on Biological Diversity and focal points for other relevant conventions, with a view to developing coherent and synergetic approaches across the conventions at national and (sub-)regional levels'. Ibid, [5].

[64] CBD, *Decision XIII/24: Cooperation with other conventions and international organizations* (2016), Annexes I and II.

[65] CBD, *Decision XIV/30: Cooperation with other conventions, international organizations and initiatives* (2018), which envisaged holding workshops to explore ways in which the Conventions can contribute to the elaboration of the post-2020 global biodiversity framework and to identify specific elements that could be included in the framework.

[66] UNEP, 'Improving the effectiveness of and cooperation among biodiversity-related conventions and exploring opportunities for further synergies.' See UNEP/CBD/SBI/1/INF/36 and UNEP/CBD/SBI/1/INF/37.

[67] CBD, *Decision 15/4: Kunming-Montreal Global Biodiversity Framework* (2022).

[68] CBD, *Decision 15/13: Cooperation with other conventions and international organizations* (2022).

[69] See Memorandum of Cooperation between The Convention on Biological Diversity and The Alpine Convention and The Carpathian Convention (2008); Memorandum of Co-operation between The Secretariat of the Convention for the Protection and Development of the Marine Environment of the Wider Caribbean Region (Cartagena, 1983) and its Protocols, including the Protocol Concerning Specially Protected Areas and Wildlife (SPAW) and the Secretariat of the Convention on Biological Diversity (CBD) (Nairobi, 1992) (1997); Memorandum

the case of the CBD-Ramsar Memorandum of Cooperation (MOC) of 2005, it initiated early institutional cooperation that intertwined their agendas and work programmes every five years, promoting the synergistic implementation of the CBD's agenda on protected areas with Ramsar's List of Wetlands of International Importance.[70] As the CBD-Ramsar MOC has expired, a new MOC and the sixth Joint Work Plan are being negotiated, with their composition to be informed by the final outcome of the Kunming-Montreal Global Biodiversity Framework.[71] On the basis of this MOC, a model could be prepared identifying cross-cutting objectives and priorities to be incorporated both in joint work plans and in national strategies and action plans. It will also be necessary to follow future developments leading to the reform or renewal of the MOC signed in 2011 between the CBD Secretariat and 27 international agencies, environmental organisations, and Conventions to achieve and implement the Strategic Plan for Biodiversity 2011–20 and its Aichi Biodiversity Targets, as the new Targets and the Kunming-Montreal Global Framework for Biodiversity may require an update of shared tasks and objectives.

3.2 Synchronising Agendas and Work Programmes

The most successful synergy practice remains the synchronisation of programmes and work plans. This has occurred between the Ramsar Convention and the CBD since the Fifth CBD-Ramsar Joint Work Plan, which covered the time period of the Strategic Plan for Biodiversity (2011–20) and operated in the context of the Ramsar Convention's implementation role for wetlands for the CBD. As an example of this intertwined cooperation, in the CBD Programme of Work on Inland Waters, Ramsar was designated as lead partner for wetlands

---

of Co-operation between the Secretariat of the Convention on International Trade in Endangered Species of Wild Fauna and Flora (Washington, DC, 1973) and the Secretariat of the Convention on Biological Diversity (Nairobi, 1992) (2001); Enhanced Memorandum of Co-operation between the Secretariat of the UN Convention on Biological Diversity (Rio De Janeiro, 1992) and the Secretariat of the Convention on the Conservation of European Wildlife and Natural Habitats (Bern, 1979) (2008), at: https://www.cbd.int/agreements/.

[70] Memorandum of Co-operation between The Convention on Wetlands (Ramsar, Iran, 1971) and The Convention on Biological Diversity (2005).

[71] Ramsar Convention, SC59 Doc 16 Rev 1, *Resolution XIV.xx: Enhancing the Convention's visibility and synergies with other multilateral environmental agreements and other international institutions* (21–25 June 2021) 7; Ramsar Convention, SC59/2022 Doc 16, *Update on Enhancing the Convention's visibility and synergies with other multilateral environmental agreements and other international institutions* (23–27 June 2022).

for the CBD.[72] The outcomes of these synchronisations were assessed by the CBD's *Global Biodiversity Outlook 5* (2020) and the preparatory work for COP15 undertaken by the CBD Subsidiary Body on Implementation.[73] They examined in depth the intersection between the Aichi Targets and the Ramsar programme's own targets for the period 2016–20. However the results were clearly insufficient. In the case of Aichi Biodiversity Target 11 on protected areas, it has been shown that the proportion of the planet's land and oceans designated as protected areas is likely to reach the targets for 2020. 'However, progress has been more modest in ensuring that protected areas safeguard the most important areas for biodiversity, are ecologically representative, connected to one another ... and are equitably and effectively managed'.[74]

For its part, Ramsar Target 12 on wetland restoration – which corresponds to Aichi Targets 14 and 15 – has been assessed by highlighting that, despite no obligation on the percentage of zones to be restored, '[r]estoration is in progress in degraded wetlands, with priority to wetlands that are relevant for biodiversity conservation, disaster risk reduction, livelihoods and/or climate change mitigation and adaptation'.[75] Complementarily, in the case of Aichi Target 15 on Ecosystem Restoration, 'progress towards the target of restoring 15% of degraded ecosystems by 2020 is limited'.[76] As was concluded in the background reports for the CBD, regarding these related targets:

> The indicators are essentially the percentage of Parties that have established restoration plans for Ramsar sites and the percentage of Parties that have implemented effective restoration or rehabilitation projects, using as baseline National Reports to

---

[72] See CBD, *Decision III/21: Relationship of the Convention with the Commission on Sustainable Development and biodiversity-related conventions, other international agreements, institutions and processes of relevance* (1996) [7], which decided '(ii) to invite the Convention on Wetlands of International Importance to cooperate as a lead partner in the implementation of activities under the Convention related to wetlands, and, in particular, requests the Executive Secretary to seek inputs from the Convention on Wetlands of International Importance, in the preparation of documentation concerning the status and trends of inland water ecosystems for the consideration of the Conference of the Parties at its fourth meeting'.

[73] CBD Subsidiary Body on Implementation, *Recommendation SBI-3/12: Cooperation with other conventions, international organizations and initiatives*, CBD/SBI/REC/3/12 (28 March 2022).

[74] *GBO-5 Inland Waters Highlights* (2021) 6, at: https://www.cbd.int/waters/doc/gbo5-inlandwaters-en.pdf.

[75] Ramsar Convention Secretariat, note 18, 5.

[76] Secretariat of the Convention on Biological Diversity, *Global Biodiversity Outlook 5: Summary for Policymakers* (2020) 10.

COP 12. Differently from [Aichi Biodiversity Target] 15, it does not have a specific percentage value in area of degraded ecosystems to be restored by the deadline.[77]

It is hoped that these gaps will be overcome in the future, thanks to the harmonisation of methodologies and indicators. As the report of the CBD Subsidiary Body on Implementation has shown, '[f]or most of the Aichi Biodiversity Targets, there has been limited progress, and, for some Targets, no overall progress' at all.[78] Thus, from the perspective of intersection with the Aichi Targets, greater efforts are needed to design and implement common projects, methodologies, and indicators. This has been insufficient to date, as wetland areas have decreased or have been degraded, which not only contributes to the loss of biodiversity but also has an impact on climate change, as pointed out in the latest Intergovernmental Panel on Climate Change (IPCC) reports.[79]

Nevertheless, the progressive synchronisation of work programmes and the division of labour according to protected interests can only be welcomed. Thus, it is suggested that the best way to join forces in achieving the objectives of Ramsar and the CBD objectives is for their design to be intertwined and their joint implementation to become part of, or complement, NBSAPs. To do so, states parties would need to make an extra effort to identify national priorities, which would inclusively reflect the interrelated interests and objectives of the biodiversity-related Conventions. Based on the selection of these

---

[77] CBD, Note by the Executive Secretary, *Considerations on Ecosystem Restoration for the Post-2020 Global Biodiversity Framework, Including on a Possible Successor to Aichi Biodiversity Target 15*, CBD/POST2020/WS/2019/11/3 (30 October 2019) 13, at: https://www.cbd.int/doc/c/fcd6/bfba/38ebc826221543e322173507/post2020-ws-2019-11-03-en.pdf.

[78] See CBD Subsidiary Body on Implementation, *Recommendation SBI-2/1: Progress in the implementation of the Convention and the Strategic Plan for Biodiversity 2011–2020 and Towards the Achievement of the Aichi Biodiversity Targets* (2018).

[79] The IPCC Report 2021 pointed out that '[d]espite the large uncertainties surrounding the quantification of the effect of additional Earth system feedback processes, such as emissions from wetlands and permafrost thaw, these feedbacks represent identified additional risk factors that scale with additional warming and mostly increase the challenge of limiting warming to specific temperature levels'. PA Arias et al, 'Technical Summary' in V Masson-Delmotte et al (eds), *Climate Change 2021: The Physical Science Basis. Contribution of Working Group I to the Sixth Assessment Report of the Intergovernmental Panel on Climate Change* (Cambridge University Press, Cambridge/New York, 2021) 99, at: https://www.ipcc.ch/report/ar6/wg1/downloads/report/IPCC_AR6_WGI_TS.pdf.

common priorities, each state party would then have to propose and adopt its own national commitment.[80]

The update of the Ramsar Convention's Strategic Plan for 2020–24 that has been carried out to take into account the new targets of the Kunming-Montreal Global Biodiversity Framework has already considered that its national application 'can be part of or supplement to the National Biodiversity Strategy Action Plan'.[81]

3.3 Upgrading Existing Institutions

In the process of adopting the new Global Biodiversity Framework, given the reluctance of the parties to progress the institutionalisation of regulatory frameworks, it will be difficult to achieve institutional reform in which new or existing bodies are given an ambitious mandate with the necessary competences and resources. It is more likely that greater visibility will be given to those already existing in the area of cooperation between the various biodiversity-related Conventions – in particular, UNEP and the Environment Management Group, and the Biodiversity Liaison Group.

3.3.1 UNEP and the Environment Management Group

Unlike most MEAs, the Ramsar Convention does not have UNEP as its Secretariat but has its own Secretariat. Previously called the Ramsar Convention Bureau, it is responsible for the day-to-day running of the Convention's activities, its administration, and the publication of periodic reports and other documents of the regime.[82] It is headed by a Secretary General and specialised staff.[83] So far, the Ramsar Convention has maintained its independence from

---

[80] See J Pittock, 'A Pale Reflection of Political Reality: Integration of Global Climate, Wetland, and Biodiversity Agreements' (2010) 1 *Climate Law* 343, 364; E Lyman, 'Rethinking International Environmental Linkages: A Functional Cohesion Agenda for Species Conservation in a Time of Climate Change' (2015) 27 *Fordham Environmental Law Review* 1, 46–47.

[81] Ramsar Convention Secretariat, note 18, 17. It states in its paragraph 36: 'Contracting Parties should implement the Strategic Plan at national and regional levels by developing national wetlands policies, strategies, action plans, projects and programmes or other appropriate ways to mobilize action and support for wetlands. This can be part of or supplement to the National Biodiversity Strategy Action Plan.' Ibid.

[82] Ramsar Convention Secretariat, *An Introduction to the Ramsar Convention on Wetlands* (Ramsar Handbooks, 5th edition, Ramsar Convention Secretariat, Gland, 2016) 20–21.

[83] In Gland, Switzerland. Ibid, 30–31.

UNEP, notwithstanding their close collaboration, as the Ramsar Convention is co-custodian with UNEP of SDG Indicator 6.6.1 on the extent of water-related ecosystems, under SDG 6 on water.[84] This independence implies that UNEP participates as a close but autonomous observer in the complex COP consultation processes of the MEAs (in particular, those related to international trade, biological diversity, wetlands, migratory species, and world heritage) and the Rio Conventions.

In addition to providing the secretariat for MEAs, UNEP has promoted greater cooperation among the various biodiversity conventions through the Environment Management Group, which has made a notable contribution by providing an analysis of UN agency activities that led to 'the identification of six main functions that UN agencies can play to directly deliver and assist Member States in implementing the post-2020 global biodiversity framework'.[85] Most notably, the Ramsar Convention acts as a co-custodian with UNEP in 'monitoring, assessment and knowledge-sharing' – in particular, the monitoring of SDG targets and indicators.[86] This role includes:

> collection and compilation [of data] as, for example, in UNEP's 'World Environment Situation Room' and the development of agreed statistical frame-works for measuring the interrelationships between ecosystems, the economy and human well-being, such as the System of Environmental-Economic Accounting (SEEA).[87]

3.3.2 The Liaison Group of the Biodiversity-related Conventions

Of particular interest is the Liaison Group of the Biodiversity-related Conventions (BLG), which brings together the heads of the Secretariats of the CBD, CITES, Ramsar, CMS, the World Heritage Convention, the IPPC, ICRW, and ITPGRFA, plus any other biodiversity-related Conventions[88] that may be adopted in the future and invited by the BLG to join as a member.[89] The BLG was established in 2004 by the CBD COP in its Decision VII/26 on Cooperation with other conventions and international organisations and initiatives that requested the CBD Executive Secretary to adopt 'a flexible frame-

---

[84] See Ramsar Convention, *Resolution XIII.7: Enhancing the Convention's visibility and synergies with other multilateral environmental agreements and other international institutions* (2018).

[85] United Nations Environment Management Group, *Supporting the Global Biodiversity Agenda: A United Nations System Commitment for Action to assist Member States delivering on the post-2020 global biodiversity framework* (2020).

[86] Ibid, 10.

[87] Ibid, 23.

[88] See note 2.

[89] See CBD, *Modus Operandi for the Liaison Group of the Biodiversity-related Conventions*, at: https://www.cbd.int/cooperation/doc/blg-modus-operandi-en.pdf.

work between all relevant actors, such as a global partnership on biodiversity, in order to enhance implementation through improved cooperation'.[90] At the following CBD COP in 2006, the proposal was further developed and the BLG was consolidated with the adoption of the Memorandum of Cooperation between Agencies to Support the Achievement of the 2010 Biodiversity Target.[91]

The BLG is chaired by rotation of the executive heads of the Conventions on an annual basis, unless otherwise agreed, and is entrusted with cooperative activities to be carried out in accordance with the following premises – although these are described on its web page as 'principles' – which define its mandate and the limited nature of its competencies, since those ultimately responsible for their fulfilment are the Secretariats of the Conventions and not their constituent parties.[92] Thus, it defines itself as 'a platform to exchange information and to enhance implementation at the national level of the objectives of each respective convention whilst also promoting synergies at the national level'.[93]

In the run-up to the 2021 and 2022 CBD COPs, the BLG considered how to enhance synergies and strengthen cooperation among the biodiversity-related Conventions, without prejudice to their specific objectives and recognising their respective mandates, and subject to the availability of resources for these Conventions, in order to improve their participation in the design of the post-2020 biodiversity framework and, in particular, to continue giving consideration to the harmonisation of national reporting.[94] However, as Karen Scott observed a decade ago:

> Although the focus on strengthening participation in existing projects is hardly surprising given the large number of joint and other collaborative programs that are currently underway involving all six conventions, the lack of progress in other areas is disappointing. Whilst the six conventions have established a joint website, there is little evidence of further administrative integration.[95]

---

[90] CBD, *Decision VII/26: Cooperation with other conventions and international organizations and initiatives* (2004).

[91] Memorandum of Cooperation Between Agencies to Support the Achievement of the 2010 Biodiversity Target (2006), at: https://www.cbd.int/doc/agreements/agmt-hoatf-unep-2006-3-27-moc-web-en.pdf.

[92] CBD, note 89, 2–3.

[93] Ibid, 2.

[94] Liaison Group of Biodiversity-related Conventions, Summary Report: Conference Call of the Liaison Group of Biodiversity-related Conventions (30 April 2020), at: https://www.cbd.int/cooperation/BLG-conference-call-30-April-2020.pdf.

[95] See KN Scott, 'International Environmental Governance: Managing Fragmentation through Institutional Connection' (2011) 12 *Melbourne Journal of*

Moreover, as pointed out by Maljean-Dubois et al, there is also evidence of an apparent level of tension within the BLG between the eight biodiversity Conventions, resulting from differences in priorities and resources.[96]

3.3.3 The Informal Advisory Group on Synergies

In 2017, the Informal Advisory Group on Synergies was appointed to 'monitor the implementation of the road map' of CBD COP15. In particular, its mandate was to prepare a report and 'provide the [CBD] Secretariat with advice on ways to optimise synergies among the biodiversity-related conventions in the development of the post-2020 biodiversity framework'.[97] Unfortunately, its work has not always been submitted in a timely manner, which has led to complaints from states parties and slowed down the process of reflection on the proposals.[98] It is expected that its mandate will be extended or integrated in the event that COP15 decides to improve the other bodies.

4 BUILDING BETTER MECHANISMS ON ASSESSMENT, MONITORING, AND REPORTING

Neither the CBD nor Ramsar envisages sanctions for a failure to adhere to its respective commitments; hence, the development of control mechanisms and the promotion of compliance are primary objectives that require greater commitment from the states parties in order to be effective. So far, as with most environmental treaties, the available soft mechanisms encourage compliance but do not punish 'non-compliance';[99] and given the outcomes, they lack the necessary teeth to improve compliance[100] – although it is a lack of resources that ultimately explains the lack of implementation. In some Conventions,

---

International Law 177, 202.

[96] Maljean-Dubois et al, note 54, 10–11.

[97] See CBD, *Decision XIII/24: Cooperation with other conventions and international organizations* [15].

[98] See EU statements and their contributions to the process in CBD, Report of the Open-Ended Working Group on the Post-2020 Global Biodiversity Framework on its Third Meeting (Part I), CBD/WG2020/3/5 (3 September 2021) 33, 44, 116, 120, at: https://www.cbd.int/doc/c/7b60/d4f7/c37d7e158e352bbc9af818a4/wg2020-03-05-en.pdf.

[99] See PGG Davies, 'Non-Compliance – A Pivotal or Secondary Function of CoP Governance?' (2013) 15 *International Common Law Review* 77, 92–93.

[100] C Redgwell, 'The Challenge of Effective Compliance and Enforcement with International Environmental Law' in Jacques Hartmann and Urfan Khaliq (eds), *The Achievements of International Law: Essays in Honour of Robin Churchill* (Hart, Oxford, 2021) 259, 275–76.

different modalities of sticks and carrots have been tested, but in no case have states allowed them to go further in terms of their policies for the conservation and sustainable use of their biodiversity. For over a decade, UNEP, the CBD, and the BLG have explored the possibilities for the harmonisation of indicators and reporting requirements between the Rio and biodiversity-related Conventions, but they have yet to reach any major agreement.[101]

The possibility of developing shared biodiversity indicators for the biodiversity-related Conventions was suggested at different stages of the negotiations leading to the adoption of the Kunming-Montreal Global Biodiversity Framework. Significantly, it was agreed that 'a globally harmonised approach to reporting, through the adoption of a standardised set of scalable global biodiversity indicators, would be beneficial'.[102] The selection of indicators became one of the most important challenges in this negotiation. The indicators respond to lessons learned from conventions and processes of particular relevance, such as those developed to assess the achievement of the SDGs and the Aichi Biodiversity Targets. The CBD website presents principles for the selection of indicators that could be useful to harmonise working methodologies and make them interoperable with other conventions.[103] Those finally incorporated in Decision 15/5 on the Monitoring Framework of the Kunming-Montreal Global Biodiversity Framework will be used on an interim basis until the final version of the monitoring mechanism is presented at the next COP16. Until then, parties will use the 'headline indicators'[104] in their national reports, supported by the 'component' and 'complementary' indicators.[105] The monitoring framework 'may be supplemented by additional

---

[101] Gardner and Grobicki, note 58, 63.

[102] CBD, Synthesis of Views of Parties and Observers on the Scope and Content of the Post-2020 Global Biodiversity Framework, CBD/POST2020/PREP/1/INF/1 (24 January 2019) 11.

[103] Source: CBD, Principles for Selecting Indicators, https://www.cbd.int/indicators/indicatorprinciples.shtml.

[104] 'Headline indicators (contained in table 1) [are] a minimum set of high-level indicators, which capture the overall scope of the goals and targets of the Kunming-Montreal Global Biodiversity Framework to be used for planning and tracking progress as set out in decision 15/6. They are nationally, regionally and globally relevant indicators validated by Parties. These indicators can also be used for communication purposes.' CBD, note 5, 3.

[105] 'Component indicators (contained in table 2) [are] a list of optional indicators that, together with the headline indicators, cover components of the goals and targets of the Kunming-Montreal Global Biodiversity Framework which may apply at the global, regional, national and subnational levels[.]' 'Complementary indicators' that are also contained in table 2, are 'a list of optional indicators for

national and subnational indicators', since in Decision 15/5, parties have acknowledged:

> the value of aligning national monitoring with the System of Environmental-Economic Accounting statistical standard in order to mainstream biodiversity in national statistical systems and to strengthen national monitoring systems and reporting as appropriate and according to their national priorities and circumstances.[106]

If role models are sought, the Ramsar Convention provides a guide for the further development of mechanisms to assist contracting parties – notably through its Montreux Record and mission system; however, in recent years, it has largely fallen into disuse[107] due to its voluntary nature and because of political and financial issues.[108] The CBD has a Subsidiary Body for Implementation to support its COP in monitoring implementation. Although this is an improvement upon previous organs – namely the Ad Hoc Open-ended Working Group on Review of Implementation of the Convention – its mandate and competences require updating in view of the challenges posed by the Kunming-Montreal Global Biodiversity Framework.

5   CONCLUSIONS

Looking ahead to the post-2020 scenario, COP15 in Kunming, at its first segment in October 2021, adopted a common commitment to integrate ecological civilisation and biodiversity conservation.[109] Most importantly from the point of view of the interaction between the Ramsar Convention and the CBD, one of the items on the agenda was the integration of biodiversity and livelihoods in wetlands.[110] This subject was discussed again at the second segment of CBD COP15 in Montreal in December 2022. The success of the CBD and the Ramsar Convention in preserving biodiversity and wetlands ultimately depends on the adoption and implementation of NBSAPs that are designed

---

thematic or in-depth analysis of each goal and target which may be applicable at global, regional, national, and subnational levels.' Ibid.

[106] Ibid, 1.

[107] See generally E Hamman, T Van Geelen, and A Akhtar-Khavari, 'Governance tools for the conservation of wetlands: the role of the Montreux Record under the Ramsar Convention' (2019) 70 *Marine and Freshwater Research* 1493.

[108] See Chapter 3 of this volume.

[109] See CBD, *Kunming Declaration 'Ecological Civilization: Building a Shared Future for All Life on Earth'* CBD/COP/15/5/Add.1 (13 October 2021).

[110] CBD, Open Ended Working Group on the Post-2020 Global Biodiversity Framework, *First Draft of the Post-2020 Global Biodiversity Framework*, CBD/WG2020/3/3 (5 July 2021) 1.

in such a way as to integrate biodiversity protection requirements, taking into account the goals of all biodiversity-related Conventions, and to achieve synergies by saving resources and duplication of efforts. This can only be achieved by articulating the necessary instances of cooperation between the different Conventions to identify common goals and priorities, which are then translated into national plans and strategies that must also be adapted to national circumstances. The proposals for the CBD Kunming-Montreal COP15 did not advance much further than the basis from which they started. As has long been criticised, the 'system lacks both an inherent hierarchy and a typical centre, but is all the same an international governing system applicable to terrestrial and marine biodiversity'.[111] It is still necessary to harmonise the methodologies for the transfer of information, reporting and control, and promotion of compliance so that the indicators are common and mutually reinforcing at the time of the evaluation of successes and shortcomings to be overcome, in accordance with a common timetable that will also make it possible to appreciate that progress is being made without regression in the commitments assumed.

It was hoped that at the end of the CBD COP15 negotiation process, a stronger institutional governance for inter-treaty cooperation would be consolidated in the long term, since it has been stated that '[o]ne suggested way of creating ownership over the post-2020 process by the biodiversity related conventions was to establish a coordination body of parties to the different biodiversity related conventions'.[112] This proposal was in line with the preparatory documents, which speak of the need to strengthen cooperation through bodies that build bridges both between the biodiversity-related Conventions and with the Rio Conventions. However, such bodies already exist. It is disturbing to consider that either they are invisible to parties or parties do not consider them to be up to the task they are expected to perform. This sense of invisibility of existing formal and informal networks is all the more worrying as one of their main missions is to reach out and increase the visibility of the work of the biodiversity-related Conventions. It is also hoped that existing synergies will be improved in order to be able to solve the problems that have been already identified, such as the lack of a common agenda for the biodiversity-related Conventions and the lack of synchronisation between their respective cycles.

---

[111] See A Johannsdottir, I Cresswell, and P Bridgewater, 'The Current Framework for International Governance of Biodiversity: Is It Doing More Harm Than Good?' (2010) 19 *Review of European, Comparative and International Environmental Law* 139, 142.

[112] CBD, *Synthesis of Views of Parties and Observers on the Scope and Content of the Post-2020 Global Biodiversity Framework*, CBD/POST2020/PREP/1/INF/1 (24 January 2019) 13.

At least the financial contribution pledged by the parties at COP15 may be provided by the new Global Biodiversity Framework Fund that was ratified eight months subsequently.[113]

In any case, the CBD already plays a fundamental role in the coordination of this constellation of conventions and institutions by taking the lead in the various networks that have made it possible for them to move (to some extent) in unison, in order to enhance cooperation and synergies among the Conventions in line with the CBD decisions. Thus, the new Kunming-Montreal Global Biodiversity Framework will achieve a further 'CBD-ification' of all biodiversity-related Conventions, including Ramsar – which, despite or because of this process, will continue to pursue its specific objectives.

---

[113] GEF, Press Release: 'New global biodiversity fund launched in Vancouver' (24 August 2023), at: https://www.thegef.org/newsroom/press-releases/new-global-biodiversity-fund-launched-vancouver.

9. The Ramsar Convention and general international water law: Complementary and mutually supportive regimes

Owen McIntyre

1 INTRODUCTION

The close interrelationship between wetlands protection and international water resources management has long been apparent. Effective conservation of wetland ecosystems depends on water resources management regimes allocating sufficient water to ensure that wetlands function sustainably. One of Hungary's key concerns in *Gabčíkovo-Nagymaros*, a leading international water law case, involved possible impacts of the dam project in question upon the Szigetkötz, an important wetland habitat.[1]

It is useful in several respects to examine the interlinkages between the regime of wetlands protection created over 50 years ago under the Ramsar Convention[2] and that of transboundary water resources management arising under general international water law. At the functional level, it is important to understand how practice under Ramsar can inform and facilitate the rapidly evolving requirement of international watercourse-related ecosystems protection, while environmental flow ('e-flow') requirements arising under international water law can play a critically important role in the maintenance of wetlands. At the structural level, while both regimes apply at different scales, there is significant overlap in relation to aquatic ecosystems management and protection. Taken together, they cover almost all terrestrial and coastal aspects

---

[1] *Case Concerning the Gabčíkovo-Nagymaros Project (Hungary/Slovakia)*, Judgment, 25 September 1997 [40].
[2] Convention on Wetlands of International Importance especially as Waterfowl Habitat, adopted 2 February 1971, entered into force 21 December 1975, 996 UNTS 245 (amended 1982 & 1987).

of the 'source-to sea' management continuum and so can facilitate integrative management of water resources, aquatic biodiversity, and other water-related environmental impacts. The evolutionary aspect of international law is also apparent in the developmental interaction of both regimes, with international water law becoming increasingly concerned with aquatic ecosystems and Ramsar becoming ever more concerned with water resources management. Indeed, at the systemic level, the tendency of each regime to permeate and influence the development and application of the other might be regarded as the epitome of the phenomenon of legal 'convergence',[3] involving 'the gradual interpenetration and cross-fertilization of previously somewhat compartmentalized areas of international law'.[4]

Though the reach and coverage of the principal instruments of international water law are improving, with the 2014 entry into force of the 1997 United Nations (UN) Watercourses Convention[5] and the global opening of the 1992 United Nations Economic Commission for Europe (UNECE) Water Convention,[6] the Ramsar Convention regime is truly global and more highly developed normatively. The Ramsar Convention benefits from over 50 years of accumulated state practice and institutional interaction, along with a wealth of technical guidance focusing on the conservation and wise use of wetland ecosystems. This Chapter examines the relationship between these two international law regimes and identifies synergies stemming from the emergence of a general requirement in international water law for states to protect natural

---

[3] See, for example, M Craven, 'Unity Diversity and the Fragmentation of International Law' (2002) XIII *Finnish Yearbook of International Law* 1–31; M Andenas, 'Reassertion and Transformation: From Fragmentation to Convergence in International Law' (2015) 46 *Georgetown Journal of International Law* 685–734; M Andenas and E Bjorge (eds), *A Farewell to Fragmentation: Reassertion and Convergence in International Law* (Cambridge University Press, Cambridge, 2015); PM Dupuy, 'The Danger of Fragmentation or Unification of the International Legal System and the International Court of Justice' (1999) 31 *New York University Journal of International Law and Politics* 791–807; G Abi-Saab, 'Fragmentation or Unification: Some Concluding Remarks' (1999) 31 *New York University Journal of International Law and Politics* 919–33.

[4] A Cassese, *International Law* (Oxford University Press, Oxford, 2001) 45.

[5] United Nations Convention on the Law of the Non-Navigational Uses of International Watercourses (New York), adopted 21 May 1997, entered into force 17 August 2014, (1997) 36 ILM 700 (UN Watercourses Convention].

[6] United Nations Economic Commission for Europe (UNECE) Convention on the Protection and Use of Transboundary Watercourses and International Lakes (Helsinki), adopted 17 March 1992, entered into force 6 October 1996 (1992) 31 ILM 1312 ('UNECE Water Convention').

ecosystems connected to international watercourses. It is increasingly apparent that normative approaches developed under either regime might serve to support effective implementation of the other, while key provisions of either treaty framework might be relevant to the interpretation and application of obligations arising under the other – although the longer-established and more technically evolved regime created under the Ramsar Convention might offer more lessons. Generally, better understanding of the relationship between both regimes can only enhance their mutual complementarity and normative coherence.

Given the interdependence of these two regimes, it is useful to examine the complementarity between the Ramsar Convention[7] – the one truly global Convention for the protection of (internationally) important wetlands[8] – and general international law relating to shared international freshwater resources, as reflected in the UN Watercourses Convention, the UNECE Water Convention, and the International Law Commission (ILC) Draft Articles on Transboundary Aquifers.[9] While numerous other environmental instruments are also relevant to the effective implementation of each regime – including, for example, the 1992 Convention on Biological Diversity[10] (CBD) – it is beyond the scope of this Chapter to examine such instruments in detail. To understand the mutually relevant normative implications of each regime, it is first necessary to examine their respective scope of application and key requirements, with a view to identifying areas of overlap and complementarity. It is also useful to outline the parallel emergence of an 'ecosystem approach' under each regime before noting specific examples of regime interaction and complementarity, including instances where inter-state cooperation on wetlands has proven easier than cooperation on shared water resources, and where detailed technical guidance developed under Ramsar informs the rather vague ecosystem protection obligations arising under international water law. The Chapter concludes with some general observations on how this close

---

[7] The preamble to the Convention expressly recognises the close inter-linkage between wetlands conservation and water resources management by noting 'the fundamental ecological functions of wetlands as regulators of water regimes'.

[8] Today the Ramsar Convention has 172 contracting parties and applies to circa 2,500 designated wetland sites covering over 250 million hectares.

[9] Draft Articles on the Law of Transboundary Aquifers, *Report of the International Law Commission on the Work of its Sixtieth Session*, UN Doc A/63/10, 2008 ('ILC Draft Articles on Transboundary Aquifers'). The UN General Assembly took note of the Draft Articles, which it annexed to Resolution 63/124, UN Doc A/RES/63/124.

[10] Convention on Biological Diversity, adopted 5 June 1992, entered into force 29 December 1993, (1992) 31 ILM 818 (CBD).

interrelationship and complementarity epitomise the increasingly apparent phenomenon of 'convergence', which counteracts that of 'fragmentation' in international law.[11]

2   RESPECTIVE SCOPE OF APPLICATION OF EACH REGIME

While there are significant areas of overlap, the scope of application of the Ramsar regime clearly does not reflect that of the key instruments of international water law. As an International Union for Conservation of Nature (IUCN) report explains, '[t]o put it simply, water basins include wetlands, but not vice versa'.[12] As regards material scope – that is, the physical elements of the natural environment as well as the activities which might impact upon such elements and which an instrument seeks to regulate –the Ramsar Convention applies to the conservation of wetlands and their flora and fauna, as broadly defined.[13] Thus, Ramsar applies to wetlands which are purely 'national' in the sense that they have no connection to shared transboundary water resources. The Convention also applies to certain estuarine and coastal areas – including coastal zones, islands, and bodies of water exceeding six metres in depth adjacent to wetlands[14] – to which the key international freshwater law instruments could not apply, at least not directly.[15] Transboundary water systems are also included under the scope of the Convention, with parties obliged to consult

---

[11] International Law Commission (ILC), Chapter X: 'Fragmentation of International Law: Difficulties arising from the Diversification and Expansion of International Law', Report of the 55th Session, UN GAOR, Supp No 3, UN Doc A/58/10 (2003) 267. See M Koskenniemi, *Fragmentation of International Law: Difficulties Arising from the Diversification and Expansion of International Law*, Report of the Study Group of the International Law Commission, UN Doc A/CN.4/L.682 (13 April 2006) 10–30.

[12] G Aguilar and A Iza, *Governance of Shared Waters: Legal and Institutional Issues* (IUCN Environmental Policy and Law Paper No 58, 2011) 101.

[13] Convention on Wetlands, note 2, Art 1.1.

[14] Ibid, Art 2.1.

[15] Article 23 of the UN Watercourses Convention provides that 'Watercourse States shall ... take all measures with respect to an international watercourse that are necessary to protect and preserve the marine environment, including estuaries, taking into account generally accepted international rules and standards'. However, this provision is primarily concerned with avoiding harm to the marine environment due to the utilisation or pollution of an international watercourse and with the protection and preservation of functionally connected marine and riverine ecosystems.

'where a water system is shared by Contracting Parties' and coordinate relevant policies and regulations.[16]

The UN Watercourses Convention applies to 'uses of international watercourses and of their waters for purposes other than navigation and to measures of protection, preservation and management related to the uses of those watercourses and their waters',[17] where a 'watercourse' is understood as 'a system of surface waters and ground waters constituting by virtue of their physical relationship a unitary whole and normally flowing into a common terminus'[18] and an 'international watercourse' is one 'parts of which are situated in different States'.[19] The UNECE Water Convention is primarily concerned with the regulation of transboundary impacts[20] and defines 'transboundary waters' to mean 'any surface or ground waters which mark, cross or are located on boundaries between two or more States'.[21] Thus, the UN Watercourses Convention has a somewhat broader scope than the UNECE Water Convention, as the former is concerned with all aspects of water utilisation, including the need to protect shared waters and international watercourse ecosystems, while the latter focuses primarily on transboundary environmental impacts.

The more recent ILC Draft Articles on Transboundary Aquifers adopt – albeit in respect of a narrower category of freshwater bodies – the broad scope of the UN Watercourses Convention and are concerned with the utilisation, protection, preservation, and management of groundwater resources,[22] while defining an 'aquifer' to mean 'a permeable water-bearing geological formation underlain by a less permeable layer and the water contained in the saturated zone'.[23] International water law is very much concerned with shared international freshwater resources, applying only simultaneously with the Ramsar Convention in respect of the protection of wetland ecosystems – especially those designated for inclusion in the Convention's List of Wetlands of International Importance that are functionally connected to a transboundary basin.

---

[16] Convention on Wetlands, note 2, Art 5.
[17] UN Watercourses Convention, note 5, Art 1(1).
[18] Ibid, Art 2(a).
[19] Ibid, Art 2(b).
[20] UNECE Water Convention, note 6, Article 1(2) defines 'transboundary impact' to mean 'any significant adverse effect on the environment resulting from a change in the conditions of transboundary waters caused by a human activity, the physical origin of which is situated wholly or in part within an area under the jurisdiction of a Party, within an area under the jurisdiction of another Party'.
[21] Ibid, Art 1(1).
[22] ILC Draft Articles on Transboundary Aquifers, note 9, Art 1(a)–(c).
[23] Ibid, Art 2(a).

However, the mutual relevance of each regime may be more extensive and pronounced than the relevant instruments would suggest. Commentators note that 'the recommendations and guidelines adopted by the [Ramsar] Conference of the Parties (COP) have expanded the spectrum of issues considered by the Convention', so that 'the key focus within the Ramsar Convention is on wetlands of *international* relevance, including the commitment to maintain ecological characteristics through the *rational use of resources*'[24] – including, most notably, water resources. In fact, commentators have recently decried Ramsar's purported evolution 'towards a "water" convention ... as a case of institutional maladaptation'.[25] The Ramsar COP recognised early on the hydrological and ecological importance of wetlands in the context of watercourses[26] and subsequently adopted detailed guidelines relevant to the management of river basins, including shared international river basins. Notable examples include 1999 guidelines[27] setting out detailed recommendations on the conservation of river basin ecosystems, including a dedicated section on 'international cooperation' concerning 'special issues related to shared river basin and wetland systems'; and 2002 guidelines[28] comprising international best practice to inform inter-state processes for allocating shared waters – at least where such waters are connected to wetlands designated as internationally important. Guidance produced under international water conventions may similarly support implementation of Ramsar objectives. For example, 2013 UNECE guidance recognises that '[i]t is important to note that water quantity is an essential element in securing the structure, function and species compositions in aquatic and water-related ecosystems'.[29]

Further extending regime overlap, Article 20 of the UN Watercourses Convention – which sets out the requirement for states to protect watercourse ecosystems – may create obligations in respect of purely internal as well as

---

[24] Aguilar and Iza, note 12, 100 (emphasis added).

[25] P Bridgewater and R Kim, '50 years on, w(h)ither the Ramsar Convention? A case of institutional drift' (2021) 30 *Biodiversity and Conservation* 3919, 3921. See also P Bridgewater and R Kim, 'The Ramsar Convention on Wetlands at 50' (2021) 4 *Nature Ecology & Evolution* 268.

[26] See, for example, *The Criteria for Identifying Wetlands of International Importance*, first endorsed by Ramsar Convention, *Recommendation IV.2* (1990).

[27] Ramsar Convention, *Resolution VII.18: Guidelines for integrating wetland conservation and wise use into river basin management* (1999).

[28] Ramsar Convention, *Resolution VIII.1: Guidelines for the allocation and management of water for maintaining the ecological functions of wetlands* (2002).

[29] UNECE, *Guide to Implementing the UNECE Water Convention* (UN, Geneva, 2013) [120].

transboundary wetlands, irrespective of any obvious transboundary impact.[30] While Article 21, regarding prevention, reduction, and control of pollution, Article 22, regarding introduction of alien or new species, and Article 23, regarding protection and preservation of the marine environment, are expressly concerned with harm caused to other states or to the marine environment and are generally understood to codify existing customary international law, Article 20 makes no mention of transboundary impact and so can be interpreted as imposing obligations on a watercourse state regarding a wetland situated entirely within its own borders, as long as that wetland is in some way connected to an international watercourse.[31] The ILC commentary to the 1994 Draft Articles, which preceded the UN Watercourses Convention, suggests that the ILC regarded Article 20 as 'laying down a general obligation to protect and preserve the ecosystems of international watercourses', regardless of any transboundary impact.[32]

This expansive view also applies to the key ecosystems provision included under the UNECE Water Convention, regarding which the implementation guide explains that '[a]lthough the Convention deals with transboundary waters, the term "ecosystems" in this provision is not necessarily limited to transboundary ecosystems nor does it exclude other than aquatic and water-related ecosystems'.[33] The same guidance further refers to UNECE recommendations on payments for ecosystem services (PES)[34] to define 'water-related ecosystems' expansively to include 'ecosystems such as forests, wetlands, grasslands, and agricultural land that play vital roles in the hydrological cycle through the services they provide'.[35] The Helsinki Declaration, adopted by the UNECE Water Convention parties at the first Meeting of the Parties in 1997, exhorts the same protection for internal and transboundary

---

[30] UN Watercourses Convention, note 5, Article 20 merely provides that 'Watercourse States shall, individually and, where appropriate, jointly, protect and preserve the ecosystems of international watercourses'.

[31] See further, O McIntyre, 'The Emergence of an "Ecosystem Approach" to the Protection of International Watercourses under International Law' (2004) 13/1 *Review of European Community and International Environmental Law* 1, 8.

[32] ILC, *Report of the International Law Commission on the Work of its Forty-Ninth Session*, UN Doc GAOR A/49/10/1994 (1994) 195, 280.

[33] UNECE, note 29, [114].

[34] UNECE, *Recommendations for Payments for Ecosystem Services in Integrated Water Resources Management*, UN Doc ECE/MP.WAT/22 (2007), at: http://www.unece.org/fileadmin/DAM/env/water/publications/documents/PES_Recommendations_web.pdf.

[35] UNECE, note 29, [114], fn 36.

water-related ecosystems in order to ensure consistency in their treatment.[36] An increasing number of international basin agreements – including those inspired by the UNECE Convention – appear to include a general obligation to protect ecosystems, whether internal or transboundary, implicitly recognising the inherent interconnectedness of the various ecological components of a drainage basin.[37]

Similarly, various regional treaties concerned with land-based sources of marine pollution tend to require the same ecosystems protection for otherwise non-international watercourses, thereby inferring the wider impacts and interests concerned.[38] Widely adopted multilateral environmental agreements (MEAs), including the CBD, already 'contribute to the protection of watercourse ecosystems even though they do not concern international watercourses *per se*', reflecting greater 'understanding of the interactions between various species and natural systems' so that 'States in their practice will recognise an expansion of both the notion of the watercourse ecosystem and the legal protection thereof'.[39]

The influential sustainable development paradigm, which correlates closely with the cardinal international water law principle of equitable and reasonable utilisation,[40] supports expansive application of the environmental protection

---

[36] UNECE, Meeting of the Parties to the Convention on the Protection and Use of Transboundary Watercourses and International Lakes, *Report of the First Meeting*, ECE/MP.WAT/2 (12 August 1997), at: https://unece.org/fileadmin/DAM/env/documents/1997/wat/ece_mp_wat2.e.pdf.

[37] See, for example, Convention of the International Commission for the Protection of the Elbe (Magdeburg, 8 October 1990), Article 1(2)(b) of which commits states to endeavour 'to achieve as natural an ecosystem as possible with a healthy diversity of species'; Convention on the Protection of the Rhine (Rotterdam, 22 January 1998), Article 3(1) of which details the means by which states are to achieve the key objective of ensuring 'sustainable development of the Rhine ecosystem'.

[38] P Birnie and A Boyle, *International Law and the Environment* (2nd edition, Oxford University Press, Oxford, 2002) 315.

[39] S McCaffrey, *The Law of International Watercourses: Non-Navigational Uses* (Oxford University Press, Oxford, 2001) 393. See also, S McCaffrey, *The Law of International Watercourses* (2nd edition, Oxford University Press, Oxford, 2007).

[40] See P Wouters and A Rieu-Clarke, 'The Role of International Water Law in Promoting Sustainable Development' (2001) 12 *Water Law* 281, 283, who conclude that '[t]he principle [of equitable and reasonable utilisation] provides, indeed requires, that states take into consideration the factors tied to sustainable development of the resource, thus providing the legal framework for *operationalising* this concept' (emphasis added). See further, O McIntyre, *Environmental Protection of*

requirements of international water law[41] so that '[r]ecourse to the notion of sustainable development might also result in the inclusion of the national environmental effects of the utilization of international watercourses as a relevant factor to consider in the application of the principle of equitable utilization'.[42] Thus, while Ramsar applies to around 2500 Wetlands of International Importance – including around 250 transboundary wetlands – the increasingly recognised legal requirement to conserve international watercourse ecosystems appears also to encompass obligations regarding purely national ecosystems and ecosystems other than those purely aquatic or water-related.

3 THE NORMATIVE REQUIREMENTS OF EACH REGIME

It is a measure of the extent to which the ILC engaged in the codification of international water law when preparing its 1994 Draft Articles that, despite slight differences in nuance and emphasis, the key global instruments of international water law each revolve around three central rules and principles: the principle of equitable and reasonable utilisation;[43] the duty to prevent significant transboundary harm;[44] and the general obligation to cooperate.[45] The former two substantive principles incorporate a wide range of social and environmental values, notably including the obligation to protect and preserve

---

International Watercourses under International Law (Ashgate, Aldershot, 2007) 246–50.

[41] See *Pulp Mills on the River Uruguay (Argentina v. Uruguay)*, Judgment of 20 April 2010, [177]. See further, O McIntyre, 'The Proceduralization and Growing Maturity of International Water Law' (2010) 22/3 *Journal of Environmental Law* 475–497; O McIntyre, 'The World Court's Ongoing Contribution to International Water Law: The Pulp Mills Case between Argentina and Uruguay' (2011) 4/2 *Water Alternatives* 124–44.

[42] X Fuentes, 'Sustainable Development and the Equitable Utilization of International Watercourses' (1998) 69 *British Yearbook of International Law* 119, 177.

[43] UN Watercourses Convention, note 5, Arts 5 and 6; UNECE Water Convention, note 6, Art 2(2)(c); ILC Draft Articles on Transboundary Aquifers, note 9, Arts 4 and 5.

[44] UN Watercourses Convention, note 5, Art 7; UNECE Water Convention, note 6, Art 2(1); ILC Draft Articles on Transboundary Aquifers, note 9, Art 6.

[45] UN Watercourses Convention, note 5, Art 8; UNECE Water Convention, note 6, Art 2(6); ILC Draft Articles on Transboundary Aquifers, note 9, Art 7.

aquatic ecosystems.[46] The latter obligation encompasses a range of primarily procedural requirements[47] regarding, most significantly, the prior notification of planned measures[48] and the routine exchange of data and information.[49] Whereas the UN Watercourses Convention appears at first reading more general in terms of its scope of application, covering all aspects of shared water utilisation, including pollution, the UNECE Water Convention focuses on the avoidance of transboundary environmental impact and the protection of international watercourses from pollution, reflecting its origins as a regional framework convention applicable to states which are predominantly economically developed and relatively abundant in water.[50] While the UNECE Water Convention also endorses the principle of equitable and reasonable utilisation, it 'covers the principle concisely in the context of activities which cause or may cause transboundary impact'[51] and includes express reference to the precautionary principle, the polluter pays principle and the principle of inter-generational equity,[52] as well as detailed provisions on the precise

---

[46] Further elaborated in UN Watercourses Convention, note 5, Art 20; UNECE Water Convention, note 6, Art 2(2)(d); ILC Draft Articles on Transboundary Aquifers, note 9, Art 10.

[47] Some commentators argue that, while the specific means of cooperating are principally procedural in nature, the general obligation to cooperate is a substantive norm. C Leb, *Cooperation in the Law of Transboundary Water Resources* (Cambridge University Press, Cambridge, 2013) 73–106.

[48] UN Watercourses Convention, note 5, Arts 11–19; UNECE Water Convention, note 6, Art 9(2)(h); ILC Draft Articles on Transboundary Aquifers, note 9, Art 15(2).

[49] UN Watercourses Convention, note 5, Art 9; UNECE Water Convention, note 6, Art 6; ILC Draft Articles on Transboundary Aquifers, note 9, Art 8.

[50] However, a closer and more holistic examination of the relevant provisions of the UNECE Water Convention suggests that it is equally concerned with the utilisation of shared water resources and with the consequences for such utilisation of transboundary environmental impacts upon shared waters. See, for example, O McIntyre, 'The UNECE Water Convention and the Principle of Equitable and Reasonable Utilisation' in A Tanzi et al (eds), *The UNECE Convention on the Protection and Use of Transboundary Watercourses and International Lakes: Its Contribution to International Water Cooperation* (Brill/Nijhoff, Leiden, 2014) 146–60. See also UNECE, note 29, 15 [80].

[51] S McCaffrey, 'The Entry into Force of the 1997 Watercourses Convention', 25 May 2014, *International Water Law Project Blog* at: http://www.internationalwaterlaw.org/blog/2014/05/25/dr-stephen-mccaffrey-the-entry-into-force-of-the-1997-watercourses-convention/.

[52] UNECE Water Convention, note 6, Art 2(5).

means of pollution prevention, control, and reduction.[53] In this manner, the two Conventions can be regarded as complementary, with the UNECE Water Convention providing normative detail on the environmental protection of shared waters and the UN Watercourses Convention elaborating more on the application of the principles of international water law to the question of quantum share of water resources in the case of overutilised watercourses in water-stressed regions. Thus far, 19 states have ratified both Conventions, thereby suggesting that they find no incongruity between their respective provisions.[54]

While the preamble to the Ramsar Convention explains that its objective is 'to stem the progressive encroachment on and loss of wetlands', the key substantive obligations imposed upon parties concern conservation and wise use of wetlands – both those included in the List of Wetlands of International Importance and others located within their territory.[55] The dual concepts of 'conservation' and 'wise use' are closely interlinked so that there is no 'presumption of a "hands-off" approach to listed wetlands'.[56] Rather, the Convention emphasises the practical benefits of wetland conservation, where appropriate, though human use and development is not promoted in the case of every wetland – each wetland should be managed having regard to its ecological character with a view to achieving its wise use.[57] The Convention's central aim of 'wise use of wetlands' appears to be the wetlands-specific version of the concept of sustainable use[58] and has been defined as 'the maintenance of their ecological character, achieved through the implementation of ecosystem approaches, within the context of sustainable development'.[59] In turn, 'ecological character' is understood as 'the combination of the ecosystem

---

[53] Ibid, Art 3.
[54] Including Chad, Denmark, Finland, France, Germany, Ghana, Guinea-Bissau, Greece, Hungary, Italy, Luxembourg, Montenegro, Netherlands, Norway, Portugal, Spain, Sweden, the United Kingdom, and Uzbekistan.
[55] See Convention on Wetlands, note 2, Arts 3.1 and 4.1
[56] M Bowman, P Davies, and C Redgwell, *Lyster's International Wildlife Law* (2nd edition, Cambridge University Press, Cambridge, 2010) 415.
[57] Convention on Wetlands, note 2, Art 3.1.
[58] Bowman et al point out that the current *Strategic Framework and Guidelines for the Future Development of the List of Wetlands of International Importance* describes the endpoint of wetland designation under Ramsar as 'achieving the long-term wise (sustainable) use of the site'. Bowman, Davies, and Redgwell, note 56, 415–16.
[59] Ramsar Convention, *Resolution IX.1, Annex A: Conceptual Framework for the Wise Use of Wetlands and the Maintenance of their Ecological Character* (2005).

components, processes and beneficial services that characterise the wetland at any given point in time'.[60] Key concepts increasingly employed in legal frameworks for ecological management – such as the ecosystem approach, as stipulated under the CBD,[61] and the notion of ecosystem services, as outlined in the 2005 Millennium Ecosystem Assessment[62] – are central to the substantive requirements of Ramsar, just as they are to key global instruments of international water law.

The Ramsar Convention requires each contracting party to 'designate suitable wetlands within its territory for inclusion in a List of Wetlands of International Importance', selected on account of 'their international significance in terms of ecology, botany, zoology, limnology or hydrology' and 'international importance to waterfowl at any season'.[63] Therefore, it allows for their unilateral designation and does not prescribe any screening procedure for the sites selected,[64] though a Ramsar Information Sheet must be completed upon designation and data revised 'at least every six years',[65] requiring the submission of detailed information concerning the Site, its legal status, and its ecological characteristics.[66] Though the requirement for parties to promote

---

[60] Ibid.

[61] CBD, note 10, citing CBD Decision V/6, which sets out 12 key principles of the ecosystem approach, together with operational guidance on their application. Bowman, Davies, and Redgwell, note 56, 417, explain that the reference to 'ecosystem approaches' in the 2005 Ramsar Conceptual Framework 'is an invocation of practice under the Biodiversity Convention, which requires the adaptive, integrated management of land, water and living resources based upon scientific methodologies encompassing all aspects of the essential interactions amongst organisms and their environment'.

[62] Millennium Ecosystem Assessment, *Ecosystems and Human Well-Being: Synthesis* (Island Press, 2005) 39–48.

[63] Convention on Wetlands, note 2, Art 2.2.

[64] Though guidance is provided regarding listing criteria: Ramsar Convention, *Recommendation 4.2: Criteria for identifying wetlands of international importance* (1990); Ramsar Convention, *Resolution 5.9: Application of the Ramsar Criteria for Identifying Wetlands of International Importance* (1993); Ramsar Convention, *Information Paper No. 5: The Criteria for Identifying Wetlands of International Importance* (2007).

[65] Ramsar Convention, *Resolution VI.13: Submission of Information on Sites Designated for the Ramsar List of Wetlands of International Importance* (1996).

[66] Ibid; Ramsar Convention, *Resolutions VIII.13: Enhancing the information on Wetlands of International Importance (Ramsar sites)* (2002) and *Resolution VIII.21: Defining Ramsar site boundaries more accurately in Ramsar Information Sheets* (2002); Ramsar Convention, *Resolution IX.1*, note 59, Annex B; and Ramsar Convention, *Resolution X.15: Describing the ecological character of wet-*

wise use of wetlands applies to both wetlands included in the List and others located within their territory, a more rigorous regime of protection applies to Sites included on the List.[67] Under Article 3.1, the requirement to promote the wise use of wetlands is qualified in relation to non-List sites in that it applies only 'as far as possible'. Also, it applies only to such non-List sites located 'within their territory', thereby suggesting that for Sites included on the List, this obligation assumes some form of collective responsibility, whereby such key Sites could be regarded as resources of common concern to the international community. Such an understanding of Article 3.1 would imply the creation of extraterritorial obligations for the contracting parties, at least involving an obligation to avoid causing significant harm to listed Sites located within the territory of neighbouring states – an obligation which would correspond neatly with those arising under general international water law regarding the prevention of significant transboundary harm and the protection of the ecosystems of international watercourses.[68] It might also require Ramsar parties to refrain from funding or otherwise assisting development projects in other states, such as dams or irrigation schemes, which impact upon the ecological character of List Sites located within the territory of such states.[69]

Ramsar also imposes a general obligation under Article 6 for contracting parties to cooperate multilaterally in the implementation of the Convention through engagement in the regular COP. Though normatively 'soft' in character, this arrangement has proven effective in ensuring the incremental

---

lands, and data needs and formats for core inventory: harmonized scientific and technical guidance (2008).

[67] Bowman, Davies, and Redgwell, note 56, 420–23.

[68] Ibid, 424–25; Ramsar Convention, *Resolution XI.9: An Integrated Framework and guidelines for avoiding, mitigating and compensating for wetland losses* (2012), Annex 3.4.

[69] Ramsar Convention, *Recommendation 3.4: Responsibility of development agencies toward wetlands* (1987) urges development agencies (including 'all banks, government institutions and international governmental agencies ... with a significant role in providing funds to countries for their development') to pursue coherent policies directed at sustainable utilisation, wise management, and conservation of wetlands, ensuring the integration of environmental aspects into all phases of the project cycle, especially prior impact assessment; while Ramsar Convention, *Recommendation 4.13: Responsibility of multilateral development banks (MDBs) towards wetlands* (1990) urges the parties to ensure that their representatives to the multilateral development banks adopt voting standards which support wetland conservation. See Bowman, Davies, and Redgwell, note 56, 425.

elaboration of a highly sophisticated Ramsar regime. More specifically, the Convention requires that:

> [t]he Contracting Parties shall consult with each other about implementing obligations arising from the Convention especially in the case of a wetland extending over the territories of more than one Contracting Party *or where a water system is shared by Contracting Parties*.[70]

This potentially overlaps significantly with corresponding obligations arising under general international water law[71] and serves to complement the cooperative procedural obligations inherent to the general legal requirement to protect and preserve the ecosystems of shared international watercourses.[72]

Though the basic obligations set out under Ramsar might at first appear modest and normatively vague, the development and operation of institutional arrangements under the Convention for the ongoing review of its implementation have been crucial to the evolution of the Ramsar regime.[73] For example, the Ramsar COP has incrementally developed guidance on practical application of the requirement for the wise use of wetlands,[74] advising the parties on, among other things, the establishment of national wetland policies, the implementation of priority measures at the national and wetland site levels, the preparation of national inventories of wetlands, environmental impact assessment (EIA) in respect of development projects potentially impacting wetland sites, and sustainable utilisation of natural elements of wetland ecosystems.[75] Such guidance is consolidated in the Ramsar *Handbooks for the Wise Use of Wetlands*.[76] Thus, Ramsar exemplifies the tendency among MEA regimes to utilise elaborate institutional arrangements, constituting 'focal points of a broad, legally significant communication process',[77] in order to address tech-

---

[70] Convention on Wetlands, note 2, Art 5 (emphasis added).

[71] See, for example, the requirements relating to consultation and negotiation concerning planned measures set out under Article 17 of the UN Watercourses Convention, note 5.

[72] See, for example, Article 20 of the UN Watercourses Convention, note 5.

[73] Bowman, Davies, and Redgwell, note 56, 428–35.

[74] Ramsar Convention, *Recommendation 3.3: Wise use of wetlands* (1987); Ramsar Convention, *Recommendation 4.10: Guidelines for the implementation of the wise use concept* (1990); Ramsar Convention, *Resolution 5.6: The wise use of wetlands* (1993); Ramsar Convention, *Resolution IX.1*, note 66, Annex A.

[75] Bowman, Davies, and Redgwell, note 56, 419.

[76] The Handbooks are available on the Ramsar Convention website at: https://www.ramsar.org/.

[77] T Gehring, 'International Environmental Regimes: Dynamic Sectoral Legal Systems' (1990) 1 *Yearbook of International Environmental Law* 35–56, 43–44.

nical, scientific, and ultimately legal questions, and thereby assist the regime's progressive evolution.[78]

4 THE 'ECOSYSTEM APPROACH' IN INTERNATIONAL WATER LAW

The principal point of intersection between general international water law and the Ramsar Convention concerns the relatively recent requirement to protect and preserve water-related ecosystems connected to shared international watercourses. The wide variety of wetland ecosystems protected under Ramsar – both transboundary wetlands and those located exclusively within the territory of one contracting party – will often constitute the most ecologically, socially, and economically significant water-related ecosystems falling within the scope of the rules of international water law. Therefore, the relatively advanced framework for ecosystems protection developed under Ramsar might be expected to play a key role in informing the normative requirements of ecosystems protection under international water law.

4.1 Ecosystem Approach

The emergence of an obligation to protect watercourse ecosystems under international law, and of the so-called 'ecosystem approach', has been well documented over the past three decades.[79] Having regard to the wealth of rele-

---

[78] See S Maljean-Dubois 'The Making of International Law Challenging Environmental Protection' in Y Kerbrat and S Maljean-Dubois (eds), *The Transformation of International Environmental Law* (A Pedone & Hart, Paris and Oxford, 2011) 25, 29–30; O McIntyre, 'Changing Patterns of International Environmental Law-Making: Addressing Normative Ineffectiveness' in S Maljean-Dubois (ed), *The Effectiveness of Environmental Law* (Intersentia, Cambridge, 2017) 187–220; O McIntyre, 'International Law-Making in the Field of Natural Resources' in C Brölmann and Y Radi (eds), *Research Handbook on the Theory and Practice of International Law-making* (Edward Elgar Publishing, Cheltenham, 2016) 442–65.

[79] See J Brunée and SJ Toope, 'Environmental Security and Freshwater Resources: A Case for International Ecosystems Law' (1994) 5 *Yearbook on International Environmental Law* 41–76; J Brunée and SJ Toope, 'Environmental Security and Freshwater Resources: Ecosystem Regime Building' (1997) 91/1 *American Journal of International Law* 26–59; AD Tarlock, 'International Water Law and the Protection of River System Ecosystem Integrity' (1996) 10/2 *Brigham Young University Journal of Public Law* 181–211; McIntyre, note 31; O McIntyre, 'The Protection of Freshwater Ecosystems Revisited: Towards

vant treaty and declarative practice lending support to the ecosystem approach[80] in the years preceding the inclusion of Article 20 in the 1997 UN Watercourses Convention, commentators noted that:

> progress made in scientific research further shows that the uses of watercourses can affect and be affected by processes related to other natural elements, such as soil degradation and desertification, deforestation and climate change ... [, which] has brought water specialists in the last decade to advocate the adoption of less economic-oriented criteria for the management of freshwater resources, following an 'ecosystem approach'.[81]

The ecosystem approach is normally understood as a spatially broad management approach which can inform the application of legal frameworks for water resources utilisation, land use planning and development control, requiring 'consideration of the whole system rather than individual components', and focusing on the interaction between '[l]iving species and their physical environments' and 'between different subsystems and their responses to stresses resulting from human activity'.[82] In the specific context of international water law, the concept has been linked to the obligation to protect international watercourses ancillary to the principle of equitable and reasonable utilisation,[83] and accordingly to the objective of sustainable development.[84] It has also been recognised – explicitly by the ILC[85] and implicitly by the International Court

---

a Common Understanding of the "Ecosystems Approach" to the Protection of Transboundary Water Resources' (2014) 23/1 *Review of European, Comparative and International Environmental Law* 88–95.

[80] The Commentary to the ILC's 1994 Draft Articles on the Non-Navigational Uses of International Watercourses, which preceded the adoption of the UN Watercourses Convention, concluded that 'there is ample precedent for the obligation contained in Article 20 in the practice of States and the work of international organizations', before proceeding to cite several leading examples. See ILC, note 32, 195, 283, 284–89.

[81] A Tanzi and M Arcari, *The United Nations Convention on the Law of International Watercourses* (Kluwer, The Hague, 1991) 8–9.

[82] Brunée and Toope, note 79, 55.

[83] According to the ILC's Commentary to Draft Article 20, note 32, 282, the 'obligation to protect the ecosystems of international watercourses is a specific application of the requirement contained in article 5 that watercourse States are to use and develop an international watercourse in a manner that is consistent with adequate protection thereof'.

[84] The ILC Commentary further explains, ibid, that Article 20 provides an 'essential basis for sustainable development'.

[85] The ILC Commentary states categorically that 'the obligation to protect the ecosystems of international watercourses is thus a general application of the prin-

of Justice (ICJ)[86] – that the ecosystems approach is inherently precautionary in nature, involving precautionary management of the impacts of activities and projects on the functioning of water-related ecosystems – for example, through the use of environmental assessment.[87] It is equally apparent that it might result in the extension of the geographical scope of international water law to the entire drainage basin,[88] even beyond the traditional obligations of international water law based on the territorial sovereignty of riparian states.[89] However, considerable uncertainty about its practical application persisted when the UN Watercourses Convention was concluded.

---

ciple of precautionary action'. Ibid. It subsequently refers to 'the principle of precautionary action [as] reflected in article 20'. Ibid, 287.

[86] The ICJ implicitly linked the precautionary principle with the ecosystem approach in the context of international watercourses in *Gabčíkovo-Nagymaros*, note 1, [140]: 'The Court is mindful that, in the field of environmental protection, vigilance and prevention are required on account of the often irreversible character of damage to the environment and of the limitations inherent in the very mechanism of reparation of this type of damage.'

[87] See the separate opinion of Judge Weeramantry in the *Gabčíkovo-Nagymaros Case*, who, describing the crucial role of (continuing) EIA, notes: 'EIA, being a specific application of the larger general principle of caution, embodies the obligation of continuing watchfulness and anticipation.' Ibid, [113]. See generally, A Trouwborst, 'The Precautionary Principle and the Ecosystem Approach in International Law: Differences, Similarities and Linkages' (2009) 18/1 *Review of European, Comparative and International Environmental Law* 26–37.

[88] For example, Birnie and Boyle, note 38, 314, doubt that the ILC's attempts to restrict the geographical scope of the requirement set out in Draft Article 20, observing: 'it is doubtful if the Commission's careful choice of terminology really does confine the potential scope of this obligation in a meaningful way. Any attempt to protect a river "ecosystem" cannot avoid affecting the surrounding land areas or their "environment".'

Similarly, McCaffrey, note 39, 393, notes that any expansive understanding of 'ecosystems' would be almost certain to include 'not only the flora and fauna in and immediately adjacent to a watercourse, but also the natural features within its catchment that have an influence on, or whose degradation could influence, the watercourse'.

[89] Article 23 of the UN Watercourses Convention extends the obligations of riparian states to protection of the marine environment, providing that: 'Watercourse States shall, individually or jointly, take all measures with respect to an international watercourse that are necessary to protect and preserve the marine environment, including estuaries, taking into account generally accepted international rules and standards.'

Despite its far-reaching implications, the normative elements of an ecosystem approach, as well as practical means for its implementation, continue to emerge. While the UN Watercourses Convention emphasises pollution control, prevention of invasive species, and protection of the marine environment, some basin-level agreements – such as the 1995 Mekong Agreement[90] – stress obligations regarding environmental flows. The ILC's Draft Articles on Transboundary Aquifers highlight 'the role of the aquifer or aquifer system in the related ecosystem'[91] and call more specifically upon states 'to ensure that the quantity and quality of water retained in an aquifer or aquifer system, as well as that discharged through its discharge zones, are sufficient to protect and preserve such ecosystems'.[92] However, while states have routinely entered into treaty commitments relating to the protection of international watercourse-related ecosystems since the early 1990s, the parameters and practical implications of a meaningful 'ecosystem approach' to the protection of such water resources are only now becoming clearer. While the 2005 Millennium Ecosystem Assessment[93] first provided detailed elaboration of the concept of 'ecosystem services', by 2019 the ICJ found that 'damage to the environment, and the consequent impairment or loss of the ability of the environment to provide goods and services, is compensable under international law'.[94] The partial award of the arbitral tribunal in the *Kishenganga Arbitration* identifies a general obligation to safeguard environmental flows for downstream states when constructing and operating major riverine infrastructure.[95]

In addition, detailed guidance on the practical requirements of an ecosystem approach has been prepared under the auspices of both water resources

---

[90] Agreement on Cooperation for the Sustainable Development of the Mekong River Basin (Chiang Mai), adopted 5 April 1995, entered into force 5 April 1995, (1995) 34 ILM 864.

[91] ILC Draft Articles on Transboundary Aquifers, note 9, Art 5(1)(i).

[92] Ibid, Art 10.

[93] Millennium Ecosystem Assessment, note 62.

[94] ICJ, *Certain Activities Carried Out by Nicaragua in the Border Area (Costa Rica v Nicaragua)*, Compensation Judgment, 2 February 2018, [42].

[95] Permanent Court of Arbitration, *Indus Waters Kishenganga Arbitration between Pakistan and India*, Partial Award, 18 February 2013, at: http://www.pca-cpa.org/showpage.asp?pag_id=1392. The arbitral tribunal was constituted pursuant to the 1960 Indus Water Treaty and operated under the auspices of the Permanent Court of Arbitration. See also, S McCaffrey, 'International Water Cooperation in the 21st Century: Recent Developments in the Law of International Watercourses' (2014) 23/1 *Review of European, Comparative and International Environmental Law* 4–14.

conventions, such as the UNECE Water Convention, and biodiversity treaties, such as the CBD. The Ramsar Convention plays a particularly notable role in this regard. Further, the procedural and institutional mechanisms required at both the international and national levels to facilitate effective implementation of a meaningful ecosystem approach to the protection of international watercourses continue to develop and strengthen. Consider, for example, the ICJ's 2010 finding in the *Pulp Mills* case that customary international law now requires transboundary EIA of major projects on international watercourses;[96] or the wealth of national water resources legislation introducing disparate regulatory measures to protect freshwater ecosystems and the services they provide.[97]

4.2 Ecosystem Services

Through the continuing evolution of the ecosystem approach, understanding of the nature and value of socially beneficial services provided by natural ecosystems has advanced considerably. 'Ecosystem services' may be understood as the benefits that people obtain from natural ecosystems, and the concept provides the basis of a methodology for economic and social valuation of natural ecosystems, thereby permitting integration of important non-marketable benefits into decision-making processes. Better understanding and assessment of the value of ecosystem services promote balanced longer-term decision-making, which inevitably favours ecosystems protection. The services provided by wetland ecosystems are essential for human wellbeing and include water purification, water storage, groundwater recharge, flood control, provision of aquatic habitats for biodiversity, nurseries for fisheries, and recreational amenities. The ecosystem services concept can raise awareness of the inherent value of services provided by wetland areas often previously regarded as unproductive and needing to be drained to make way for development. In this way, the concept can assist riparian states to mobilise public support for measures required to meet their obligations regarding ecosystems conservation, as codified under Article 20 of the UN Watercourses Convention and Articles 2 and 3 of the UNECE Water Convention.

---

[96] *Pulp Mills*, note 41.

[97] Including, for example, the designation of a water 'reserve' for ecosystems or the imposition of ecological controls on water trading mechanisms. See S Burchi, 'A Comparative Review of Contemporary Water Resources Legislation: Trends, Developments and an Agenda for Reform' (2012) 37/6 *Water International* 613–27.

Detailed guidance has emerged on the application of the ecosystem services concept in policy-making. The World Resources Institute (WRI) has produced recommendations on methodologies for review of ecosystem services in impact assessment processes,[98] which might prove particularly effective in ensuring the protection of transboundary watercourse ecosystems as EIA is now recognised as a requirement of general international law.[99] The ecosystem services concept can significantly assist transboundary water cooperation where co-riparian states must communicate and agree upon equitable measures for the utilisation and protection of shared watercourses based on a common understanding of their cost and benefit implications for each state. Thus, generally accepted methodologies for the valuation of benefits provided by related natural ecosystems can facilitate effective cooperative engagement between co-basin states. Once again, the Ramsar regime plays a central role – notably through its work on the economics of ecosystems and biodiversity for water and wetlands, highlighting the economic value of aquatic ecosystems to society and providing a structured means for policymakers to take these into account.[100] Other MEA Secretariats have also contributed, with a 2008 CBD report including an Annex I which examines in detail 'ecosystem services by inland waters which can be affected by inappropriate water allocations and unsustainable water use'.[101]

Such ongoing elaboration of the ecosystem services concept in the context of shared water resources further clarifies the standards underlying ecological obligations in international water law. Commentators suggest that 'water and wetland ecosystems are perhaps among the best studied of habitats in terms of ecosystem services', while also noting a 'lack of attention to ecosystem services within the context of transboundary freshwater ecosystems and law'.[102]

---

[98] F Landsberg et al, *Ecosystem Services Review for Impact Assessment: Introduction and Guide to Scoping* (WRI Working Paper, Washington DC, November 2011).

[99] *Pulp Mills*, note 41, [204].

[100] D Russi et al, *The Economics of Ecosystems and Biodiversity (TEEB) for Water and Wetlands* (IEEP, London and Brussels/Ramsar Secretariat, Gland, 2013), at: http://www.teebweb.org/publication/the-economics-of-ecosystems-and-biodiversity-teeb-for-water-and-wetlands/.

[101] S Brels, D Coates, and F Loures, *Transboundary Water Resources Management: The Role of International Watercourse Agreements in Implementation of the CBD*, CBD Technical Series No. 40 (Secretariat of the Convention on Biological Diversity, Montreal, 2008) 37–42.

[102] A Rieu-Clarke and C Spray, 'Ecosystem Services and International Water Law: Towards a More Effective Determination and Implementation of Equity?' (2013) 16/2 *Potchefstroom Electronic Law Journal* 12, 13.

The discourse on ecosystem services often includes discussion of the potential role of PES arrangements, which 'are widely promoted to secure ecosystem services through incentives to the owners of land from which they are derived' and 'to foster conservation and poverty alleviation in the global south'.[103] Key international actors have once again developed guidance on how such payments systems might work.[104]

Rieu-Clarke and Spray conclude that 'at least the basis of an ecosystem approach can be found in the key provisions of international water law' and regard the ecosystem services paradigm as 'a new conceptual framework for the detailed assessment and management of ecosystems in terms of the values and services that flow from ecosystems to humans'.[105] However, they also warn of a range of pitfalls 'when considering the linkages between ecosystem services, law and transboundary freshwaters', and recognise 'the need for further research through case studies at the transboundary basin level'.[106] Therefore, while it is clear that an ecosystem services framework for decision-making can function 'to highlight the multiplicity of the benefits that freshwater ecosystems and their *collective* protection provide',[107] considerable work needs to be done on the development of appropriate legal mechanisms. Clearly, practice under Ramsar may hold important lessons in this regard for international water law.

4.3 Environmental Flows

Technical guidance plays a critical role in the calculation and maintenance of e-flows intended to provide 'a methodological approach that incorporates environmental concerns into the process of allocating water rights among different users'.[108] This is now central to implementation of an ecosystem approach, particularly in the case of transboundary basins where states compete for quantum share of scarce water resources. The 'e-flows' concept is defined as 'the quantity, timing and quality of water flows required to sustain freshwater and estuarine ecosystems and the human livelihoods and well-being that

---

[103] R de Francisco et al, 'Payment for Environmental Services and Unequal Resource Control in Pimampiro, Ecuador' (2013) 26 *Society and Natural Resources* 1217–33.

[104] Including, for example, IUCN, *PAY – Establishing Payments for Watershed Services* (IUCN, Gland, 2006) and UNECE, *Guidance on Water and Adaptation to Climate Change* (UNECE, Geneva, 2009) 101.

[105] Rieu-Clarke and Spray, note 102, 23–24.

[106] Ibid, 29.

[107] Ibid, 31.

[108] Brels, Coates, and Loures, note 101, 13.

depend on these ecosystems',[109] where '[t]he overriding objective of e-flows is to modify the magnitude and timing of flow releases from water infrastructure (e.g. dams) to restore natural or *normative* flow regimes that benefit downstream river reaches and their riparian ecosystems'.[110] The Ramsar regime – including the influential Scientific and Technical Review Panel (STRP) – is a key institutional actor to recognise the importance of e-flows and engage in preparing technical studies and reports on different aspects of the concept, including the ecological vulnerability of wetlands and reviews of e-flow methodologies.[111] Typifying the collaborative aspect of normative convergence, in December 2012 the Ramsar and CBD Secretariats jointly published an influential technical report on methodologies applicable to calculating the e-flows required for the maintenance of healthy estuarine ecosystems.[112]

The e-flows concept has received significant judicial endorsement. In the *Kishenganga Arbitration*, the 2013 partial award established that 'hydro-electric projects ... must be planned, built and operated with environmental sustainability [and minimum environmental flow in particular] in mind'.[113] The tribunal found that '[i]t is established that principles of international environmental law must be taken into account even when ... interpreting treaties concluded before the development of that body of law',[114] and based an obligation to maintain minimum e-flows on 'the "principle of general international law" that States have "a duty to prevent, or at least mitigate" significant harm to the environment when pursuing large-scale construction activities'.[115] The tribunal further suggested that obligations regarding e-flows could arise under the 'requirement under general international law to undertake an environmental impact assessment where there is a risk that the proposed industrial activity may have a significant adverse impact in a transboundary context, in particular, on a shared resource', as recognised in the *Pulp Mills* case.[116] In its

---

[109] 2007 Brisbane Declaration, at: https://riverfoundation.org.au/wp-content/uploads/2017/02/THE-BRISBANE-DECLARATION.pdf.

[110] N LeRoy Poff and JH Matthews, 'Environmental Flows in the Anthropocene: Past Progress and Future Prospects' (2013) 5 *Current Opinion in Environmental Sustainability* 667, 667 (emphasis added).

[111] See Brels, Coates, and Loures, note 101, 13.

[112] J Adams, *Determination and Implementation of Environmental Water Requirements for Estuaries*, Ramsar Technical Report No 9/CBD Technical Series No 69 (2012).

[113] *Indus Waters Kishenganga Arbitration between Pakistan and India*, note 95, [454].

[114] Ibid, [452].

[115] Ibid, [451].

[116] Ibid, [450]. See *Pulp Mills*, note 41, [204].

final award, the arbitral tribunal determined the precise minimum e-flow to be maintained downstream of the dam.[117] The ICJ has subsequently confirmed that impairment of minimum flow may amount to significant transboundary harm.[118] In assessing state and treaty practice, commentators note that '[t]he need to provide environmental flows in order to conserve ecological integrity of water basins is becoming more and more important'.[119]

4.4 General Technical Guidance

Ecosystems protection and the ecosystem approach have enjoyed considerable international attention in recent years, with a broad range of actors working to develop technical guidance for implementation of this complex paradigm.[120] International water Convention Secretariats have been at the forefront of developing guidance to promote common understanding and uniform implementation of ecosystems obligations set out therein. In 1993, the Secretariat to the UNECE Water Convention issued ground-breaking Guidelines on the Ecosystem Approach in Water Management[121] and subsequently produced more detailed guidance on specific aspects thereof, including the 2007 Recommendations on Payments for Ecosystem Services in Integrated Water Resources Management.[122] Similarly, recognising the key significance of international water resources for achieving their own ecological objectives, biodiversity Conventions have issued guidance informing the ecosystems protection obligations set out under Articles 2 and 3 of the UNECE Water Convention and Article 20 of the UN Watercourses Convention. For example, a 2008 study conducted by the CBD Secretariat recognised that '[t]he equitable and sustainable allocation and management of water are crucial for maintaining the ecological function of freshwater ecosystems', before exploring in

---

[117] *Indus Waters Kishenganga Arbitration between Pakistan and India*, Final Award, 20 December 2013. See BH Desai and BK Sidhu, 'Making sense of the Kishenganga final award', *The Tribune,* Chandigarh, 9 January 2014, 15.

[118] ICJ, *Certain Activities Carried Out by Nicaragua in the Border Area (Costa Rica v Nicaragua)*, Judgment of 16 December 2015, [119].

[119] Aguilar and Iza, note 12, 99.

[120] In the marine context, for example, see R Long, 'Legal Aspects of Ecosystem-Based Marine Management in Europe' (2012) 26 *Ocean Yearbook* 417–84.

[121] Source: http://www.unece.org/fileadmin/DAM/env/water/publications/documents/Part%20One_WaterSeries1.pdf. These Guidelines contained detailed guidance on 'Legal, Planning and Institutional Measures' and on 'Regulatory Framework', 5–7.

[122] UNECE, note 34.

detail the synergies between the CBD and the UN Watercourses Convention and UNECE Water Convention.[123] Recognition of the overlap between these instruments aids development of a coherent understanding of the requirements of an ecosystem approach to the management of international water resources, considering the universal nature of the CBD[124] and the fact that the ecosystem approach has principally emerged under the auspices of this instrument.[125]

5 THE 'ECOSYSTEM APPROACH' UNDER RAMSAR

The ecosystem approach is a central and defining feature of the Ramsar regime. The Convention was primarily concerned, *ab initio*, with the protection of wetland ecosystems, initially as waterfowl habitat, as noted in the 2005 Conceptual Framework for the Wise Use of Wetlands and the Maintenance of their Ecological Character.[126] Its scope of concern has extended to providing more generally 'the framework for national action and international cooperation for the conservation and wise use of wetlands and their resources'.[127] Each of the developments outlined above as facilitating an ecosystem approach in respect of shared international freshwater resources is addressed in technical guidance issued under Ramsar. Updated 2008 guidance on EIA and strategic environmental assessment provides a clear example.[128]

---

[123] Brels, Coates, and Loures, note 101, 5.

[124] With 193 parties, one commentator suggests that universally adopted conventions such as the CBD have attained what 'can be called quasi-constitutional status ... in the metaphysical sense of a deeply influential norm-creating and value-ordering document'. See MA Drumble, 'Actors and Law-making in International Environmental Law' in M Fitzmaurice, DM Ong, and P Merkouris (eds), *Research Handbook in International Environmental Law* (Edward Elgar Publishing, Cheltenham, 2010) 16.

[125] The conservation of ecosystems is one of the CBD's three main objectives set out under Article 1; while Article 8(f) obliges state parties to 'rehabilitate and restore degraded ecosystems'. See CBD, *Decision V/6: Ecosystem Approach*, UN Doc UNEP/CBD/COP/5/23 (22 June 2000); CBD, *Decision VII/11: Ecosystem approach*, UN Doc UNEP/CBD/COP/7/21 (13 April 2004).

[126] Ramsar Convention, *Resolution IX.1*, note 59, Annex A.

[127] See Bridgewater and Kim, '50 years on,' note 25, 3921, citing the landing page of the Convention's website. See further CM Finlayson et al, 'The Ramsar Convention and ecosystem-based approaches to the wise use and sustainable development of wetlands' (2011) 14 *Journal of International Wildlife Law and Policy* 176–98.

[128] Ramsar Convention, *Resolution X.15*, note 66; Ramsar Convention, *Resolution X.16: A Framework for processes of detecting, reporting and respond-*

Most significantly, the Ramsar regime has developed extensive technical guidance elaborating key obligations inherent to the ecosystem approach, published as a series of *Handbooks for the Wise Use of Wetlands*.[129] This guidance sets out best practice on the management of wetlands at the national level and notably includes *Handbook 3 (Laws and Institutions)* and *Handbook 8 (Water-related Guidance)*. Continuing elaboration of practical, sector-specific technical guidance under the auspices of such a widely ratified Convention functions to clarify the due diligence requirements of the ecosystem approach in this field. *Handbook 20 (International Cooperation)* links Ramsar guidance, which focuses on ecosystems protection at the national level, with the management of international water resources. The role of Ramsar guidance illustrates how the technical complexity of issues underlying international environmental law often dictates that scientific expertise plays a central role in its implementation, requiring intense interaction between environmental scientists and environmental lawyers at every stage in its development and application.[130] In terms of the process of international law-making, the Ramsar STRP is one of those vital institutions which constitute the 'focal points of a broad, legally significant communication process'.[131] Such institutions play a critical role in promoting convergence, at least as regards primary international rules, and thereby counter the risk of 'normative incompatibility between various sectors of international law' due to sectoral fragmentation.[132]

Ramsar has played a leading role regarding ecosystem services in particular, with the Convention preamble describing wetlands as 'a resource of great economic, cultural, scientific, and recreational value', and institutional communications under Ramsar regularly highlighting the role of wetland ecosystems in the provision of ecosystem services. Recognition of the value of ecosystem services is implicit in the overarching objective of 'wise use' and the discourse under Ramsar has long emphasised the practical human benefits of pursuing sustainable exploitation. The 2009–15 Strategic Plan advised that the wise use of wetlands requires that conservation decisions be made with an awareness of the ecosystem services provided by wetlands;[133] while the 2009 Strategic Framework and Guidelines listed as their objective the development and

---

ing to change in wetland ecological character (2008); Ramsar Convention, *Resolution X.17: Environmental Impact Assessment and Strategic Environmental Assessment: updated scientific and technical guidance* (2008).

[129] *Ramsar Convention Handbooks for the Wise Use of Wetlands*, note 76.
[130] See McIntyre (2016), note 78.
[131] Gehring, note 77, 44.
[132] Craven, note 3, 3.
[133] Ramsar Convention, *Resolution X.1: The Ramsar Strategic Plan 2009–2015* (2008) 6.

maintenance of an international network of wetlands 'important for the conservation of global biological diversity and for sustaining human life through the maintenance of their ecosystem components, processes and benefits/services'.[134] The 2012 Report of the Secretary-General on the implementation of the Convention at the global level highlights the increasingly important role of wetlands in providing a range of ecosystem services relating, among other things, to climate change adaptation and mitigation, water and food security, water quality, and sustainable tourism.[135] This focus is maintained in the 2016–24 Strategic Plan, which calls, in respect of Strategic Goal 1: *Addressing the Drivers of Wetland Loss and Degradation*, for parties to ensure that '[w]ater use respects wetland ecosystem needs for them to fulfil their functions and provide services at the appropriate scale'.[136] Similarly, Target 11, listed under Strategic Goal 3: *Wisely Using All Wetlands*, commits the parties to ensuring that '[w]etland functions, services and benefits are widely demonstrated, documented and disseminated'. It appears, therefore, that there exists greater awareness under the Ramsar regime of the value and relevance of services provided by natural ecosystems than is currently recognised – at least formally – in the practice of international water law. The practice developed under Ramsar can inform emerging practice regarding the ecosystem approach in international water law and thereby assist mutually supportive implementation of both the international wetlands and water resources regimes.

6   REGIME INTERACTION AND COMPLEMENTARITY

The two regimes complement and support each other, with Ramsar promoting greater inter-state cooperation over shared water resources and transboundary water cooperation facilitating implementation of Ramsar. Highly developed practice under Ramsar intended to meet obligations regarding wetland ecosystems can facilitate the obligation to protect international watercourse ecosystems that has emerged under conventional and customary international water law. In turn, river basin organisations and other cooperative institutional arrangements established under watercourse agreements can ensure that e-flows are maintained for the purposes of conserving wetland sites protected under Ramsar. Cooperative ecosystem protection measures may be adopted

---

[134] Ibid, 2, [6].

[135] Ramsar Convention, *COP11 DOC 7: Report of the Secretary General on the implementation of the Convention at the global level* (2012) [5].

[136] Ramsar Convention, *Resolution XII.2: The Ramsar Strategic Plan 2016–2024* (2015).

under either regime, having regard to the priority concerns of the states involved or the relative sensitivity of either field of international cooperation. Such arrangements might also arise independently, with either regime providing scope for alternate means of pursuing ecosystems conservation objectives.

Though China and Russia did not conclude their bilateral Treaty on Good Neighbourliness until 2001[137] and the related Agreement on the Rational Utilisation and Protection of Transboundary Waters until 2008,[138] they had concluded a 1996 Agreement regarding two Ramsar listed nature reserves on either side of the Chinese-Russian boundary[139] seeking to give effect to Article 5 of Ramsar, requiring that the states concerned consult 'in the case of wetlands extending over the territories of more than one Contracting State'.[140] Lake Khanka/Xingkai is the source of the Songcha River, part of the transboundary Amur drainage basin, and the Agreement includes among its objectives the protection of ecosystems within the conservation area and the facilitation of bilateral cooperation in the rational utilisation of natural resources, for which a Joint Commission has been established.[141] Thus, as parties to the Ramsar Convention, China and Russia could cooperate on conservation of shared watercourse wetland ecosystems under the auspices of Ramsar before they could do so by means of international water law; while China was one of only three states that voted against the UN General Assembly Resolution adopting the UN Watercourses Convention, reflecting its view that the Convention did not adequately reflect the territorial sovereignty of watercourse states.[142]

---

[137] Treaty of Good Neighbourliness, Friendship and Cooperation between the Russian Federation and the People's Republic of China, adopted 16 July 2001, Article 19 of which provides: 'the Contracting Parties shall cooperate in the protection and improvement of the environment, prevention of transboundary pollution, equitable and reasonable utilisation of the boundary watercourses and the living resources in the Northern Pacific and the basins of the boundary rivers; undertake joint efforts in protecting rare species of flora, fauna and the natural ecosystems in the border areas ...'

[138] Agreement between the Government of the Russian Federation and the Government of the People's Republic of China on the Rational Utilisation and Protection of Transboundary Waters, adopted 29 January 2008.

[139] Agreement on the Natural Reserve 'Lake Khanka/Xinkai', adopted 29 January 2008.

[140] See S Vinogradov, 'Can the Dragon and Bear Drink from the Same Well? Examining Sino-Russian Cooperation on Transboundary Rivers through a Legal Lens' [Part I] (2013) 23/3 *Journal of Water Law* 95, 103.

[141] Agreement, note 139, Art 2.

[142] UNGA Resolution 51/229, UN Doc A/RES/51/229 (1997). See, Vinogradov, note 140, 101. See DJ Devlaeminck and X Huang, 'China and the global water

Sino-Russian practice under Ramsar also demonstrates that the Convention can assist with effective implementation of watercourse-related ecological requirements once these have been established under international water law agreements. Pursuant to the 2001 Treaty and 2008 Water Agreement outlined above, in 2008 Russia proposed the Amur Regional Initiative under the auspices of Ramsar in order to promote international cooperation in conservation and sustainable use of the Amur basin ecosystems[143] in line with the requirements of Article 5 of the Ramsar Convention and related guidance on the wise use of wetlands,[144] and with the ecosystems protection requirements arising under international water law. It is noteworthy that the proposed Initiative involves Russia, China, and Mongolia, and could possibly include Korea and Japan as interested parties, when one considers 'China's aversion to a multilateral approach to managing international watercourses where more than just two states are concerned, and its clear preference for bilateral interactions and arrangements'.[145] Thus, in certain situations, commitments under Ramsar may encourage broader inter-state cooperative engagement than corresponding requirements of international water law alone. Similarly, specific international

---

conventions in light of recent developments: Time to take a second look?' (2020) 29/3 *Review of European, Comparative and International Environmental Law* 395–405.

[143] Vinogradov, note 140, 103. See further, T Minaeva, 'Cooperation prospects of the Russian Federation, China and Mongolia under the Ramsar Convention' in *Status and Prospects of the Russian-Chinese Cooperation in Environment Conservation and Water Management* (Materials of the International Conference, Moscow, 27–28 September 2007).

[144] For example, *Ramsar Handbook 20*, on *International Cooperation: Guidelines and other support for international cooperation under the Ramsar Convention on Wetlands*, advises that: 'In the area of shared river basins Contracting Parties should, where appropriate, seek to harmonise their implementation of Article 5 of the Ramsar Convention with obligations arising from any watercourse agreements to which they may also be signatories.'

[145] Vinogradov, note 140, 100, points out that 'China has about 50 agreements regarding its shared waters, all of which are bilateral, despite the fact that many of them relate to multi-state basins'. See further, P Wouters and H Chen, 'China's "soft-path" to transboundary water cooperation examined in the lights of two global UN water conventions – exploring the "Chinese way"' (2013) 22 *Journal of Water Law* 229, 232. It should be pointed out, however, that China has been a 'dialogue partner' in the Mekong River Commission (MRC) since the mid-1990s and shares water data and information with the MRC. See Y Feng et al, 'Water Cooperation Priorities in the Lancang-Mekong River Basin Based on Cooperative Events Since the Mekong River Commission Establishment' (2019) 29 *Chinese Geographical Science* 58–69.

water agreements can facilitate implementation of requirements arising under Ramsar.

Indicating the close inter-linkages between these two bodies of law, the ICJ attached great significance in relation to environmental damage of a shared watercourse to the fact that territory disputed between the parties contained a wetland site designated under Ramsar.[146] Judge Sepulveda-Amor's separate opinion stressed the risks posed by Nicaragua's activities of habitat loss and changes to groundwater beneath the wetland, as had been pointed out in a report submitted to the ICJ by the Ramsar Secretariat.[147] While requiring both parties to refrain from sending military or civilian personnel to the disputed area, the ICJ identified a necessary exception whereby Costa Rica could dispatch civilian personnel for the purposes of environmental protection of the wetland, 'in respect of which Costa Rica bears obligations under the Ramsar Convention', provided that 'Costa Rica shall consult with the Secretariat of the Ramsar Convention in regard to these actions'.[148] In another separate opinion, Judge Greenwood emphasised that 'the disputed area falls within a wetland notified by Costa Rica under the Ramsar Convention', and that Costa Rica had 'assumed responsibilities under the terms of the Convention for the protection of the environment in the disputed area'.[149] Judge Greenwood appears to have regarded environmental management of international water resources and related ecosystems as dependent upon meaningful inter-state cooperation,[150] as is explicit under the Ramsar Convention[151] and the key multilateral instruments of international water law.[152]

---

[146] ICJ, *Certain Activities Carried Out by Nicaragua in the Border Area (Costa Rica v Nicaragua)*, Request for Provisional Measures, Order of 8 March 2011.

[147] Ibid, Separate Opinion of Judge Sepulveda-Amor, [25]. See J Harrison, 'Significant International Environmental Cases: 2011–12' (2012) 24/3 *Journal of Environmental Law* 559, 560.

[148] *Costa Rica v Nicaragua*, note 146, [80].

[149] Ibid, Separate Opinion of Judge Greenwood, [14] and [15].

[150] Ibid, [15]. See also, Separate Opinion of Judge Sepulveda-Amor, note 147, [34]. See further, Harrison, note 147, 561.

[151] Convention on Wetlands, note 2, Art 5.

[152] UNECE Water Convention, note 6, Arts 2(2)(d) and 2(6); UN Watercourses Convention, note 5, Art 8; and ILC Draft Articles on Transboundary Aquifers, note 9, Art 7. For example, the UNECE Guide to Implementing the Water Convention, note 29, 26 [113], advises that Article 2(2)(d), requiring States to 'take all appropriate measures … [t]o ensure conservation and, where necessary, restoration of ecosystems … implies, for example, *close cooperation among those who establish these measures*, including consultations with local populations' (emphasis added).

Disputes concerning utilisation and environmental protection of transboundary watercourses have been raised within the Ramsar institutional framework on numerous occasions. At the tenth COP, Iraq alleged that dams constructed upstream in Turkey and Iran operated to restrict water flows into the Hawizeh Marsh.[153] Notably, Ukraine's project to construct the deep-water Bystroe navigation canal through the Danube Delta[154] resulted in successive Ramsar COP resolutions calling for a range of measures to ensure effective inter-state cooperation in mitigating project impacts.[155] Such steps largely reflected normative requirements also arising under general international water law, including the generally applicable customary requirement for EIA incorporating assessment of transboundary impacts of any major project potentially affecting a transboundary watercourse.[156] The ICJ found that, while EIA is an autonomous requirement under customary international law, it is also necessary to ensure effective notification of the project to potentially affected states and to meet certain due diligence requirements under the general duty to prevent significant transboundary harm.[157]

Detailed technical guidance and normative practice developed under Ramsar can directly inform interpretation and practical application of the general ecosystems obligation set out under Article 20 of the UN Watercourses Convention and corresponding provisions of international basin agreements. Under Article 31(3)(c) of the Vienna Convention on the Law of Treaties (VCLT),[158] in interpreting a treaty provision, '[t]here shall be taken into account, together with the context ... any relevant rules of international law applicable in the relations between the parties'. This provision, contained in one of the key constitutional instruments of public international law, provides the basis for the principle of

---

[153] Ramsar Convention, *COP10 DOC 7: Report of the Secretary General pursuant to Article 8.2 concerning the List of Wetlands of International Importance* 2008), [20]. For the response of Turkey and Iran, see Ramsar Convention, COP10, *Report of the Meeting* (2008) [91]–[92]. See generally, Bowman, Davies, and Redgwell, note 56, 424–25.

[154] See M Koyano, 'The significance of the Convention on Environmental Impact Assessment in a Transboundary Context (Espoo Convention) in international environmental law: examining the implications of the Danube Delta case' (2008) 26/4 *Impact Assessment and Project Appraisal* 299–314.

[155] Ramsar Convention, *Resolution IX.15: The status of sites in the Ramsar List of Wetlands of International Importance* (2005), [27(iv)]; Ramsar Convention, *Resolution 10.13: The status of sites in the Ramsar List of Wetlands of International Importance* (2008) [27(II)].

[156] *Pulp Mills*, note 41, [204].

[157] Ibid, [121] and [204].

[158] 1969 Vienna Convention on the Law of Treaties, 1155 UNTS 331 [VCLT].

'systemic integration',[159] which can address negative effects of 'fragmentation' of international law[160] by functioning 'to limit the potential risk of norms, regimes or tribunals entering into conflict with each other'.[161]

With over 400 basin agreements in force – which can vary considerably in terms of their normative treatment of key issues, including ecosystems conservation – as a body of rules, international water law faces significant risk of such fragmentation.[162] In this regard, practice under Ramsar is noted for the continuing development and effective application of the ecosystems obligations inherent to international water law.[163] Focusing particularly on the relevance of Articles 2.1, 3, and 4.1 of Ramsar for the interpretation of Article 20 of the UN Watercourses Convention, Lee determines that these satisfy the three essential components of an interactional framework of analysis – 'namely shared understanding; criteria of legality; and practice of legality' – in order to be included among the 'rules of international law applicable in

---

[159] See C McLachlan, 'The Principle of Systemic Integration and Article 31(3)(c) of the Vienna Convention' (2005) 54 *International and Comparative Law Quarterly* 279–320; J Kammerhofer, 'Systemic Integration, Legal Theory and the ILC' (2008) 19 *Finnish Yearbook of International Law* 157–82; P Sands, 'Treaty, Custom and the Cross-Fertilisation of International Law' (1998) 1 *Yale Human Rights & Development Law Journal* 85–106. For a judicial application of the principle of systemic integration, see *Oil Platforms (Islamic Republic of Iran v United States of America), Merits, Judgment of 6 November 2003*, ICJ Reports 2003, 161, [40–41].

[160] For a full discussion of the problem of fragmentation of international law, see the seminal Final Report of the ILC; Koskenniemi, note 11. See also, G Hafner, 'Pros and Cons Ensuing from Fragmentation of International Law' (2003–4) 25 *Michigan Journal of International Law* 849–63.

[161] Kammerhofer, note 159, 280. See further, D French, 'Treaty Interpretation and the Incorporation of Extraneous Rules' (2006) 55/2 *International and Comparative Law Quarterly* 281–314. While there is no universally accepted definition of 'fragmentation' in international law, it is normally understood to consist of the segregation of international law into highly specialised fields, governed by their own autonomous rules and principles, resulting in disunity and incoherence in the development and application of international law. See R Deplano, 'Fragmentation and Constitutionalisation of International Law' (2013) 6/1 *European Journal of Legal Studies* 67, 69.

[162] See, for example, NA Zawahri and S Mitchell, 'Fragmented Governance of International Rivers: Negotiating Bilateral versus Multilateral Treaties' (2011) 55/3 *International Studies Quarterly* 835–58.

[163] J Lee, *Preservation of Ecosystems of International Watercourses and the Integration of Relevant Rules: An Interpretive Mechanism to Address the Fragmentation of International Law* (Brill, Leiden, 2014).

the relations between the parties' pursuant to Article 31(3)(c) of the VCLT.[164] Focusing on the 2002 Ramsar Guidelines[165] and the 2012 Strategic Framework and Guidelines,[166] Lee highlights that 'the scope of "conservation" is informed by the recognition of the ecosystem and the ecosystem approach',[167] and that 'conservation decisions are made with an awareness of the importance of the ecosystem services provided by wetlands'.[168] The interpretive implications for international watercourse ecosystem obligations are obvious. Lee's analysis further highlights that the Ramsar Strategic Framework and Guidelines 'elaborate the processes and procedural mechanisms required for the conservation of wetlands',[169] requiring 'a continuous, long-term process that is adaptable and dynamic, where a management plan will grow as information becomes available', and which 'necessitates a review of all existing Ramsar sites to determine the effectiveness of management arrangements'.[170] Generally, in addition to promoting the normative development of the Ramsar regime beyond an abstract objective of ecosystems conservation, the intense institutional engagement required of the contracting parties under Ramsar in order to elaborate subsidiary rules and to identify and reference complementary rules constitutes an effective process of inter-state communication which produces shared understanding and enhances the legitimacy of the entire regime.[171]

In identifying the 'parties' in respect of which Ramsar practice might inform interpretation of relevant rules of international water law, Article 31(3)(c) of

---

[164] Ibid, 214–58.

[165] Ramsar Convention, *Resolution VIII.14: New Guidelines for Management Planning for Ramsar Sites and Other Wetlands* (2002).

[166] Ramsar Convention, *Resolution XI.8, Annex 2: Strategic Framework and Guidelines for the Future Development of the List of Wetlands of International Importance* (2012).

[167] Lee, note 163, 220.

[168] Ibid, 229.

[169] Ibid, 223. On the importance of the elaboration of procedural rules of international water law, see O McIntyre, 'The Contribution of Procedural Rules to the Environmental Protection of Transboundary Rivers' in L Boisson de Chazournes, C Leb, and M Tignino (eds), *Freshwater and International Law: The Multiple Challenges* (Edward Elgar Publishing, Cheltenham, 2013) 239–65; O McIntyre, 'The Proceduralization and Growing Maturity of International Water Law' (2010) 22/3 *Journal of Environmental Law* 475–97.

[170] Lee, note 163, 230.

[171] J Brunée and SJ Toope, *Legitimacy and Legality in International Law: An Interactional Account* (Cambridge University Press, Cambridge, 2010) 179. See Lee, note 163, 220 and 237–38.

the VCLT does not require that the parties to the interpreted treaty be party to the influencing external treaty, as long as the rule:

> can be said to be implicitly accepted or tolerated by all parties to the treaty under interpretation in the sense that it can reasonably be considered to express the common intentions or understanding of all members as to the meaning of the term concerned.[172]

Therefore, the interpretive guidance provided by the wealth of Ramsar normative practice might inform interpretation of the international watercourse ecosystems obligations not only of the 172 contracting parties to the Ramsar Convention, but also of non-party states which have tacitly accepted Ramsar standards – for example, by cooperating with Ramsar states in the management of wetlands in a manner consistent with the requirements of the Ramsar regime.

7 CONCLUSION

In a range of ways – including operation of the principle of systemic integration – Ramsar can significantly influence the continuing evolution of ecosystems obligations under international water law and thereby address the risk of fragmentation of such obligations due to the relative proliferation of global, regional, basin-level, and bilateral water instruments. It seems likely that practice established under Ramsar will promote convergence around coherent standards and procedures for ecosystems protection under these two related regimes. Likewise, rules of international water law must be implemented to guarantee allocation of sufficient water to ensure that wetlands function sustainably. While employment of e-flow methodologies in transboundary watercourse regimes can help to identify the quantity, timing, and nature of the flow required, the ecosystem services concept can provide a framework for valuing the human benefits derived from wetlands and for equitably sharing the benefits and burdens involved in the conservation of wetland ecosystems providing such services.

This intense interrelationship typifies the phenomenon of 'convergence' in international law, which can occur by a variety of means. For example, rules developed under either regime might be taken into account, pursuant to Article 31(3)(c) of the VCLT, in the interpretation of key conventional provisions of the other regime. Similarly, the Ramsar and CBD COPs and Secretariats provide the 'machinery for institutional dialogue', which plays an increasingly important role in ensuring the 'interpenetration' and 'cross-fertilization' indic-

---

[172] McLachlan, note 159, 314–15; Lee, note 163, 255–56.

ative of the integrative process of 'convergence', which counteracts international law's fragmentation.[173] For example, they continue jointly to elaborate detailed guidance on the integration into international water law of ecosystems obligations arising under each of these flagship global environmental conventions.[174] Such actions further enhance the normative coherence of both bodies of law.

---

[173] Cassese, note 4.

[174] See, for example, Ramsar Convention, *Resolution IX.3: Engagement of the Ramsar Convention on Wetlands in Ongoing Multilateral Processes Dealing with Water* (2005); Brels, Coates, and Loures, note 101; Adams, note 112; Russi et al, note 100.

10. The cultural values and services of wetlands: Evolution and obligations under the Ramsar Convention
Evan Hamman[1]

1 INTRODUCTION

Wetlands are places of tremendous social, cultural, and ecological significance. They produce numerous and varied benefits to humans, including flood control, sanitation, water supply, storm protection, recreation, and tourism.[2] Nevertheless, wetlands have often been undervalued in planning and land use decision-making.[3] In many parts of the world, the drainage or conversion of wetlands for human exploitation – particularly agriculture – has been actively encouraged. In the past, such actions were considered 'civilised';[4] yet paradoxically, they have had a negative impact on society and the economy.[5]

---

[1] The author is indebted to Professor Lucas Lixinski for earlier comments on a draft of this Chapter.
[2] Millennium Ecosystem Assessment, *Ecosystems and Human Well-Being: Wetlands and Water Synthesis* (World Resources Institute, 2005); N Davidson et al, 'Worth of wetlands: revised global monetary values of coastal and inland wetland ecosystem services' (2019) 70(8) *Marine and Freshwater Research* 1189.
[3] I-M Gren et al, 'Primary and secondary values of wetland ecosystems' (1994) 4 *Environmental and Resource Economics* 55.
[4] GVT Matthews, *The Ramsar Convention on Wetlands: Its History and Development* (Ramsar Convention Secretariat, 2013) 2. On the negative perceptions of wetlands in the United States, see RC Gardner, *Lawyers, Swamps, and Money: U.S. Wetland Law, Policy, and Politics* (Island Press, Washington DC, 2011), Chapter 1.
[5] See, eg, KD Schuyt, 'Economic consequences of wetland degradation for local populations in Africa' (2005) 53(2) *Ecological Economics* 177.

Over the last five decades, wetlands have been increasingly recognised for their cultural values.[6] Many wetlands exist on customary or traditional lands, and important artefacts and spiritual places can be found within and surrounding these environments. There is also a growing body of evidence that the use of traditional ecological knowledge and customary practices can promote the wise use of wetlands.[7] The relationship between humans and wetlands is thus very much a two-way street.[8] As Papayannis and Pritchard put it:

> Wetlands and culture coexist. Wetland-related cultures and their diversity can support sustainable livelihoods and the well-being of human societies ... Loss of wetland-related culture is a threatening sign of wetland loss, and loss of wetlands often results in unsustainable livelihoods.[9]

Despite the increasing recognition of this under the Ramsar Convention,[10] recent studies have critiqued the development of a top-down and technocratic approach to implementing the Convention.[11] The path to increased recognition

---

[6] For an early analysis of the concept, see RJ Reimold, JH Phillips, and MA Hardisky, 'Socio-Cultural Values of Wetlands' in V Kennedy (ed), *Estuarine Perspectives* (Academic Press, New York, United States, 1980).

[7] G Oviedo and M Kenza Ali, *Indigenous peoples, local communities and wetland conservation* (Ramsar Convention Secretariat, 2018).

[8] See, eg, Ramsar Convention, *Resolution IX.21: Taking into account the cultural value of wetlands* (2005), which recognised that 'cultural actions may be determined by ecological processes and vice versa'. See also Ramsar Convention, *Resolution XI.8, Annex 2 (Rev. COP13): Strategic Framework and guidelines for the future development of the List of Wetlands of International Importance* (2018) Annex 2, 67, which noted that 'wetlands exist within landscapes in which people's activities are influenced by the wetlands and the delivery of their ecosystem services, and in which the wetlands themselves are influenced by the use of such services by dependent local communities (e.g., by forms of traditional management). There are many examples where the ecosystem structure and functioning of the wetland have developed as a result of cultural features or legacies. There are also many examples where the maintenance of the ecosystem structure and functioning of wetlands depends upon the interaction between human activities and the wetland's biological, chemical, and physical components'.

[9] T Papayannis and D Pritchard, *Culture and wetlands – a Ramsar guidance document* (Ramsar Convention, 2008) 20.

[10] Convention on Wetlands of International Importance especially as Waterfowl Habitat, adopted 2 February 1971, entered into force 21 December 1975, 996 UNTS 245 (amended 1982 & 1987).

[11] D Joshi et al, 'Ramsar Convention and the wise use of wetlands: rethinking inclusion' (2021) 39(1–2) *Ecological Restoration* 36. On the under-involvement of Indigenous Peoples and local communities in the Convention, see P Bridgewater

of culture has by no means been straightforward, with several parties objecting to the draft text of culture-related resolutions on the basis of either incongruence with trade principles or inconsistency across multilateral environmental agreements (MEAs).[12]

This Chapter charts the evolution of culture within Ramsar, with a particular focus on commitments to identify and protect culture.[13] It examines the influence that certain discourses, resolutions, international agreements, and declarations have played – and continue to play – in establishing a link between wetlands and culture. Special attention is given to the phrases 'cultural value' and 'cultural services', tracing them to the conservation literature[14] and to the key commitment to maintain the ecological character of Ramsar Sites. Towards the end of the Chapter, an argument is developed that Ramsar will be pulled closer towards debates which currently permeate cultural heritage discourses – in particular, questions surrounding who owns, values, and makes decisions about culture. Another possibility is that Ramsar will retreat to its original objective – that of the conservation of wetlands as habitat for migratory waterfowl,[15] and possibly other wetland-dependent species.

Either way, any continued focus on the cultural aspects of wetlands ought to occur within a legitimate and defendable framing. For this to happen, the politics and power relations embedded in wetland-culture linkages need to be better understood and, where possible, confronted.[16] Navigating the question

---

and RE Kim, '50 years on, w(h)ither the Ramsar convention? A case of institutional drift' (2021) 30(13) *Biodiversity and Conservation* 3919.

[12] On the challenges at the intersection between cultural values and international trade, see V Cable, 'The New Trade Agenda: Universal Rules Amid Cultural Diversity' (1996) 72(2) *International Affairs* 227; T Voon, 'UNESCO and the WTO: A Clash of Cultures?' (2006) 55(3) *International and Comparative Law Quarterly* 635.

[13] For an earlier evolution of culture-nature linkages under Ramsar, less related to legal obligations, see D Pritchard, 'Culture and Nature: The case of the Ramsar Convention' in B Verschuuren and S Brown (eds), *Cultural and Spiritual Significance of Nature in Protected Areas* (Routledge, Abingdon, 2019).

[14] On the link between cultural values and biodiversity, see in DA Posey (ed), *Cultural and Spiritual Values of Biodiversity* (UNEP and Intermediate Technology Publications, London, 1999). For the link with protected areas and the related question of 'significance', see B Verschuuren et al, *Cultural and spiritual significance of nature. Guidance for protected and conserved area governance and management* (IUCN, Gland, 2021).

[15] Bridgewater and Kim, note 11. See also Chapter 7 in this volume.

[16] Joshi et al, note 11.

of 'value' will be no easy task. As Boer and Gruber point out, in the context of cultural heritage discourses:

> Decisions about what to conserve and legally protect as heritage are often bound up with issues of cultural identity, which are inherently political and the subject of continuously changing discourses about what is of value at a particular time, and what can be ignored or discarded.[17]

Indeed, scholars in the cultural heritage literature have noted the emergence of an 'authorized heritage discourse',[18] whereby certain 'experts' define and control what 'heritage' is and how it should be protected – often to the exclusion of local communities.[19] It may be that Ramsar will not suffer the same fate, with only a comparatively tangential focus on culture, 'pulling it in' – as it seems – with the overarching goal of more effective 'site management'.[20] On the other hand, there are clearly instances where Ramsar has 'pushed towards culture', recognising wetlands as sites of exceptional cultural value or having heritage features necessitating valuation and protection in their own right.[21] Ultimately, any legal obligations to identify and protect 'culture' still reside

---

[17] B Boer and S Gruber, 'Heritage Discourses' in K Rubenstein and B Jessup (eds), *Environmental Discourses in International and Public Law* (Cambridge University Press, Cambridge, 2012) 10.

[18] L Smith, *Uses of Heritage* (Routledge, Abingdon, 2006).

[19] L Lixinski, *International Heritage Law for Communities: Exclusion and Re-imagination* (Oxford University Press, Oxford, 2019). See also Chapter 6 in this volume.

[20] See Ramsar Convention, *Resolution VIII.19: Guiding Principles for taking into account the cultural values of wetlands for the effective management of sites* (2002). See in particular, Guiding Principle 21 – To incorporate the cultural aspects of wetlands in management planning. See also D Pritchard (with M Ali and T Papayannis), *Guidance: Rapid Cultural Inventories for Wetlands* (Ramsar Culture Network, 2016) 19. See also Question 11.4 of Ramsar Convention, *National Report Form (COP14 format)*, at: https://www.ramsar.org/document/national-report-form-for-cop14-offline-version, which asks parties specifically whether cultural values of wetlands have been included in the management planning for Ramsar Sites and other wetlands.

[21] See the way some of the 'characteristics' are worded in Ramsar Convention, *Resolution IX.21*, note 8, 2, although they all do link to the concept of ecological character in some way. See also the explanation of culture-nature linkages under Ramsar in Pritchard, note 13.

within the broader commitment of parties to maintain ecological character through the prism of wise use (ie, Article 3.1).[22]

2 CULTURE AND THE RAMSAR REGIME

2.1 Coming to Terms with Culture

In the 1960s, the meaning, identification, and protection of culture – or, more specifically, 'cultural heritage' – was evolving under the auspices of the United Nations Educational, Scientific and Cultural Organisation (UNESCO) – notably through the creation of the World Heritage Convention (WHC).[23] Concurrently, negotiations were occurring for the protection of wetland environments, with a primary focus on waterfowl. Unsurprisingly, perhaps, the original text of Ramsar does not define or explicitly provide any clear obligations regarding culture. That said, there is clear evidence in the recitals that the drafters were 'convinced' that wetlands were 'a resource of great economic, cultural, scientific, and recreational value'. Pritchard attributes this recognition to an early desire to combine conservation with sustainable use of resources – something of a precursor to sustainable development,[24] which would later envelop Ramsar and many other MEAs.[25]

As the notion of culture took on greater prominence within Ramsar, it was perhaps inevitable that the regime came to terms with a more precise definition of 'culture' – or, at the very least, a more detailed explanation of what it means in the context of wetlands. A 2011 publication noted the enormity of this task and (quite reasonably) adopted UNESCO's definition: '[Culture is] a set of distinctive spiritual, material, intellectual and emotional features of society or

---

[22] D Pritchard, 'The "ecological character" of wetlands: a foundational concept in the Ramsar Convention, yet still cause for debate 50 years later' (2021) 73(10) *Marine and Freshwater Research* 1127; Pritchard, note 20, 2.

[23] Convention Concerning the Protection of the World Cultural and Natural Heritage, adopted 16 November 1972, entered into force 17 December 1975, 1037 UNTS 151.

[24] D Pritchard, 'Wise Use Concept of the Ramsar Convention' in CM Finlayson et al (eds), *The Wetland Book* (Springer, Dordrecht, 2018).

[25] Bridgewater and Kim, note 11. In the case of sustainable development under the World Heritage Convention, see *Policy for the Integration of a Sustainable Development Perspective into the Processes of the World Heritage Convention*, adopted by the General Assembly of the States Parties to the Convention at its 20 Session (Paris, 2015), by its Resolution 20 GA 13.

a social group, that encompasses, in addition to art and literature, lifestyles, ways of living together, value systems, traditions and beliefs.'[26]

In 2016, the Ramsar Culture Network (RCN) produced guidance on *Rapid Cultural Inventories for Wetlands*, which defined 'culture' for the purposes of Ramsar as:

> a property of human groups or societies which expresses aspects of their identity, shared values, attitudes, beliefs, knowledge systems, creativity and other practices. It conditions the ways in which people interact with each other and with their environment. Culture can be exhibited in both material and non-material ways, and it is constantly evolving.[27]

An almost identical conception was put forward two years later at a Ramsar culture workshop on the Isle of Vilm (Germany), with the notable inclusion of 'tangible and intangible heritage'.[28]

The 2018 definition is attractive for a variety of reasons – not least because of its breadth in encapsulating 'beliefs, values and attitudes', as well as the inclusion of both tangible and intangible heritage.[29] Moreover, the acknowledgement that culture is dynamic (ie, 'constantly evolving') is reflective of the myriad ways in which human beings interact with their environment, including where the base conditions of the ecosystem are themselves changing.

Of course, any attempt to define something as unwieldy and contentious as 'culture' invites rigorous debate – for example, that the definition is biased at the hands of the designer.[30] The same is true for the related concept of

---

[26] T Papayannis and D Pritchard, *Culture and Wetlands in the Mediterranean: An Evolving Story* (Med-INA, Athens, 2011) 15. This definition has similarities to the concept of intangible cultural heritage in heritage discourses and instruments.

[27] Pritchard, note 20, 2.

[28] German Federal Agency for Nature Conservation, International Academy for Nature Conservation and Ramsar Convention Secretariat, 'The cultural and spiritual significance of wetlands – supporting the integration of nature and culture in their governance and management Workshop Report' (26 February to 2 March 2018).

[29] Early attempts at protecting heritage under international law focused on tangible (eg, built) aspects of heritage. See, for example, the Hague Convention for the Protection of Cultural Property in the Event of Armed Conflict adopted 14 May 1954, entered into force on 7 August 1956, 249 UNTS 216. More recent efforts have targeted intangible components as well. See, in particular, the Convention for the Safeguarding of the Intangible Cultural Heritage, a UNESCO treaty adopted 17 October 2003, entering into force in 2006, 2368 UNTS 3.

[30] M Cocks, 'Biocultural diversity: moving beyond the realm of "indigenous" and "local" people' (2006) 34(2) *Human Ecology* 185.

'spirituality'.[31] On one level, a definition is important as it draws the bounds around what is, and what is not, within Ramsar's 'cultural' ambit. It can help to distinguish culture-related goals from other narrower regime objectives such as migratory waterbird conservation. On the other hand, any formal attempt to unravel culture (eg, through Conference of the Parties (COP) Resolutions) risks further subjecting the regime to dominant (Western) political discourses – or, perhaps more likely, a failure to garner any form of consensus at all.[32]

To some extent, Ramsar has managed to sidestep these difficult definitional challenges by homing in on a narrower set of wording within its governance architecture – that is, the 'cultural value' and 'services' of wetlands.

2.2 The Evolution of Culture under Ramsar

The original focus of Ramsar was on the protection of wetlands primarily for waterfowl habitat. Since the 1990s, however, there has been a noticeable shift – or rather, 'drift', as Bridgewater and Kim put it[33] – towards the social, cultural, and economic benefits that wetlands provide. As the RCN has highlighted, although the preamble to the Convention recognised the value of wetlands, 'serious attention to the incorporation of cultural aspects into the work of the Ramsar Convention began only in the late 1990s'.[34]

Since that time, it seems that a greater effort has been made by the Convention's parties and institutions to recognise culture-wetland linkages through resolutions, guidelines, technical publications, and other policy actions (Table 10.1). A focus on Indigenous peoples and the cultural values of wetlands seemed to be particularly strong in the so-called 'building phase' of the Convention between 1996 and 2005 (COPs 6, 7, 8, and 9).[35]

The two most important initiatives relating to culture during this period were Resolution VIII.19 (COP8, 2002) and Resolution IX.21 (COP9, 2005), both of which focused specifically on the 'cultural values' of wetlands. Through both resolutions, attempts were made to establish a closer link with Indigenous

---

[31] N Cooper et al, 'Aesthetic and spiritual values of ecosystems: Recognising the ontological and axiological plurality of cultural ecosystem "services"' (2016) 21 *Ecosystem Services* 218.

[32] See the comments of Brazil, for example, that examination of culture in 2002 had been extended beyond its original mandate. Ramsar Convention, *Report of the 8th meeting of the Conference of the Parties* (2002) 18.

[33] Bridgewater and Kim, note 11.

[34] Ramsar Culture Network, *Ramsar and Culture: Incorporating cultural aspects in the Ramsar Convention* (2015), at: https://wli.wwt.org.uk/wp-content/uploads/2020/12/culture_and_ramsar_150323.pdf.

[35] Bridgewater and Kim, note 11.

Table 10.1 *Historical Events Relating to Recognition of 'Culture' Under the Ramsar Convention*

Year	Event
1967	Early discussions recommend treaty should include reference in the preamble to wetlands being very important from cultural, scientific, economic, and social points of view.[1]
1971	Convention text adopted, with a statement in the preamble that wetlands constitute 'a resource of great economic, cultural, scientific, and recreational value, the loss of which would be irreparable'.
1987 (COP3)	Links strengthened between Ramsar, the Man and Biosphere Programme, and the WHC parties (eg, Brazil) suggest expanding listing criteria to include cultural factors. Suggested existing criteria are insufficient to motivate developing countries to participate in Ramsar and the expense of implementing activities 'was not justified' by referring only to waterfowl.[2]
1990	New datasheet for Ramsar Sites developed with a space to include information on social and cultural values.
1990 (COP4)	New criteria for identifying wetlands of international importance adopted, including noting that a wetland could also be of substantial value in supporting human communities by, among other things, maintaining cultural values.[3]
1996 (COP6)	Resolution VI.1 working definition of 'ecological character' referenced the phrase 'attributes' of a wetland, including biological diversity and unique cultural and heritage features. These attributes may lead to certain uses or the derivation of particular products, but they may also have intrinsic, unquantifiable importance.
1999 (COP7)	Resolution VII.8 adopted guidelines for establishing and strengthening local community and Indigenous participation in the management of wetlands. Resolution acknowledged the Convention's commitment to wetlands' cultural value.
2000	Mediterranean Wetlands Committee developed guiding principles for the inclusion of cultural values in wetland sustainable use in the Mediterranean region.
2001	Ramsar Secretariat establishes a programme of work with the aim of establishing a formal position on the cultural values of wetlands under the Convention.
2002	The theme for the 2002 World Wetlands Day focuses on 'Wetlands: Water, Life, and Culture' including highlighting the 'cultural heritage of wetlands'.
2002 (COP8)	Resolution VIII.19 adopted guiding principles for taking into account the 'cultural values' of wetlands for the effective management of sites.
2003	Ramsar Strategic Plan 2003–08 recognises 'cultural issues' among other aspects as requiring further work and highlighting the maintenance of cultural values by local communities and Indigenous peoples as part of one of the three pillars of action.[4]

Year	Event
2005 (COP9)	Resolution IX.21 on taking into account the cultural values of wetlands adopted, including establishment of the Culture Working Group.
2008 (COP10)	Culture and Wetlands – Ramsar Guidance Document produced and launched.
2012	New Ramsar Information Sheet format released, including specific information on cultural ecosystem services.
2013	Culture Working Group evolved into the RCN.
2016	Rapid Cultural Inventories for Wetlands released.
2016	Ramsar Strategic Plan 2016–24 adopted. Wetlands deliver a wide range of ecosystem services, including spiritual and cultural inspiration.
2017	Report by the Ramsar Convention Secretariat released investigating the role of cultural values, practices, and traditions in World Heritage/Ramsar Sites.[5]
2018 (COP13)	Resolution on cultural values and practices of Indigenous peoples and local communities and their contribution to climate change mitigation and adaptation in wetlands is adopted.

Notes
1 Matthews, note 4, 16.
2 Ramsar Convention, 23.
3 Matthews, note 4, 46.
4 Ramsar Convention, –(2002) 8.
5 R McInnes, M Ali, and D Pritchard, (Ramsar Convention Secretariat, Gland, 2017).

peoples and local communities, though reportedly with limited success.[36] At the same time, both resolutions were drafted in such a way as to respond to earlier requests to designate wetlands of international importance on the basis of cultural as well as ecological criteria. As the (then) Culture Working Group noted, Resolution IX.21 'dealt with [this] in a general manner, by advising Contracting Parties to *consider* cultural values as well as ecological values in the process of site designation'.[37]

Under Resolution VIII.19, the parties adopted non-binding 'Guiding Principles' for taking into account the cultural values of wetlands. The Resolution annexed certain principles for identifying, preserving, and reinforcing those values. For example, Principle 21 provided for the incorporation of cultural values into management planning, and the most recent National Report Form now includes a specific question relating to this objective.[38] The same template also asks the party to provide information on whether Resolution

---

[36] Bridgewater and Kim, note 11.
[37] Papayannis and Pritchard, note 9, 7 (emphasis added).
[38] Ramsar Convention, *National Report Form (COP14 format)*, note 20, Question 11.4.

VIII.19 has been 'used or applied' and whether case studies, participation in projects, or successful experiences on cultural aspects of wetlands have been undertaken.[39]

At the time, Resolution VIII.19 was hailed by some as 'a pioneering statement on cultural values in wetlands', and as having 'broader significance for the conservation of all types of ecosystems'.[40] However, the extent to which Resolution VIII.19 influenced other MEAs is debatable, with delegates to COP9 in Uganda 2005 arguing that cultural measures should be consistent with parties' rights and obligations under other MEAs (rather than the other way around).[41] Accordingly, in Resolution IX.21, the most the parties were willing to agree to was that:

> in the application of the existing criteria for identifying Wetlands of International Importance, a wetland *may also be considered* of international importance when, *in addition to relevant ecological values*, it holds examples of significant cultural values, whether material or non-material, linked to its origin, conservation and/or ecological functioning.[42]

Resolution IX.21 further recognised four potentially relevant 'characteristics', which may comprise part of the ecological character of a wetland:

(i) sites which provide a model of wetland wise use, demonstrating the application of traditional knowledge and methods of management and use that maintain the ecological character of the wetland;
(ii) sites which have exceptional cultural traditions or records of former civilisations that have influenced the ecological character of the wetland;
(iii) sites where the ecological character of the wetland depends on the interaction with local communities or indigenous peoples; and
(iv) sites where relevant non-material values such as sacred sites are present and their existence is strongly linked with the maintenance of the ecological character of the wetland.[43]

These characteristics would become known as 'social or cultural values' in the current Ramsar Information Sheet (RIS) template and the 'push-pull' culture dynamic mentioned earlier is evident in the approach. Whereas characteristics (ii) and (iv) attempt to 'push' Ramsar outwards to acknowledge cultural

---

[39] Ibid, Questions 10.1 and 10.2.
[40] Oviedo and Ali, note 7, 17.
[41] International Institute for Sustainable Development (IISD), *Summary of the Ninth Meeting of the Conference of the Parties to the Ramsar Convention on Wetlands* (8–15 November 2005) 12.
[42] Ramsar Convention, *Resolution IX.21*, note 8, 2.
[43] Ibid.

aspects of wetlands, characteristics (i) and (iii) seem to 'pull culture in' by reinforcing the underlying narrative of more effective site management (for conservation purposes).

This dynamic is also apparent in various other resolutions and recommendations over the years – in particular, surrounding the involvement of Indigenous peoples and public participation by:

- strengthening the role of Indigenous peoples in wetland management;[44]
- enhancing public awareness of wetland values;[45] and
- promoting participatory management as a tool for wise use.[46]

2.3 Roadblocks to the Recognition of Culture

The two culture-related Resolutions (above) served different purposes, though were reportedly linked through a desire for 'implementation' of the former by the latter.[47] In this regard, COP9 was seen as something of a turning point for Ramsar, representing the first COP 'where everyone agreed that culture must be discussed'.[48] While that may indeed be true, the official record shows that several parties opposed an expanded focus on culture. Concerns arose, specifically, that the Ramsar Convention was heading into 'dangerous waters' and away from its 'ecologically-based' roots[49] (concerns which would later be realised).[50] Papayannis and Pritchard have noted the 'fears' that certain parties held that the incorporation of culture into Ramsar might be used to protect economic activities in ways which might frustrate free trade objectives.[51]

Although many of these concerns were evident at COP9 (2005), three years prior, at COP8, Australia and New Zealand had proposed that any traditional activities related to wetlands ought to be consistent with World Trade Organization requirements.[52] At the same COP, Brazil formally objected to an

---

[44] Ramsar Convention, *Recommendation 6.3: Involving local and indigenous people in the management of Ramsar wetlands* (1996).
[45] Ramsar Convention, *Recommendation 5.8: Measures to promote public awareness of wetland values in wetland reserves* (1993).
[46] Ramsar Convention, *Resolution VIII.36: Participatory Environmental Management (PEM) as a tool for management and wise use of wetlands* (2002).
[47] Ramsar Convention *Resolution IX.21*, note 8, 1.
[48] IISD, note 41, 15.
[49] Ibid.
[50] Bridgewater and Kim, note 11.
[51] Papayannis and Pritchard, note 9, 7.
[52] International Institute for Sustainable Development (IISD), *Summary of the Eighth Meeting of the Conference of the Parties to the Ramsar Convention on Wetlands* (18–26 November 2002) 9.

expanded notion of culture, arguing that 'the mandate' given for discussion of culture had 'been exceeded'.[53] Brazil's main concern surrounded 'the [proposed] multifunctional character of agriculture', which would keep developing countries (such as itself) out of international agricultural markets and contribute to distorted prices and production levels.[54] Many parties also opposed adding a new criterion for listing Ramsar Sites based solely on cultural values.[55]

The culture-related sensitivities evident at COPs 8 and 9 carried forward into other work of Ramsar's technical representatives – for example, in 2008, guidance on cultural values of wetlands, where the authors took 'great care' in drafting to avoid 'problematic implications for the realm of international trade or other areas of political, legal or technical sensitivity'.[56] As a further reflection of the sensitivities around culture, the guidance was not proposed for adoption within a COP resolution, but rather made available to the parties as a technical resource.[57]

3 IDENTIFYING AND PROTECTING VALUES AND SERVICES

Today, Ramsar makes a slight, though not unimportant, distinction between the concepts of 'cultural values' and 'cultural services'. As revealed below, the two terms are not synonymous, although their conflated usage in certain forums may suggest otherwise. To confuse matters further, it is theoretically possible to calculate both the 'value of cultural services' and the 'cultural value of wetlands'.[58] Such a distinction is made clear in the current RIS, where parties are to describe the ecological character of their site.[59] Figure 10.1 shows the relationship of social and cultural values and cultural services under Ramsar to the concept of ecological character. Of course, many other 'non-culture related' values and services may comprise the ecological character of a site.

---

[53] Ramsar Convention, note 32, 18.
[54] Ibid.
[55] See, eg, IISD, note 52.
[56] Papayannis and Pritchard, note 26, 44.
[57] Papayannis and Pritchard, note 9, 9.
[58] See also the language of the socio-cultural value of wetland services used in R de Groot et al, *Valuing wetlands: guidance for valuing the benefits derived from wetland ecosystem services*, Ramsar Technical Report No 3/CBD Technical Series No 27 (Ramsar Convention Secretariat, Gland, 2006) 20.
[59] Ramsar Convention, *Resolution XI.8, Annex 2*, note 8, Section 7.3.

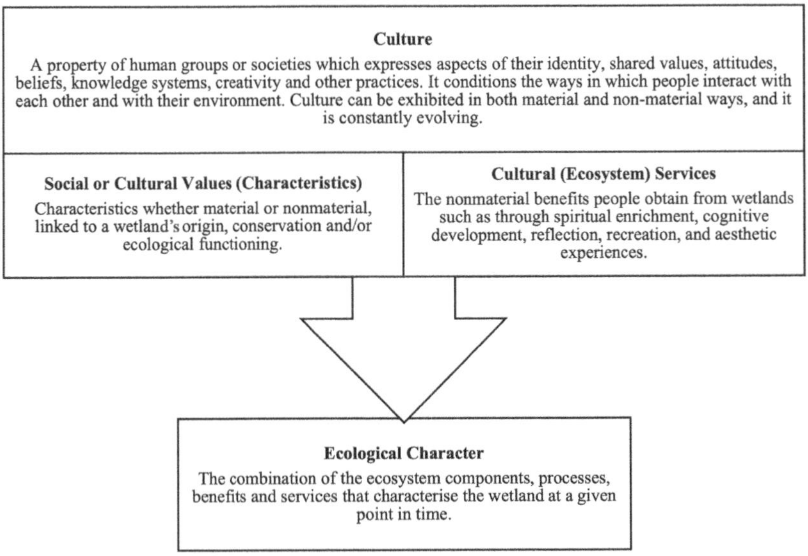

```
┌─────────────────────────────────────────────────────────────────────┐
│                              Culture                                │
│ A property of human groups or societies which expresses aspects of  │
│ their identity, shared values, attitudes, beliefs, knowledge        │
│ systems, creativity and other practices. It conditions the ways in  │
│ which people interact with each other and with their environment.   │
│ Culture can be exhibited in both material and non-material ways,    │
│ and it is constantly evolving.                                      │
├──────────────────────────────────┬──────────────────────────────────┤
│ Social or Cultural Values        │ Cultural (Ecosystem) Services    │
│ (Characteristics)                │ The nonmaterial benefits people  │
│ Characteristics whether material │ obtain from wetlands such as     │
│ or nonmaterial, linked to a      │ through spiritual enrichment,    │
│ wetland's origin, conservation   │ cognitive development,           │
│ and/or ecological functioning.   │ reflection, recreation, and      │
│                                  │ aesthetic experiences.           │
└──────────────────────────────────┴──────────────────────────────────┘
                                  ↓
                    ┌─────────────────────────────┐
                    │    Ecological Character     │
                    │ The combination of the      │
                    │ ecosystem components,       │
                    │ processes, benefits and     │
                    │ services that characterise  │
                    │ the wetland at a given      │
                    │ point in time.              │
                    └─────────────────────────────┘
```

Figure 10.1   The relationship of social and cultural values (characteristics) and cultural (ecosystem) services under Ramsar to the concept of ecological character

3.1   The Recognition of Cultural Values

Just as the recitals to the Convention highlighted the importance of the cultural values of wetlands, so too has Ramsar's framework embraced the evolving terminology of 'values'. At COP6 in Brisbane (1996), for instance, a working definition of 'ecological character' was put forward as follows: 'the structure and inter-relationships between the biological, chemical, and physical components of the wetland. These derive from the interactions of individual processes, functions, *attributes and values* of the ecosystem(s).'[60]

The term 'value' was proposedly defined as 'the perceived benefits to society, either direct or indirect, that result from wetland functions'.[61] Notably, cultural values were excluded, although the 'unique cultural and heritage

---

[60] Ramsar Convention, *Resolution VI.1: Working Definitions of Ecological Character, Guidelines for Describing and Maintaining the Ecological Character of Listed Sites, and Guidelines for Operation of the Montreux Record* (1996) (emphasis added).

[61] Ibid, 3.

features' of wetlands were included in the definition of 'attributes'.[62] The proposed definition of 'attributes' also noted that such features 'may lead to certain uses or the derivation of particular products, but they may also have intrinsic, unquantifiable importance'.[63] An alternative definition of 'ecological character' was eventually accepted in 1999 (which did not reference the word 'values')[64] and revised again in 2005 (see Section 3.3 below).

'Cultural values' were again referenced in 1999, when the parties adopted Resolution VII.8, which annexed guidelines for establishing and strengthening local community and Indigenous participation in the management of wetlands.[65] The term was further utilised in 2002 and 2005 in Resolutions VIII.19 and IX.21, respectively.[66] Problematically, the term 'cultural value' was not defined in any of these resolutions and, confusingly, was used alongside other related terms that were also not defined, such as 'cultural significance',[67] 'cultural and heritage values',[68] 'cultural aspects',[69] 'cultural concern',[70] 'cultural elements',[71] and 'cultural heritage'.[72] Despite the increasing use of the term 'cultural values' under Ramsar, therefore, its exact meaning has remained unclear and no agreed definition is readily available. Although it may be the case that defining 'cultural value' is less important than understanding its application, a few points of distinction may prove useful for context.

First, it seems clear that cultural values are intended to be distinguished from ecological values[73] and further separated from economic values.[74] The natural capital of a wetland can theoretically produce all three of these values. Second, cultural values may be distinguishable from 'cultural aspects', with

---

[62] Ibid.

[63] Ibid.

[64] Ramsar Convention, *Resolution VII.10, Wetland Risk Assessment Framework* (1999).

[65] Ramsar Convention, *Resolution VII.8, Guidelines for establishing and strengthening local communities' and indigenous people's participation in the management of wetlands* (1999) 7.

[66] See the discussion in Section 2.2 of this Chapter above.

[67] Ramsar Convention, *Resolution VIII.19*, note 20, 3.

[68] Ibid, 4.

[69] Ibid, 6.

[70] Ibid, 3.

[71] Ibid.

[72] Ramsar Convention, *Resolution IX.21*, note 8, 1 and 2; Ramsar Convention, *Resolution VII.8*, note 65, 5.

[73] Ramsar Convention, *Resolution XI.8, Annex 1: Ramsar Site Information Sheet (RIS) – 2012 revision* (2012) 38.

[74] de Groot et al, note 58, 18.

some parties, for example, suggesting that not all 'aspects' of wetlands could or should be considered 'values'.[75] Third, the language of cultural values conforms broadly with that of the recitals to the Convention (ie, cultural values of wetlands as a resource), which subsequently aligned with the emergence of economic approaches to nature conservation in the 1980s and 1990s.[76] Although influential organisations such as the International Union for Conservation of Nature (IUCN) utilised the phrase 'cultural resources'[77] in the mid-1990s, by the early 2000s, it had clearly shifted to the language of cultural values.[78]

Finally, the use of the term 'value' is a key component of heritage discourses and many cultural heritage frameworks at the international and domestic level turn on the importance of identifying and protecting heritage value. Consider, for example, the practice of the WHC in protecting the Outstanding Universal Value (OUV) of World Heritage properties. As with the notion of ecological character under Ramsar, OUV becomes the baseline for the future management of sites, with a Statement of OUV now required at the time of listing.[79] In a cultural heritage sense, given the breadth of discourses and instruments, 'values' may have a variety of meanings – although one popular conception sees them as 'the qualities and characteristics assigned by people to an object, a feature or a place, be it a building, a landscape, a forest, or a mountain'.[80]

---

[75] Papayannis and Pritchard, note 9, 3. Note that 'cultural aspects' is used on occasion in this Chapter as a catchall term to refer broadly to cultural services, values, and/or characteristics.

[76] See, generally, R de Groot, 'Environmental functions as a unifying concept for ecology and economics' (1987) 7(2) *Environmentalist* 105; C Folke and T Kåberger (eds), *Linking the Natural Environment and the Economy: Essays from the Eco-Eco Group* (Springer, Dordrecht, 1991).

[77] IUCN, *Guidelines for Protected Area Management Categories* (IUCN, Gland, 1994) 7.

[78] IUCN 'Protected Areas – About', at: https://www.iucn.org/theme/protected-areas/about.

[79] World Heritage Convention, *Operational Guidelines for the Implementation of the World Heritage Convention* (2021 edition) [51], at: https://whc.unesco.org/en/guidelines/.

[80] Citing M De la Torre (ed), *Assessing the Values of Cultural Heritage* (The Getty Conservation Institute, Los Angeles, CA, 2002).

3.2 The Recognition of Cultural Services

During the early 2000s, the 'socio-cultural' values of wetlands were seen as connected to the broader 'ecosystem services-ecosystem functions' paradigm.[81] Conceptually, functions are distinct from services, with the latter understood as those functions that are being 'actively' used (Figure 10.2).[82]

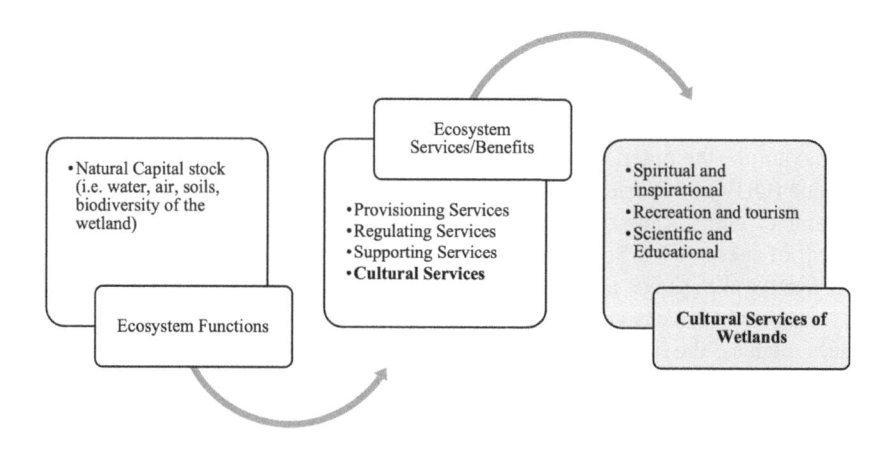

Figure 10.2   Derivation of 'cultural services' of wetlands under Ramsar

By 2005, the language of ecosystem services had become popularised across MEAs – Ramsar included – through the publication of the Millennium Ecosystem Assessment,[83] which, among other things, recognised 'cultural services' as a specific category of ecosystem services.[84] The Assessment's classification of 'ecosystem services' was swiftly adopted by Ramsar in the

---

[81] de Groot et al, note 58.

[82] EG Baggethun and R de Groot, 'Natural Capital and Ecosystem Services: The Ecological Foundation of Human Society' in RM Harrison and RE Hester (eds), *Ecosystem Services* (RSC Publishing, Cambridge, 2010) 111.

[83] P Bridgewater, 'A new context for the Ramsar Convention: Wetlands in a changing world' (2008) 17(1) *Review of European Community & International Environmental Law* 100.

[84] Millennium Ecosystem Assessment, note 2.

guidance for ecological character descriptions in 2008,[85] and an updated RIS defined 'cultural services' as 'the nonmaterial benefits people obtain from ecosystems such as through spiritual enrichment, cognitive development, reflection, recreation, and aesthetic experiences'.[86]

Following their classification in the Millennium Ecosystem Assessment, cultural services were differentiated from other services, including supporting, provisioning, and regulating services (Figure 10.3). The distinction is largely premised on the notion that cultural services are 'primarily driven by human experience [of nature]'[87] and may include, for example, the following:

- recreational hunting and fishing;
- water sports, recreation, and tourism;
- nature study pursuits;
- educational values;
- cultural heritage;
- contemporary cultural significance, including for arts and creative inspiration, and including existence values;
- aesthetic and 'sense of place' values;
- spiritual and religious values; and
- important knowledge systems and importance for research.[88]

---

[85] Ramsar Convention, *Resolution X.15: Describing the ecological character of wetlands, and data needs and formats for core inventory: harmonized scientific and technical guidance* (2008).

[86] Ramsar Convention, *Resolution XI.8, Annex* 1, note 73, 36.

[87] The URBES Project, *Cultural Ecosystem Services – A gateway to raising awareness for the importance of nature for urban life* 2, at: https://oppla.eu/sites/default/files/uploads/urbesfactsheet08web2.pdf.

[88] Ramsar Convention, *Resolution X.15*, note 85, 11.

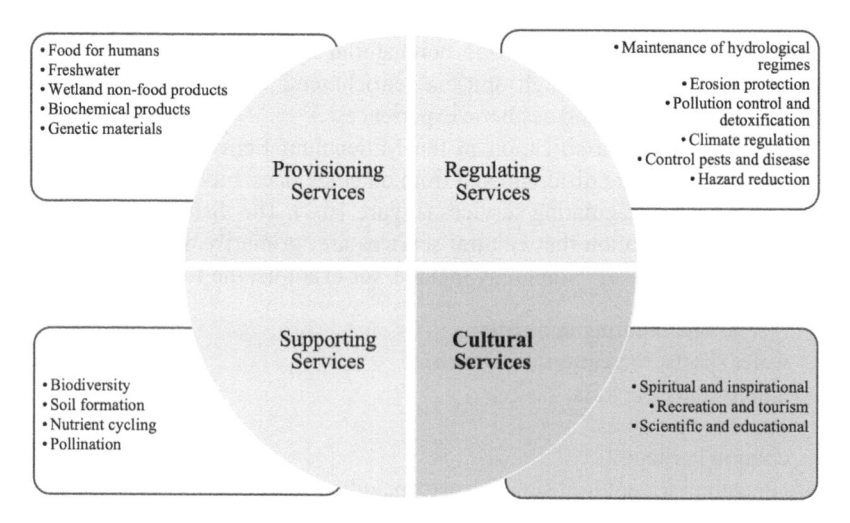

```
┌─────────────────────────┐ ┌─────────────────────────────┐
│ • Food for humans       │ │ • Maintenance of hydrological│
│ • Freshwater            │ │                      regimes │
│ • Wetland non-food products│ • Erosion protection       │
│ • Biochemical products  │ │ • Pollution control and     │
│ • Genetic materials     │ │             detoxification  │
│                         │ │ • Climate regulation        │
│          Provisioning   │ │ • Control pests and disease │
│          Services       │ │   Regulating  • Hazard reduction │
│                         │ │   Services                  │
└─────────────────────────┘ └─────────────────────────────┘

┌─────────────────────────┐ ┌─────────────────────────────┐
│          Supporting     │ │   Cultural                  │
│          Services       │ │   Services                  │
│ • Biodiversity          │ │ • Spiritual and inspirational│
│ • Soil formation        │ │ • Recreation and tourism    │
│ • Nutrient cycling      │ │ • Scientific and educational│
│ • Pollination           │ │                             │
└─────────────────────────┘ └─────────────────────────────┘
```

Figure 10.3   *The different classifications of ecosystem services provided by wetlands under Ramsar*

3.3   Obligations Relating to Values and Services

It is accordingly clear that both values and services are capable of being recognised by Ramsar's governance approach, yet the extent to which such recognition shades into obligation is not settled. While Ramsar adopts a non-regulatory approach,[89] there are certainly key provisions which give rise to obligations.[90] This Chapter is not the place to debate exactly what the obligations are or might be under the Ramsar – either procedural or substantive. However, two narrow lines of legal academic enquiry are pursued:

- Is there an obligation on parties to identify cultural services and/or values (hereinafter 'cultural aspects') at the time of designation of a Ramsar Site?

---

[89] RC Gardner and KD Connolly, 'The Ramsar Convention on Wetlands: Assessment of International Designations within the United States' (2007) 29(2) *National Wetlands Newsletter* 6–15; E Gallo-Cajiao, *Review of the international policy framework for conserving migratory shorebirds in the East Asian-Australasian Flyway* (EAAFP, Brisbane, 2014).

[90] V Koester, *The Ramsar Convention on the Conservation of Wetlands: A Legal Analysis of the Adoption and Implementation of the Convention in Denmark*, IUCN Environmental Policy and Law Paper No 23 (1989); Joshi et al, note 11; and O Ferrajolo, 'State obligations and non-compliance in the Ramsar system' (2011) 14(3–4) *Journal of International Wildlife Law & Policy* 243.

- Notwithstanding the answer to the question above, what, if any, are the obligations on parties to protect cultural aspects of a Ramsar Site?[91]

In relation to the first question above, it is broadly accepted that a wetland may not be listed as a wetland of international importance unless it meets one of the nine criteria for inclusion.[92] Moreover, it has been a longstanding position of the Convention that the ecological character of a site be described by the contracting party at the time of designation through completion of a RIS.[93] As noted above, despite arguments for cultural criteria to be included as a reason for designation, this has not yet transpired. The most the parties have agreed to, in the form of a non-binding resolution, is that 'significant' cultural and social values ('characteristics') may be referenced at the time of designation. The 'right' to identify these values is recognised in Resolution IX.21:

> [I]n the application of the existing criteria for identifying Wetlands of International Importance, a wetland *may also be considered* of international importance when, in addition to relevant ecological values, it holds examples of significant cultural values, whether material or non-material, linked to its origin, conservation and/or ecological functioning.[94]

The language is clearly discretionary and not obligatory. Moreover, in terms of legal requirements at the point of designation, the Ramsar treaty text remains authoritative. Article 2 sets out the 'mandatory' requirements for designating Ramsar Sites, including that parties must describe the boundaries of designated wetlands, including delimiting their area on a map. Moreover, designated wetlands should be selected on account of their international significance in terms of ecology, botany, zoology, limnology, or hydrology. Finally, under Article 2.6, each party has an obligation – denoted by the word 'shall' – to consider its international responsibilities for the conservation, management, and wise use of migratory stocks of waterfowl at the point of Ramsar designation.

On this basis, it may be reasonably concluded there is no obligation to identify cultural aspects of a site at the time of designation. It is therefore likely that parties have something less than an obligation – probably a dis-

---

[91] Although it is acknowledged Ramsar's 'wise use' ambit seeks to extend to all wetlands within a party's jurisdiction (see Article 3.1), the focus here is on designated sites of international importance (ie, Ramsar Sites).

[92] Ramsar Convention Secretariat, *Ramsar Sites Criteria*, at: www.ramsar.org/sites/default/files/documents/library/ramsarsites_criteria_eng.pdf.

[93] See, eg, Ramsar Convention, *Recommendation 4.7: Mechanisms for improved application of the Ramsar Convention* (1990); Ramsar Convention, *Resolution VI.1*, note 60.

[94] Ramsar Convention, *Resolution IX.21*, note 8, 2 (emphasis added).

cretion – to include such information where they consider it relevant. One argument against this conclusion is the inclusion of the word 'ecology' under Article 2.2, which necessarily includes human ecology – that is, the study of the relationships between humans and their environment, potentially including the cultural aspects of wetlands. Even if this were the case, the accompanying words in Article 2.2 'should be selected ... on account of their international significance in terms of ecology' would need to be read to mean 'must be selected' for the obligation to exist. Such a conclusion is unlikely, given that the sciences listed in Article 2.2 are 'or' options and not 'ands'. Moreover, the designation of only 'suitable wetlands' in Article 2.1 and the 'right' for parties to add to the list in Article 2.5 strongly suggest that the designation process is discretionary on the part of the party – in terms of both which sites they put forward and the basis on which they do.[95]

In relation to the second question above, Ramsar clearly imposes a commitment under Article 3.1 to maintain the ecological character of listed sites. Services and values are 'protected' in an obligatory sense through the related concepts of ecological character and wise use. The 'wise use of wetlands' is the maintenance of ecological character, 'achieved through the implementation of ecosystem approaches, within the context of sustainable development'.[96] The definition of 'ecological character' under the Convention is 'the combination of the ecosystem components, processes, benefits and services that characterise the wetland at a given point in time'.[97] That point in time is at designation, when the ecological character of the site is described by the contracting party.[98]

Accordingly, any requirement to 'protect' cultural aspects of wetlands is best viewed – in an obligatory sense – through the action of maintaining the ecosystem components, processes, benefits, and services that characterise the wetland, including the cultural services as identified at the point of designation. It follows that, as there is currently no hard requirement to identify cultural services at the point of designation, the obligation to maintain such services is enlivened only to the extent to which the description of the site documents such values. Finally, the 'maintenance of ecological character' must be pursued in the context of sustainable development, meaning (arguably) that

---

[95] This conclusion is further confirmed by a resolution in 1993, which noted that 'any decision on designation remains the prerogative of the Contracting Party in whose territory the wetland is situated'. See Ramsar Convention, *Resolution 5.9: Application of the Ramsar Criteria for Identifying Wetlands of International Importance* (1993).

[96] Ramsar Convention, *Resolution IX.1 Annex A: A Conceptual Framework for the wise use of wetlands and the maintenance of their ecological character* (2005).

[97] Ibid.

[98] Ibid.

only those cultural aspects of wetlands which are not inconsistent with the pursuit of sustainable development can and must be maintained. This conclusion would address the perverse result, for example, where cultural services or values are identified in the ecological character description, yet somehow have a negative impact on the overall functioning of the wetland in terms of sustainable development.[99]

4 CHALLENGES ON THE HORIZON

4.1 Neutrality, Inclusion, and Objectivity in Valuation

One of the foremost challenges that Ramsar faces with respect to the identification and protection of the cultural aspects of wetlands – whether or not the end goal is effective site management – is how best to ensure that characteristics and services are objectively valued and facilitate the meaningful inclusion of local knowledge holders. Overcoming the 'long-standing intellectual conflicts regarding the value of nature' will be no easy feat in this regard.[100] As the 2018 Isle of Vilm workshop into the cultural significance of wetlands acknowledged, 'Ramsar reflects many of the typical biases in global environmental governance towards (for example) particular dominant languages, forms of science, world-views, belief systems, norms, orthodoxies and paradigms'.[101]

Moreover, as Joshi and others have observed, the Ramsar Convention seems to have been heavily influenced by 'external, mostly Northern biophysical scientists, conservationists, and [male] representatives from technical national government institutions. This has resulted in a top-down, technocratic vision and meaning of the wise use framework, which focuses on implementation and management through governmental agencies'.[102]

Similar problems have been faced by the WHC – the only other site-based conservation treaty that seeks to ascribe a 'global value' to socio-ecological systems. The WHC has been critiqued for, among other things, its ignorance of Indigenous world views, its increasing politicisation of decision-making,

---

[99] It would seem the definition of 'Sustainable Development' of the 1987 Brundtland Commission is to be utilised in this regard (and applied to a wetland context). See Ramsar Convention, *Resolution IX.1 Annex A*, note 96, 6.

[100] S Stålhammar and H Thorén, 'Three perspectives on relational values of nature' (2019) 14 *Sustainability Science* 1201.

[101] German Federal Agency for Nature Conservation et al, note 28, 3.

[102] Joshi et al, note 11.

and an unbalanced list of World Heritage Sites biased towards those 'valued' by the Global North.[103]

It should therefore be remembered that 'assigning values [to ecosystems] is in itself not a neutral or objective exercise'.[104] Any attempt to identify, document, qualify, or quantify the 'worth' of cultural values or services of wetlands (on their own or in comparison to other cultural attributes) is likely to be met with conflict. In identifying 'significant' cultural values[105] and assigning value to cultural services, states are faced with complex methodological tasks, many of which cannot be 'captured by economic valuation methods'.[106] While there are available methods for valuing cultural services from ecosystems (eg, through participant interviews, questionnaires, behavioural observations),[107] they take considerable time, effort, and expertise, which states may be unable or unwilling to invest in. By the same token, conflicting or politicised cultural claims to the same wetland may mean that some states are unlikely to objectively account and defend all of the values that a wetland provides. In such circumstances, the state may be prone to overemphasising (or underemphasising) certain services or values, potentially to the detriment of others.

4.2 Navigating the Complexities of Scale

Allied to the above challenge is the issue of how Ramsar navigates the complex question of value at scale. The Convention's 2016 Guidance for Rapid Cultural Inventories of Wetlands hints at some of the practical challenges in this regard:

> Particular values may be held in common by a particular community of stakeholders or a social group who derive benefits (tangible or intangible) from the wetland systems (or stories) concerned. These people may live in the area concerned or they may not ... For values that are of global significance as the common heritage of humanity, it may be the global community as a whole who should be mentioned. In the case of cultural practices, there may be details to record concerning roles played by particular members or groups in the community, such as elders, leaders, nominated celebrants, men, women, children, etc ...[108]

---

[103] See L Meskell, *A Future in Ruins* (Oxford University Press, Oxford, 2018).

[104] Verschuuren et al, note 14, 22 (citing P Jepson and S Canney, 'Values-led conservation' (2003) 12(4) *Global Ecology and Biogeography* 271).

[105] Parties are asked in the RIS to identify 'significant cultural values'. See Ramsar Convention, *Resolution XI.8, Annex 2*, note 8, 68.

[106] de Groot et al, note 58, 20.

[107] Ibid, 22.

[108] Pritchard et al, note 20, 16.

Parties may struggle to identify the scale at which cultural value is exhibited, or should be exhibited. Comments in a New Zealand context, for instance, noted this in the context of fishing activities at potential Ramsar Sites: 'It is difficult to consider differing degrees of cultural importance because such sites are used by local communities and are therefore highly important at a local level.'[109]

The recognition of 'internationally significant values' of a wetland may also trigger the expansion of domestic regulatory obligations for site management depending on the nature of the legal system.[110]

Again, the question of competing value scales has plagued the WHC. The 'universality' of OUV does not always align with the day-to-day values of local communities (which may oppose the listing in the first place). This is routinely seen in the controversies surrounding highly urban World Heritage properties, listed for OUV, but which succumb to local developmental agendas.[111] It may also be the case that local communities practise their heritage in a way that makes it unsuitable for 'international' designation in the first place – for example, in the case of the World Heritage Site of Bagan (Myanmar), where the local community's religious practices conflicted with the state's narrative of OUV.[112] Accordingly, as Pritchard points out in the Ramsar context, '[t]here is a particularly difficult balance to strike, in the context of a Convention, when pronouncing at international level on issues concerning

---

[109] K Denyer and H Robertson, *National guidelines for the assessment of potential Ramsar wetlands in New Zealand* (New Zealand Department of Conservation, Wellington, 2016) 32.

[110] Australia, for example, has provisions under its national environmental legislation that any significant impacts on Ramsar Sites must not be 'inconsistent' with obligations under the Ramsar Convention. See Australia's Environment Protection and Biodiversity Conservation Act 1999 (Cth) s 138. Schedule 6 of the Environment Protection and Biodiversity Conservation Regulations 2000 (Cth) also sets out general principles for the management of Ramsar Sites (most of which are managed by state and territory governments).

[111] The delisting of Dresden Elbe Valley (Germany) from the World Heritage List in 2009 shows the extent to which global, national, and local agendas can conflict. The case of Vienna recently listed as 'In Danger' to local conflicting development is another.

[112] As Kraak has reported, 'with the aim of a World Heritage status, the government's approach to its heritage shifted'. See AL Kraak, 'Ruins, Rituals and Sunset Sacrifice: The contesting values of Bagan in Myanmar' Proceedings of the Australia International Council on Monuments and Sites Conference, 5–8 November 2015, at: http://www.aicomos.com/wp-content/uploads/Ruins-Rituals-and-Sunset-Sacrifice-The-contesting-values-of-Bagan-in-Myanmar-full-paper.pdf.

human society and environment – which are by definition rooted in a variety of locally distinctive approaches'.[113]

4.3 Classifying Cultural Services

The approach of seeing culture through the prism of an ecosystem services paradigm may be criticised as 'utilitarian and neoliberal', and as potentially having 'very serious consequences' for particular cultural groups.[114] Indeed, it is tempting to call out the binary categories of services which Ramsar deploys as feeding into a system which ultimately fails to address 'the power and politics of how access to, use, and control of wetlands resources are determined'.[115] In other words, a desire to document and categorise distinct kinds of wetland services (in line with the Millennium Ecosystem Assessment) may mean the emergence of categories which do not fully explain the complexities of human-wetland interactions.[116]

For example, sustenance for humans in the form of fish, molluscs, grains, and so on is represented as a 'provisioning service' under the RIS, as are the non-food products that may be extracted from a wetland (eg, timber, reeds, and fibre).[117] In traditional communities and Indigenous societies, such services may be intimately connected to cultural services, appropriately classified under Ramsar as 'cultural heritage', 'inspiration', or 'sense of place' values. Accordingly, there are likely to be instances where the non-material benefits of wetlands are inseparable from the material. As Cooper and others write, 'cultural services are often not discrete, but bundled up with others'.[118] This may lead to double-counting, under-counting or other perverse information-gathering results which do not accurately reflect the service being provided or utilised.[119] This has ramifications for the legal obligation to maintain the 'described' ecological character of Ramsar Sites discussed earlier.

---

[113] Pritchard, note 13, 192.

[114] Verschuuren et al, note 14.

[115] Joshi et al, note 11, 39. Note that this argument was not in the context of ecosystem services.

[116] B Fisher, RK Turner, and P Morling, 'Defining and classifying ecosystem services for decision-making' (2009) 68 *Ecological Economics* 643; P Horwitz and CM Finlayson, 'Wetlands as Settings for Human Health: Incorporating Ecosystem Services and Health Impact Assessment into Water Resource Management' (2011) 61(9) *BioScience* 678.

[117] Ramsar Convention, *Resolution XI.8, Annex 1*, note 73, 35–36.

[118] Cooper et al, note 31.

[119] As McInnes and others have reported, some ecosystem services are more frequently reported than others. See RJ McInnes et al, 'Wetland Ecosystem Services and the Ramsar Convention: An Assessment of Needs' (2017) 37 *Wetlands* 123.

Such potential criticisms are by no means specific to Ramsar and may apply more broadly to the practicalities around identifying and valuing cultural ecosystem services. As Chan and others point out:

> The values that conform least well to economic assumptions – variously lumped together with/as cultural services – have proven elusive in part because valuation is complicated by the properties of intangibility and incommensurability, which has in turn led to their exclusion from economic valuation.[120]

Under Ramsar, the 'lumping together' of services, or seeing culture as a 'residual category' of services,[121] could explain an over-representation of documented cultural services in the Ramsar Sites Information Service relative to other reported services (provisioning, regulating, and supporting).[122]

Moreover, although the ecosystem services-cultural services framing attempts to consider 'multiple scales, from local to global',[123] it is arguably unable to mediate conflicts within the same classification of services. Navigating 'trade-offs' between ecosystem services has been acknowledged before – for example, an increase in the supply of 'provisioning ecosystem services such as food and water [may] result in reductions in the supply of supporting, regulating, and cultural services ...'[124] But what of the trade-offs within the same classification of cultural service – for example, between competing claims to cultural heritage, recreational pursuits, or spiritual and religious values? The current approach places the decision concerning trade-offs on the party at the time of designation, and the RIS requires no explicit justification of the methodological basis for how such trade-offs were or are to be resolved in the context of a proposed listing.[125]

---

[120] KMA Chan, T Satterfield, and J Goldstein, 'Rethinking ecosystem services to better address and navigate cultural values' (2012) 74 *Ecological Economics* 8.

[121] T Daniel et al, 'Contributions of cultural services to the ecosystem services agenda' (2012) 109 (23) *Proceedings of the National Academy of Sciences* 8812.

[122] Oviedo and Ali, note 7, 25.

[123] Millennium Ecosystem Assessment, note 2, 19.

[124] Ibid, 53.

[125] In addition, although parties may theoretically have access to improved methods to value ecosystem services, including cultural services, the recognition of such approaches, domestically, is likely to still be informal and not enshrined in legislation or incorporated into management planning. See RJ McInnes, 'Recognizing Ecosystem Services from Wetlands of International Importance: An Example from Sussex, UK' (2013) 33 *Wetlands* 1001; J Bell-James et al, 'Can't see the (mangrove) forest for the trees: Trends in the legal and policy recognition of mangrove and coastal wetland ecosystem services in Australia' (2020) 45 *Ecosystem Services* 101148.

5 CONCLUSION

In 1971, the recitals to the Ramsar Convention explicitly acknowledged the cultural value of wetlands, yet few would have predicted at the time the regime would move as far as it has towards governing cultural aspects. Although the Convention has increasingly recognised wetland-culture linkages, it is clear that Ramsar is not – and was never intended to be – a cultural heritage regime. In a sense, Ramsar has been drawn towards culture through its broader journey examining the intersection of humans, wetlands, and water. Since the late 1990s, Ramsar – like other MEAs – has been enveloped by a sustainable development agenda and responded by massaging key concepts such as wise use and ecological character. This move has paralleled the increasing popularity of the ecological economics movement – in particular, the embrace of ecosystem services from which cultural services are born. This Chapter has traced that evolution and examined the extent to which the 'recognition of culture' may crystallise into obligation. It has suggested that while there is arguably no hard requirement to identify cultural aspects at the point of designation, there is a commitment, pursuant to Article 3.1, to maintain cultural aspects to the extent to which they:

- comprise part of the documented ecological character description of the site in the RIS; and
- are consistent with the broader context of sustainable development.

This Chapter predicted that the further Ramsar moves towards the cultural aspects of wetlands, the more it will likely find itself intertwined with debates affecting neighbouring discourses such as cultural heritage. One of the strongest critiques permeating cultural heritage studies is the contention that only certain 'experts' have been able to define – or are 'qualified' to define – what culture is and how it should be protected. These critiques have focused on the emergence, for example, of a discourse which 'privileges expert values and knowledge about the past and its material manifestations, and dominates and regulates professional heritage practices'.[126] The (hitherto) under-integration of local and Indigenous communities,[127] combined with the capacity of parties to control the cultural narrative, only feeds further into these debates. Ramsar may be able to sidestep some of these concerns, given its reluctance to promulgate specific cultural criteria for which a site could be listed;[128] yet

---

[126] Smith, note 18, 4.

[127] Bridgewater and Kim, note 11; Joshi et al, note 11.

[128] D Stroud and N Davidson, 'Fifty years of criteria development for selecting wetlands of international importance (2022) 73(10) *Marine and Freshwater Research* 1134.

its encouragement of parties to identify culture-wetland linkages necessarily invites a methodological melting pot which exposes parties – and the regime as a whole – to similar criticisms.

11. Wetlands, climate change, and international law

An Cliquet

1 INTRODUCTION

'How peat could protect the planet.'[1] 'What the world needs now to fight climate change: More swamps.'[2] 'Swamp power: how the world's wetlands can help stop climate change.'[3] These are just a few titles of scientific or newspaper articles that point to the tremendous importance of wetlands in the fight against climate change.

The relationship between wetlands and climate change is a double-edged sword: on the one hand, wetlands play a crucial role in both mitigation of and adaptation to climate change; while on the other hand, wetlands are increasingly under pressure from the effects of climate change, which will necessitate additional efforts to protect and restore them. Both aspects are interrelated: in order to fulfil their functions for mitigation and adaptation, wetlands need to be resilient; but climate change in many instances aggravates the already dire state that many wetlands are in, reducing their capacity to fulfil their role in climate change mitigation and adaptation.

This Chapter examines the extent to which international law addresses this twofold relationship between wetlands and climate change (Section 2 on mitigation and adaptation and wetlands, and Section 3 on the protection

---

[1] V Gewin, 'How peat could protect the planet' (2020) 578 *Nature* 204.

[2] W Moomaw, G Davies, and M Finlayson, 'What the world needs now to fight climate change: More swamps', *The Conversation*, 12 September 2018, at: https://theconversation.com/what-the-world-needs-now-to-fight-climate-change-more-swamps-99198.

[3] A Neslen, 'Swamp power: how the world's wetlands can help stop climate change', *The Guardian*, 20 July 2015, at: https://www.theguardian.com/environment/2015/jul/20/swamp-power-how-the-worlds-wetlands-can-help-stop-climate-change.

of wetlands under climate change). The focus is on the Ramsar Convention,[4] but the international climate regime is also briefly dealt with, as well as the Convention on Biological Diversity (CBD)[5] and EU law as key instruments to protect ecosystems and biodiversity.

Unsurprisingly, the Ramsar Convention treaty text from 1971 itself does not explicitly refer to climate change, as this was not yet a broad policy issue at the time of the adoption of the Convention. This has not prevented the conference bodies from dealing extensively with climate change and wetlands in different respects, covering mitigation, adaptation, and the protection of wetlands under pressure from climate change. This has been done through Conference of the Parties (COP) resolutions: since 2002, several have been adopted that explicitly address climate change.[6]

Climate change has also been addressed in various other documents within the framework of the Ramsar Convention, including briefing notes,[7] policy

---

[4] Convention on Wetlands of International Importance especially as Waterfowl Habitat, adopted 2 February 1971, entered into force 21 December 1975, 996 UNTS 245 (amended 1982 & 1987).

[5] Convention on Biological Diversity, adopted 5 June 1992, entered into force 29 December 1993, 1760 UNTS 79.

[6] Ramsar Convention COP resolutions relating to climate change include: COP8, *Resolution VIII.3: Climate change and wetlands: impacts, adaptation, and mitigation* (2002); COP10, *Resolution X.24: Climate change and wetlands* (2008); COP11, *Resolution XI.14: Climate change and wetlands: implications for the Ramsar Convention on Wetlands* (2012); COP12, *Resolution XII.11: Peatlands, climate change and wise use: Implications for the Ramsar Convention* (2015); COP13, *Resolution XIII.12: Guidance on identifying peatlands as Wetlands of International Importance (Ramsar Sites) for global climate change regulation as an additional argument to existing Ramsar criteria* (2018); COP13, *Resolution XIII.13: Restoration of degraded peatlands to mitigate and adapt to climate change and enhance biodiversity and disaster risk reduction* (2018); COP13, *Resolution XIII.15: Cultural values and practices of indigenous peoples and local communities and their contribution to climate-change mitigation and adaptation in wetlands* (2018); COP13, *Resolution XIII.16: Sustainable urbanization, climate change and wetlands* (2018); COP14, *Resolution XIV.17: The protection, conservation, restoration, sustainable use and management of wetland ecosystems in addressing climate change* (2022).

[7] See, eg, SM Fennessy and G Lei, 'Wetland Restoration for Climate Change Resilience' *Ramsar Briefing Note 10* (Ramsar Convention Secretariat, Gland, 2018).

briefs,[8] technical reports,[9] and the *Global Wetland Outlook* reports.[10] Some of these documents are dealt with more extensively below.

2 WETLANDS AND CLIMATE CHANGE MITIGATION AND ADAPTATION

2.1 The Role of Wetlands in Climate Change Mitigation and Adaptation

Wetlands occupy only 5%–8% of the land surface of the Earth but hold 20%–30% of the estimated global soil carbon,[11] making them a vital tool in decreasing and halting emissions of greenhouse gases. Coastal marshes and mangroves capture an average of 6–8 tonnes of carbon dioxide equivalent (CO_2e) per hectare per year – about two to four times greater than the global rates observed in mature tropical forests. Peatlands are particularly important: they cover 3% of the global land area but contain approximately 30% of all the carbon on land, equivalent to 75% of all atmospheric carbon and twice the carbon stock in the global forest biomass.[12] Around 50 million hectares of peatlands are currently drained globally and this is responsible for approximately 4% of all anthropogenic greenhouse gas emissions. At least half of

---

[8] L Dinesen et al, 'Restoring drained peatlands: A necessary step to achieve global climate goals', *Ramsar Policy Brief 5* (Ramsar Convention Secretariat, Gland, 2021).

[9] See, eg, H Gitay, CM Finlayson, and NC Davidson, 'A Framework for assessing the vulnerability of wetlands to climate change' *Ramsar Technical Report No 5/CBD Technical Series No 57* (Ramsar Convention Secretariat, Gland, & Secretariat of the Convention on Biological Diversity, Montreal, 2011).

[10] RC Gardner and CM Finlayson, *Global Wetland Outlook: State of the world's wetlands and their services to people 2018* (Ramsar Convention Secretariat, Gland, 2018); M Courouble et al, *Global Wetland Outlook: Special Edition 2021* (Ramsar Convention Secretariat, Gland, 2021).

[11] On wetlands as carbon pool in the European Union, see D Abdul Malak et al, *Carbon pools and sequestration potential of wetlands in the European Union* (European Topic Centre on Urban, Land and Soil Systems, Vienna and Malaga, 2021) 11; see also L Kopsieker, G Costa Domingo, and E Underwood, *Climate mitigation potential of large-scale restoration in Europe: Analysis of the climate mitigation potential of restoring habitats listed in Annex I of the Habitats Directive* (Institute for European Environmental Policy, Brussels, 2021) 8.

[12] C Perceval and R Cadmus, 'Wetlands: The Hidden Resource for Climate Mitigation and Adaptation', 1 December 2015, at: https://www.ramsar.org/news/wetlands-the-hidden-resource-for-climate-mitigation-and-adaptation.

these should be restored by 2030 to enable global warming to remain below 1.5–2.0°C.[13]

According to the *Global Wetland Outlook 2018*, wetlands are the world's largest carbon stores, but the benefits are partially counteracted by the release of methane from wetlands. Rising temperatures due to climate change can lead to additional release of greenhouse gases from wetlands, particularly as a consequence of permafrost melting.[14] This points to the importance of conservation and restoration of wetlands, so that they can continue to serve as carbon sinks. The *Global Wetland Outlook 2021* observes that undisturbed peatlands as well as coastal blue carbon ecosystems (mangroves, seagrass beds, salt marshes) are exceptionally powerful carbon sinks but can also be significant sources of greenhouse gases if degraded or converted.[15] Also, the April 2022 Intergovernmental Panel on Climate Change (IPCC) Report points to the mitigation potential of conservation, improved management, and restoration of forests and other ecosystems (coastal wetlands, peatlands, savannas, and grasslands).[16]

Wetlands can also play a crucial role in protecting humans from the effects of climate change, including disaster risk reduction in case of extreme weather events such as flooding or droughts.[17] According to a Ramsar Briefing Note, '[p]rioritizing wetland protection and restoration can enhance climate adaptation and resilience' against extreme weather events such as storms and flooding.[18] However, due to the loss and degradation of wetlands, their role for adaption to climate change – as well as many other ecosystem services that they provide – will be lost or diminished.[19]

---

[13] Ramsar Convention, *Harnessing Wetland Wise Use, Protection and Restoration in Delivering Climate Change Outcomes: Brief on the Ramsar Convention on Wetlands for UNFCCC COP 26*, at: https://www.ramsar.org/sites/default/files/unfccc_cop26_briefing_e_0.pdf.

[14] Gardner and Finlayson, note 10, 36.

[15] Courouble et al, note 10, 36.

[16] IPCC, *Climate Change 2022: Mitigation of Climate Change, Summary for Policymakers* (2022).

[17] Gardner and Finlayson, note 10, 40; Courouble et al, note 10, 23–24; Wetlands International, *Wetlands and climate change adaptation* (2012), at: https://www.wetlands.org/publications/wetlands-and-climate-change-adaptation-brochure-2/; see also R Kumar et al, 'Wetlands for disaster risk reduction: Effective choices for resilient communities' *Ramsar Policy Brief 1* (Ramsar Convention Secretariat, Gland, 2017).

[18] Fennessy and Lei, note 7.

[19] RC Gardner et al, 'State of the World's Wetlands and their Services to People: A compilation of recent analyses' *Ramsar Briefing Note 7* (Ramsar Convention Secretariat, Gland, 2015).

2.2 International Climate Change Law and Wetlands

The role of ecosystems as carbon sinks for greenhouse gas emissions has been recognised in international climate change law. The United Nations Framework Convention on Climate Change of 1992 (UNFCCC)[20] recognised the role of ecosystems in Article 4 by obliging state parties 'to promote and cooperate in the conservation and enhancement of sinks and reservoirs of greenhouse gases, including biomass, forests and oceans as well as other terrestrial, coastal and marine ecosystems'. The 2015 Paris Agreement[21] also refers to the role of both ecosystems and biodiversity in its preamble, '[r]ecognizing the importance of the conservation and enhancement, as appropriate, of sinks and reservoirs of the greenhouse gases referred to in the Convention', and '[n]oting the importance of ensuring the integrity of all ecosystems, including oceans, and the protection of biodiversity, recognized by some cultures as Mother Earth'.

Article 5 of the Paris Agreement provides that sinks and reservoirs of greenhouse gases should be conserved and enhanced. This article also establishes a formal basis for the decisions that have already been taken through the framework of the UNFCCC. REDD+[22] is an important mechanism that was created under the UNFCCC to promote the protection of forests as carbon sinks. REDD+ provides funding to developing countries for climate mitigation activities and sustainable forest management.[23] Its initial focus was on reducing emissions from deforestation and forest degradation, but this was subsequently broadened to include the role of conservation, sustainable forest management, and enhancement of forest carbon stocks in developing countries. The initiative was first introduced at the UNFCCC COP of 2005 and continued to be discussed at later COPs.[24] Despite the absence of a formal legal agreement before 2015 on REDD+, the COP of 2007 asked countries to

---

[20] United Nations Framework Convention on Climate Change, adopted 9 May 1992, entered into force 21 March 1994, 1771 UNTS 107.

[21] Paris Agreement, adopted 12 December 2015, entered into force 4 November 2016, 3156 UNTS 79.

[22] 'REDD' stands for 'Reducing Emissions from Deforestation and Forest Degradation in Developing Countries'.

[23] See the REDD+ Web platform at UNFCCC: http://redd.unfccc.int/. The REDD Web platform was established by UNFCCC COP *Decision 2/CP.13: Reducing emissions from deforestation in developing countries: approaches to stimulate action*, FCCC/CP/2007/6/Add.1 (2007) [10].

[24] For an overview of the relevant COP decisions, see http://unfccc.int/files/land_use_and_climate_change/redd/application/pdf/compilation_redd_decision_booklet_v1.1.pdf.

undertake REDD+ action on a voluntary basis.[25] In order to guide this work, the United Nations (UN) developed the UN-REDD programme in 2008.[26] The programme is supported by, among others, the Forest Carbon Partnership Facility[27] at the World Bank. The Ramsar COP has noted the importance of REDD+ in helping to achieve the objectives of the Ramsar Convention and has encouraged Ramsar parties to promote the importance of wetlands in ongoing discussions on this issue.[28] Although the REDD+ mechanism is certainly valuable, it applies only to forests and thus only covers forested wetlands protected under the Ramsar Convention, and not other wetland types.

The Paris Agreement primarily seeks to mitigate greenhouse gases through Nationally Determined Contributions (NDCs), meant as a bottom-up approach for the parties to the Paris Agreement to formulate national plans to mitigate greenhouse gases and thus contribute to the overall binding goals to stay well below 2°C and keep the temperature increase to 1.5°C.[29] Greenhouse gases can be mitigated by reducing emissions from various human activities or through 'nature-based solutions'. A November 2021 Report[30] from the World Wide Fund for Nature (WWF) assessed the integration of nature-based solutions in updated or revised NDCs from 114 countries. The Report found a significant increase in the number of NDCs that mentioned wetlands and mangroves compared to the previous versions. Additionally, 51 updated NDCs mentioned wetlands compared to 32 previous NDCs and 43 mentioned mangroves compared to 29 previous NDCs. Although this is an improvement, the Report advocated the upscaling of nature-based solutions in the NDCs. A 2023 study reached a similar conclusion.[31]

Under the UNFCCC, greater attention is also gradually being paid to adaption to the effects of climate change. Within the Cancun Adaptation Framework of 2010, parties were encouraged to increase the resilience of socio-economic and ecological systems, including through sustainable management of natural

---

[25] UNFCCC COP, *Decision 2/CP.13: Reducing emissions from deforestation in developing countries: approaches to stimulate action*, FCCC/CP/2007/6/Add.1 (2007) [1].
[26] Source: http://www.un-redd.org/.
[27] Source: https://www.forestcarbonpartnership.org/.
[28] Ramsar Convention, *Resolution XI.14*, note 6, [14].
[29] See Paris Agreement, note 21, Art 2(1)(a).
[30] WWF, *NDCs – A Force for Nature*, 4th edition, Nature in Enhanced NDCs (November 2021), at: https://wwfint.awsassets.panda.org/downloads/wwf_ndcs_for_nature_4th_edition.pdf.
[31] H Zhai, B Gu, and Y Wang, 'Evaluation of policies and actions for nature-based solutions in nationally determined contributions' (2023) 131 *Land Use Policy* 106710.

resources.[32] The 2015 Paris Agreement repeats and formally adopts the Cancun Adaptation Framework in Article 7(9):

> Each Party shall, as appropriate, engage in adaptation planning processes and the implementation of actions, including the development or enhancement of relevant plans, policies and/or contributions, which may include: ... (e) Building the resilience of socioeconomic and ecological systems, including through economic diversification and sustainable management of natural resources.

2.3 International Biodiversity Law and Climate Change Mitigation and Adaptation

The Ramsar COP has also addressed the issue of mitigation and adaptation in several resolutions. Resolution VIII.3 on climate change and wetlands from 2002, among other things, called on 'relevant countries to take action to minimise the degradation, as well as promote restoration, and improve management practices of those peatlands and other wetland types that are significant carbon stores, or have the ability to sequester carbon'.[33] The Resolution also exhorts parties 'to manage wetlands so as to increase their resilience to climate change and extreme climatic events, and to reduce the risk of flooding and drought in vulnerable countries by, *inter alia*, promoting wetland and watershed protection and restoration.'[34]

In Resolution X.24 of 2008, the parties acknowledged the 'increasing evidence that some types of wetlands play important roles as carbon stores', but were also 'concerned that this is not yet fully recognised by international and national climate change response strategies, processes, and mechanisms'. The parties recognised the significant progress made since Ramsar COP8 (2002) 'with respect to peatland inventory and awareness of the carbon storage function of wetlands such as peatlands'.[35] The Resolution urged contracting parties to:

> take immediate action to reduce the degradation, promote restoration, improve management practices of peatlands and other wetland types that are significant greenhouse gas sinks, and to encourage expansion of demonstration sites on peat-

---

[32] UNFCCC COP, *Decision 1/CP.16: The Cancun Agreements: Outcome of the work of the Ad Hoc Working Group on Long-term Cooperative Action under the Convention*, FCCC/CP/2010/7/Add.1 (2010) [14, d].
[33] Ramsar Convention, *Resolution VIII.3*, note 6, [15].
[34] Ibid, [14].
[35] Ramsar Convention, *Resolution X.24*, note 6, [8]–[9].

land restoration and wise use management in relation to climate change mitigation and adaptation activities.[36]

In the same Resolution, attention is also paid to adaptation. The parties noted that:

> the wise use and restoration of wetlands contributes to building the resilience of human populations to climate change impacts and can attenuate natural disasters expected with climate change, such as the use of restored floodplain wetlands to reduce risks from flooding.[37]

The Resolution further recommends that parties promote restoration for several wetland types, including rivers, lakes, coastal wetlands, and peatlands.[38]

In 2012, Resolution XI.14 highlighted that the continuing degradation and loss of some types of wetlands cause the release of large amounts of stored carbon and thus exacerbate climate change.[39] The parties recognised that:

> scientific reports indicate that degradation and loss of many types of wetlands is occurring more rapidly than in other ecosystems and that climate change is likely to exacerbate this trend, which will further reduce the mitigation and adaptation capacity of wetlands, and, since the conservation and wise use of wetlands has the potential to halt this degradation, the designation of Ramsar Sites, together with their effective management, as well as that of other wetlands, can in some regions play a vital role in carbon sequestration and storage and therefore in the mitigation of climate change.[40]

Parties are urged to sequester and store carbon for climate change mitigation, on the one hand, by maintaining and enhancing the ecological functions of wetlands; and on the other hand, by reducing or halting the release of stored carbon from the degradation and loss of wetlands. Parties should also promote the ability of wetlands to contribute to nature-based climate change adaptation.[41]

The importance of nature-based solutions delivered by wetland ecosystems to address climate change is also recognised and confirmed by Resolution XIV.17.[42] In order to fulfil these functions, parties are 'strongly' encouraged

---

[36] Ibid, [32].
[37] Ibid, [18].
[38] Ibid, [30]–[32].
[39] Ramsar Convention, *Resolution XI.14*, note 6, [26].
[40] Ibid, [13].
[41] Ibid, [26].
[42] Ramsar Convention, *Resolution XIV.17*, note 6.

both to phase out or modify policies that contribute to wetlands loss and degradation and to take action to conserve and restore wetlands.[43]

Several Ramsar resolutions have specifically addressed peatlands. Resolution XII.11 of 2015 encouraged parties, as appropriate, 'to consider limiting activities that lead to drainage of peatlands and may cause subsidence, flooding and the emission of greenhouse gases and urged greater international cooperation, technical assistance and capacity building to address this'.[44] Resolution XIII.13 from 2018 is specifically dedicated to the restoration of peatlands for climate mitigation and adaptation. The Resolution encourages parties to 'develop or improve legislation on restoration and rewetting of degraded peatlands, as well as on the protection and sustainable use of peatlands in general'; and to 'conserve existing peatlands and to restore degraded peatlands in their territory, as one means to contribute to climate-change mitigation, adaptation, biodiversity conservation, and disaster risk reduction'.[45] Peatland conservation and restoration measures are seen as a way to contribute to NDCs under the Paris Agreement.[46]

In addition to the resolutions, the Ramsar Fourth Strategic Plan 2016–24[47] sees the importance of wetlands for climate change mitigation and adaptation as one of its priority areas. Target 12 of the Strategic Plan states: 'Restoration is in progress in degraded wetlands, with priority to wetlands that are relevant for biodiversity conservation, disaster risk reduction, livelihoods and/or climate change mitigation and adaptation.' Annex 1 to the Strategic Plan includes baselines and indicators for the various targets. The baseline for Target 12 shows that 68% of parties have identified priority sites for restoration and 70% have implemented restoration or rehabilitation programmes. Indicators to assess the implementation of Target 12 are twofold:

- [Percentage] of parties that have established restoration plans [or activities] for sites. (Data source: National Reports).
- [Percentage] of parties that have implemented effective restoration or rehabilitation projects. (Data source: National Reports).[48]

---

[43] Ibid, [11].
[44] Ramsar Convention, *Resolution XII.11*, note 6, [21].
[45] Ramsar Convention, *Resolution XIII.13*, note 6, [23]–[24].
[46] Ibid, [32].
[47] Ramsar Convention, *Resolution XII.2: The Ramsar Strategic Plan 2016–2024* (2015).
[48] Ibid, Annex 1.

The Ramsar Strategic Plan refers to synergies with the CBD Aichi Targets. Target 12 of the Ramsar Strategic Plan corresponds to Aichi Target 15,[49] stating that 15% of degraded ecosystems should be restored by 2020, thereby contributing to climate change mitigation and adaptation. According to *Global Biodiversity Outlook 5*, this target has not been met.[50] Aichi Target 10 is also relevant for wetlands: according to this target, by 2015, the multiple anthropogenic pressures on coral reefs and other vulnerable ecosystems impacted by climate change or ocean acidification should be minimised so as to maintain their integrity and functioning. This target has also not been achieved.[51]

As a follow-up to the Aichi Biodiversity Targets, the Kunming-Montreal Global Biodiversity Framework was adopted at CBD COP15 in December 2022.[52] On mitigation and adaptation, Target 8 is important:

> Minimize the impact of climate change and ocean acidification on biodiversity and increase its resilience through mitigation, adaptation, and disaster risk reduction actions, including through nature-based solutions and/or ecosystem-based approaches, while minimizing negative and fostering positive impacts of climate action on biodiversity.

Other targets can also contribute to climate mitigation and adaptation through the protection and restoration of biodiversity. These include Target 2, which calls for at least 30% of areas of degraded ecosystems to be under effective restoration by 2030; and Target 3, which calls for 30% protected areas by 2030.[53]

As the Global Biodiversity Framework is yet another (non-binding) COP decision, the question is to what extent these new targets will be more successful than their predecessors, such as the Aichi Targets. Also, a specific focus on wetlands is absent. Given their important role for both biodiversity and climate

---

[49] CBD COP, *Decision X/2: The Strategic Plan for Biodiversity 2011–2020 and the Aichi Biodiversity Targets* (UNEP/CBD/COP/DEC/X/2, 2010); Target 15: 'By 2020, ecosystem resilience and the contribution of biodiversity to carbon stocks has been enhanced, through conservation and restoration, including restoration of at least 15% of degraded ecosystems, thereby contributing to climate change mitigation and adaptation and to combating desertification.' On interactions with the CBD, see further Chapter 8 in this volume.

[50] Secretariat of the Convention on Biological Diversity, *Global Biodiversity Outlook 5* (Montreal, 2020) 100.

[51] Ibid, 78.

[52] The Post-2020 Biodiversity Framework should have been adopted in 2020 at COP15 but was postponed to 2022 because of the COVID-19 pandemic.

[53] Annex to Decision 15/4. Kunming-Montreal Global Biodiversity Framework, CBD/COP/DEC/15/4 (2022).

change mitigation and adaptation, wetlands deserve specific attention in the Global Biodiversity Framework.[54]

Another example of an initiative that points to the importance of wetlands for climate mitigation and adaptation is the UN Decade on Ecosystem Restoration 2021–30,[55] described as a 'rallying call for the protection and revival of ecosystems all around the world, for the benefit of people and nature. It aims to halt the degradation of ecosystems, and restore them to achieve global goals'.[56] The Ramsar Convention, as a global partner to the UN Decade on Restoration, has produced three factsheets to assist restoration efforts by contracting parties, policymakers, and practitioners.[57] In the framework of the Decade, several international non-governmental organisations (NGOs) have specifically advocated a 'Wetland Decade under the UN Decade on Ecosystem Restoration (2021–2030)'.[58]

2.4 Concerns for Wetlands and Climate Change Mitigation and Adaptation

While it is clear that wetlands play a very important role in climate change mitigation and adaptation, and that this role has increasingly been recognised in both international climate change law and international biodiversity law, particular concerns warrant further consideration – namely, the focus on forests; the focus on carbon storage and sequestration; and the lack of binding and/or concrete obligations.

---

[54] See also Wetlands International, *Policy Brief. Ensuring the Global Biodiversity Framework prioritises measures to safeguard wetlands and wetland biodiversity*, at: https://www.wetlands.org/publications/ensuring-the-global-biodiversity-framework-prioritises-measures-to-safeguard-wetlands-and-wetland-biodiversity/.

[55] United Nations Decade on Ecosystem Restoration (2021–2030), A/RES/73/284 (6 March 2019).

[56] UN Environment Programme, 'About the UN Decade', at: https://www.decadeonrestoration.org/about-un-decade.

[57] Ramsar Convention, *Wetlands restoration: unlocking the untapped potential of the Earth's most valuable ecosystem* (2021); Ramsar Convention, *Realizing the full potential of marine and coastal wetlands: why their restoration matters* (2021); Ramsar Convention, *Restoring drained peatlands: now an environmental imperative* (2021).

[58] International Organization Partners of the Ramsar Convention on Wetlands, 'Call for Wetland Decade under the UN Decade on Ecosystem Restoration (2021–2030)', 22 March 2019, at: https://www.iucn.org/news/water/201903/call-wetland-decade-under-un-decade-ecosystem-restoration-2021-2030.

Although both climate change law and biodiversity law recognise a variety of ecosystems as carbon sinks, which play an important role in climate change mitigation, the primary focus in international law and policy is often on the role of forests. This is also apparent from non-binding commitments such as the Bonn Challenge and the New York Declaration on Forests.[59] Although the increased attention on forest ecosystems certainly has its merits, other ecosystems such as non-forested peatlands can also play a major role in carbon sequestration.[60] If mitigation policy is too narrowly focused on forests without acknowledging the importance of other ecosystems, there is a danger that policymakers may pursue ecologically perverse programmes by, for example, planting trees in grassland ecosystems.[61] Scientists are thus calling for the scope of initiatives such as the Bonn Challenge to be broadened to include other ecosystems.[62] Also, recent policy initiatives, such as the EU Biodiversity Strategy to 2030, include tree planting schemes, raising concerns that this might lead to inappropriate afforestation.[63]

Another potentially perverse outcome is that the conservation and restoration of forests are seen exclusively as a way to enhance carbon stocks, disregarding biodiversity and other ecosystem services that are provided by forests and other ecosystems.[64] The REDD+ mechanism, for example, has been criticised for its focus on enhancing forest carbon stocks, as it is possible that other

---

[59] Sources: https://www.bonnchallenge.org/ and https://forestdeclaration.org/; see also A Cliquet, 'International law and policy on restoration' in S Allison and S Murphy (eds), *Routledge Handbook of Ecological and Environmental Restoration* (Routledge, Abingdon, 2017) 387–400.

[60] See M Strack (ed), *Peatlands and Climate Change* (International Peatland Society, 2nd edition, Jyväskylä, 2023).

[61] JW Veldman et al, 'Tyranny of trees in grassy biomes' (2015) 347 *Science* 484; JW Veldman et al, 'Where Tree Planting and Forest Expansion are Bad for Biodiversity and Ecosystem Services' (2015) 65 *BioScience* 1011.

[62] VM Temperton et al, 'Step back from the forest and step up to the Bonn Challenge: how a broad ecological perspective can promote successful landscape restoration' (2019) 27 *Restoration Ecology* 705.

[63] C Tölgyesi et al, 'Urgent need for updating the slogan of global climate actions from "tree planting" to "restore native vegetation"' (2021) 30 *Restoration Ecology* 3, e13594.

[64] A Telesetsky, A Cliquet, and A Akhtar-Khavari, *Ecological Restoration in International Environmental Law* (Routledge, Abingdon, 2017) 270.

services and social issues could be adversely affected;[65] and it is also clear that monocultures provide far fewer benefits than restored natural forests.[66]

The role of wetlands in climate mitigation is increasingly recognised in international law and policy, but many initiatives are soft law or are voluntary or political commitments. The pronouncements of multilateral agreements, such as the Ramsar Convention and the CBD, are non-binding decisions or recommendations. Also, the language of the different Ramsar Resolutions remains soft, 'urging' or 'encouraging' states to protect and restore wetlands. It may be asked whether the Ramsar Convention framework has sufficient teeth to stop further degradation and restore wetlands, and thus contribute to climate change mitigation and adaptation.

Much will depend on the willingness of states to take action at the national level. Many initiatives exist, such as the Scottish government's Climate Change Plan 2018–32 update, which aims to restore at least 250 000 hectares of degraded peatland by 2030,[67] and the National Peatland Action Programme in Wales.[68] But here again, the question arises as to whether the national initiatives taken thus far are adequate and whether the challenges are being sufficiently addressed.[69] Voluntary initiatives – mostly driven by NGOs – also exist, including major programmes such as the Global Peatlands Initiative,[70]

---

[65] JM Bullock et al, 'Restoration of ecosystem services and biodiversity: conflicts and opportunities' (2011) 10 *Trends in Ecology and Evolution* 541.

[66] S Alexander et al, 'Opportunities and Challenges for Ecological Restoration within REDD+' (2011) 19 *Restoration Ecology* 683; see also A Di Sacco et al, 'Ten golden rules for reforestation to optimize carbon sequestration, biodiversity recovery and livelihood benefits' (2021) 27 *Global Change Biology* 1328.

[67] Scottish Government, *Securing a green recovery on a path to net zero: Climate change plan 2018–2032 – update*, 16 December 2020, at: https://www.gov.scot/publications/securing-green-recovery-path-net-zero-update-climate-change-plan-20182032/pages/12/.

[68] Natural Resources Wales, 'The National Peatland Action Programme', 23 August 2024, at: https://naturalresources.wales/evidence-and-data/maps/the-national-peatland-action-programme/?lang=en; see also V Jenkins and J Walker, 'Maintaining, Enhancing and Restoring the Peatlands of Wales: Unearthing the Challenges of Law and Sustainable Land Management' (2022) 34 *Journal of Environmental Law* 163.

[69] R Andersen et al, 'An overview of the progress and challenges of peatland restoration in Western Europe' (2017) 25 *Restoration Ecology* 271; I Brown, 'Challenges in delivering climate change policy through land use targets for afforestation and peatland restoration' (2020) 107 *Environmental Science & Policy* 36.

[70] Source: https://www.globalpeatlands.org/. The Global Peatlands Initiative is an effort to save peatlands as the world's largest terrestrial organic carbon stock

the Global Mangrove Alliance,[71] and the Freshwater Challenge.[72] However, the lack of concrete and/or binding obligations for restoration, the often predominantly quantitative targets, and the lack of guidance create a need for better policies, laws, and standards.[73]

Given the urgency of both the climate and biodiversity crises, there is a pressing need for urgent and strong protection. First, remaining wetlands such as peatlands should be protected though strong, legally binding norms. An immediate non-deterioration obligation for all remaining wetlands is vital. Examples of such clauses for protected areas can be found in EU law – most significantly, in Article 6(2) of the Habitats Directive.[74]

Second, further commitments to protect remaining wetlands are required, as are binding targets on restoration. Again, we can look to EU law and policy for inspiration. The EU Biodiversity Strategy to 2030[75] commits to protect at least 30% of land and sea area by 2030, of which 10% should be strictly protected. According to the Strategy, all remaining primary and old-growth forests in the European Union should be strictly protected; but '[s]ignificant areas of other carbon-rich ecosystems, such as peatlands, grasslands, wetlands, mangroves and seagrass meadows should also be strictly protected, taking into account projected shifts in vegetation zones'.[76] Member States have until the end of 2023 to demonstrate significant progress in designating new protected areas.

---

and to prevent carbon from being emitted into the atmosphere. It was formed by experts and institutions at the UNFCCC COP in Marrakech in 2016.

[71] The Global Mangrove Alliance seeks to increase global mangrove cover by 20% by 2030. It is a collaboration between NGOs, governments, industry, local communities, and funders, and was launched in 2017. See https://www.mangrovealliance.org/.

[72] 'The Freshwater Challenge (FWC) is a country-led initiative, launched at the UN Water Conference in March 2023. Its aim is to restore 300,000 km of degraded rivers and 350 million hectares of degraded wetlands by 2030. See https://www.freshwaterchallenge.org/about-the-challenge.

[73] See, eg, SY Lee et al, 'Better restoration policies are needed to conserve mangrove ecosystems' (2019) 3 *Nature Ecology and Evolution* 870; see also A Cliquet et al, 'Upscaling ecological restoration: toward a new legal principle and protocol on ecological restoration in international law' (2021) 30 *Restoration Ecology* e13560.

[74] See Chapter 12 in this volume.

[75] Communication from the Commission to the European Parliament, the Council, the European Economic and Social Committee and the Committee of the Regions, *EU Biodiversity Strategy for 2030: Bringing nature back into our lives*, COM(2020) 380 final (Brussels, 20 May 2020).

[76] Ibid, Section 2.1.

The European Commission will assess by 2024 whether stronger measures, including further legislation, are needed.[77]

Another important part of the EU Biodiversity Strategy is the EU nature restoration plan.[78] This advances the commitment to raise the level of implementation of existing law, including the request that EU Member States ensure there is no deterioration in conservation trends and the status of all protected habitats and species by 2030. The European Commission has also been tasked with developing a proposal for binding restoration targets. The Strategy further includes specific commitments for different ecosystems, including freshwater ecosystems – for example, at least 25 000 kilometres of rivers will be restored into free-flowing rivers by 2030 through the removal of primarily obsolete barriers and the restoration of floodplains and wetlands.

A European Parliament Resolution of 2021[79] warmly welcomed the commitment to draw up a legislative proposal on the EU nature restoration plan, including binding restoration targets. According to the Resolution, the proposal should include ecosystem, habitat, and species-specific targets at the EU and Member State levels on the basis of their ecosystems, 'with a particular emphasis on ecosystems for the dual purposes of biodiversity restoration and climate change mitigation and adaptation', stressing that these should include forests, grasslands, wetlands, peatlands, pollinators, free-flowing rivers, coastal areas, and marine ecosystems; and that 'after restoration, no ecosystem degradation should be allowed'.[80] A strong restoration law has also been advocated by restoration scientists in their Declaration in support of an ambitious restoration law,[81] which specifically mentioned the importance of restoration of wetlands.

The European Commission proposed a Nature Restoration Law on 22 June 2022.[82] While this Proposal was one of the most promising initiatives

---

[77] Ibid.

[78] Ibid, Section 2.2.

[79] European Parliament Resolution on the EU Biodiversity Strategy for 2030: Bringing nature back into our lives (2020/2273(INI)), 9 June 2021, *OJ C* 67, 8 February 2022, 25–55.

[80] Ibid, [36]–[37].

[81] Declaration – Scientists in Support for an Ambitious EU Nature Restoration Law, approved during the 12th European Conference on Ecological Restoration, organised by the European Chapter of the Society for Ecological Restoration, 7–10 September 2021.

[82] European Commission, Proposal for a Regulation of the European Parliament and of the Council on nature restoration, COM(2022) 304 final, 2022/0195 (COD), Brussels, 22 June 2022, at: https://environment.ec.europa.eu/publications/nature-restoration-law_en. For an analysis of this law proposal, see several notes from SERE Legal Working Group, at: https://Chapter.ser.org/europe/.

for wetland restoration – not just for the European Union, but possibly also setting a precedent for other regions – it suffered from misinformation and opposition by certain lobby groups and political parties.[83] As a result, the Proposal was weakened during the subsequent legislative procedure.[84] A political compromise on an amended and weakened text was reached between the European Commission, the European Council, and the European Parliament in November 2023. The law was finally adopted in June 2024 and entered into force on 18 August 2024.[85] The law was adopted in the form of a regulation, which means that it is directly applicable in EU Member States and need not be transposed into national law.

The purpose of this law is to contribute to the recovery of nature, to achieve climate mitigation and adaptation objectives, to enhance food security, and to meet international commitments.[86] The preamble of the law refers on several occasions to the importance of wetlands for carbon storage and climate adaptation. However, to function in this regard, wetlands must be in good condition; hence, restoration measures are required. The law is not limited to the restoration of wetland ecosystems, but wetlands, peatlands, rivers, and coastal ecosystems are all explicitly addressed. The law sets out restoration obligations for different terrestrial ecosystems listed in Annex I, including several coastal and inland wetland types.[87] The added value of the law is that the restoration obligations impose quantitative targets and deadlines for taking restoration measures. It also includes non-deterioration clauses and obligations for continuous improvement of the listed habitat types. Aside from these obligations for listed habitat types, the law sets out specific obligations for the restoration of the natural connectivity of rivers and the natural functions of related floodplains.[88] For peatlands in agricultural use, quantitative targets for restoration and rewetting have been included,[89] in addition to the restoration obligations for peatlands under Annex I. Contrary to scientific insights on the importance of peatlands, the quantitative targets for restoring and rewetting peatlands in

---

[83] A Cliquet et al, 'The negotiation process of the EU Nature Restoration Law Proposal: Bringing nature back in Europe against the backdrop of political turmoil?' (2024) 32 *Restoration Ecology* e14158.

[84] In the European Union, a law proposed by the European Commission must be adopted by both the European Parliament and the European Council.

[85] Regulation (EU) 2024/1991 of the European Parliament and of the Council of 24 June 2024 on nature restoration and amending Regulation (EU) 2022/869, *OJ L*, 2024/1991, 29 July 2024.

[86] Ibid, Art 1.
[87] Ibid, Art 4.
[88] Ibid, Art 9.
[89] Ibid, Art 11.

agricultural use were dramatically lowered in ambition in the finally adopted law. Additionally, the restoration obligations for marine ecosystems, forests, and even urban ecosystems may also be of relevance to wetlands.[90] The law must be implemented by EU Member States through national restoration plans that are due within two years of its adoption.[91] Despite being weakened, the law still has the potential to substantially upscale the restoration of wetlands within the European Union.

Another EU legislative initiative that may have an impact on the protection and restoration of wetlands is the amendment of the Land-Use, Land-Use Change and Forestry Regulation.[92] The Regulation sets a binding commitment for each Member State to ensure that accounted emissions from land use are entirely compensated by an equivalent accounted removal of CO_2 from the atmosphere through action in the sector (no debit rule). From 2026 to 2030, the scope will be extended to all managed land, including wetlands. Furthermore, the amended Regulation moves away from the no debit rule, setting a net carbon removal target of 310 million tonnes of CO_2e by 2030. According to the European Environment Agency, the European Union is not on track to meet the 2030 net removal target of 310 metric tons of carbon dioxide equivalent.[93]

A third EU legislative initiative with relevance to the restoration of wetlands in the context of climate change is the so-called 'Taxonomy Regulation', which includes a classification system for sustainable activities in order to allow for sustainable investment.[94] It applies to activities such as renewable

---

[90] Ibid, respectively Arts 5, 12, and 8.

[91] Ibid, Arts 14–19.

[92] Regulation (EU) 2018/841 of the European Parliament and of the Council of 30 May 2018 on the inclusion of greenhouse gas emissions and removals from land use, land use change and forestry in the 2030 climate and energy framework, and amending Regulation (EU) No 525/2013 and Decision No 529/2013/EU, *OJ L* 156, 19 June 2018; amended by Regulation (EU) 2023/839 of the European Parliament and of the Council of 19 April 2023 amending Regulation (EU) 2018/841 as regards the scope, simplifying the reporting and compliance rules, and setting out the targets of the Member States for 2030, and Regulation (EU) 2018/1999 as regards improvement in monitoring, reporting, tracking of progress and review, *OJ L* 107, 21 April 2023.

[93] European Environment Agency, 'Greenhouse gas emissions from land use, land use change and forestry in Europe', 24 October 2023, at: https://www.eea.europa.eu/en/analysis/indicators/greenhouse-gas-emissions-from-land#:~:text=Removals%20are%20estimated%20to%20increase,50%20Mt%20CO2e.

[94] Regulation (EU) 2020/852 of the European Parliament and of the Council of 18 June 2020 on the establishment of a framework to facilitate sustainable investment, and amending Regulation (EU) 2019/2088, *OJ L* 198, 22 June 2020;

energy and energy efficiency, but – interestingly – also applies to wetland restoration. The considerations of Regulation 2021/2139 highlight the role of wetland restoration in climate mitigation and adaptation.[95] Annex I sets out the technical screening criteria for determining the conditions under which an economic activity qualifies as contributing substantially to climate change mitigation and whether that economic activity causes no significant harm to any of the other environmental objectives. This Annex contains a chapter on wetland restoration. 'Wetlands' are defined with references to the definitions of 'wetland' and 'peatland' in the Ramsar Convention.[96] The technical criteria include requirements for a restoration plan (in relation to which the Regulation also refers to Ramsar guidelines), a climate benefit analysis and a guarantee of permanence. The latter requires that the wetland status of the area in which the activity takes place be guaranteed by one of the following measures:

- The area is designated to be retained as wetland and may not be converted to other land use;
- The area is classified as a protected area; or
- The area is the subject of a legal or contractual guarantee ensuring that it will remain a wetland.[97]

All of these initiatives confirm that increasing attention is being paid to wetlands in the context of climate change. Whereas in the past, forests were the primary focus, recent legislative initiatives demonstrate a clear policy recognition of the role that wetlands play in climate change mitigation and adaptation. However, in order to fulfil this role, wetlands must have a good conservation status. The next section of this Chapter elaborates on the conservation of wetlands under pressure from climate change.

Commission Delegated Regulation (EU) 2021/2139 of 4 June 2021 supplementing Regulation (EU) 2020/852 of the European Parliament and of the Council by establishing the technical screening criteria for determining the conditions under which an economic activity qualifies as contributing substantially to climate change mitigation or climate change adaptation and for determining whether that economic activity causes no significant harm to any of the other environmental objectives, C/2021/2800, *OJ L* 442, 9 December 2021.

[95] Regulation 2021/2139, consideration 18.
[96] Ibid, Annex I, 2.1, footnotes 64 and 65.
[97] Ibid, Annex I, 2.1. [4.1].

3 THE PROTECTION OF WETLANDS IN A CHANGING CLIMATE

3.1 Effects of Climate Change on Wetlands

Natural wetlands have been disappearing at an alarming rate, with an estimated loss of up to 87% since the start of the eighteenth century.[98] According to the *Global Wetland Outlook 2018*, between 1970 and 2015, inland and marine/coastal wetlands declined by approximately 35% (where data is available) – three times the rate of forest loss.[99] It further noted that '[s]ince 1970, 81% of inland wetland species populations and 36% of coastal and marine species have declined'.[100]

There are several drivers of wetland loss, including land use change, water abstraction, the introduction or removal of species, and resource consumption.[101] Climate change represents another threat to wetlands and may influence water volumes, flows, temperature, invasive species, nutrient balance, and fire regimes.[102] Climate change is expected to act in conjunction with other existing pressures to wetlands, many of which are already a cause for concern. Therefore, it is important to prevent or reduce additional stressors that can reduce the ability of wetlands to respond to climate change.[103]

A common recommendation by scientists to climate-proof ecosystems is to increase their resistance and resilience.[104] 'Resistance' means the ability of a system to remain unchanged in the face of external forces; whereas 'resilience' means the ability of a system to recover from perturbations.[105]

---

[98] NC Davidson, 'How much wetland has the world lost? Long-term and recent trends in global wetland area' (2014) 65 *Marine and Freshwater Research* 936.

[99] Gardner and Finlayson, note 10, 5.

[100] Ibid.

[101] Ibid, 45.

[102] Ibid, 52; Courouble, note 10, 18.

[103] KL Erwin, 'Wetlands and global climate change: the role of wetland restoration in a changing world' (2009) *Wetlands Ecology and Management* 17, 71.

[104] See, eg, N Heller and E Zavaleta, 'Biodiversity management in the face of climate change: a review of 22 years of recommendations' (2009) 142 *Biological Conservation* 4; A Trouwborst, 'International nature conservation law and the adaptation of biodiversity to climate change: a mismatch?' (2009) 21 *Journal of Environmental Law* 419; A Cliquet, 'International and European law on protected areas and climate change: need for adaptation or implementation?' (2014) 54 *Environmental Management* 720.

[105] See further J Lawler, 'Climate Change Adaptation Strategies for Resource Management and Conservation Planning' (2009) 1162 *The Year in Ecology and Conservation Biology* 81.

Ecosystems should be made more resilient in order to be able to withstand the additional pressure from climate change, as resilient systems will most likely be able to continue to function. The restoration of ecosystems and ecosystem functions and the recovery of species are seen as important strategies for increasing the resilience of ecosystems to recover from the negative effects of climate change.[106] Building resilience in wetlands will help them not only to cope with the effects of climate change, but also to fulfil their capacity for climate change mitigation and adaptation and the provision of other ecosystem services.[107]

Several Ramsar resolutions seek to enhance the resilience of wetlands to cope with climate change. Resolution VIII.3 of 2002 called on contracting parties to 'manage wetlands so as to increase their resilience to climate change and extreme climatic events'.[108] Resolution X.24 (2008) urged contracting parties to:

> manage wetlands wisely to reduce the multiple pressures they face and thereby increase their resilience to climate change and to take advantage of the significant opportunities to use wetlands wisely as a response option to reduce the impacts of climate change.[109]

Resolution XI.14 (2012) moved from mere 'management' to enhancing and restoring wetlands by exhorting contracting parties:

> to maintain or improve the ecological character of wetlands, including their ecosystem services, to enhance the resilience of wetlands as far as possible in the face of climate-driven ecological changes including, where necessary, to promote the restoration of degraded wetlands.[110]

3.2 Designation and Management of Protected Areas

One important measure that can increase the resilience of wetlands is the designation and management of protected areas. This is recognised in Resolution XI.14 of 2012, which highlighted the role that the designation and

---

[106] J Mawdsley, R O'Malley, and D Ojima, 'A Review of Climate-Change Adaptation Strategies for Wildlife Management and Biodiversity Conservation' (2009) 23 *Conservation Biology* 1080, 1083.

[107] Gardner et al, note 19.

[108] Ramsar Convention, *Resolution VIII.3*, note 6, [14].

[109] Ramsar Convention, *Resolution X.24*, note 6, [28].

[110] Ramsar Convention, *Resolution XI.14*, note 6, [26].

effective management of Ramsar Sites can play in promoting adaptation and resilience to climate change.[111]

The criteria for the designation of Ramsar Sites can also take climate change into account. Criterion 2 provides that '[a] wetland should be considered internationally important if it supports vulnerable, endangered, or critically endangered species or threatened ecological communities'.[112] The guidelines for the application of Criterion 2 specifically mention climate change, as follows:

> [When] identifying sites with threatened ecological communities, greatest conservation value will be achieved through the selection of sites with ecological communities that have one or more of the following characteristics. They: i) are globally threatened communities or communities at risk from direct or indirect drivers of change, particularly where these are of high quality or particularly typical of the biogeographic region; and/or ... iv) can no longer develop under contemporary conditions (because of climate change or anthropogenic interference for example).[113]

A 2018 Resolution added peatlands to the criteria for the designation of Ramsar Sites by replacing and superseding Appendix E2 of the Strategic Framework and guidelines for the future development of the List of Wetlands of International Importance of the Convention on Wetlands.[114] Peatlands considered for designation under Criterion 1[115] include:

> pristine, peat-forming peatlands, some human-modified and naturally degrading peatlands that are no longer forming peat, and restored or rehabilitated peatlands that meet the criteria. They may consist of a mosaic of different peatland types with various levels of human impact.[116]

---

[111] Ibid, [8].

[112] Ramsar Convention, *The Ramsar Sites Criteria*, at: https://www.ramsar.org/sites/default/files/documents/library/ramsarsites_criteria_eng.pdf.

[113] Ramsar Convention, *Resolution IX.1, Annex B. Revised Strategic Framework and guidelines for the future development of the List of Wetlands of International Importance* (2005).

[114] Ramsar Convention, *Resolution XIII.12*, note 6; for the strategic framework, see Ramsar Convention, *Resolution XI.8 Annex 2 (Rev COP13): Strategic Framework and guidelines for the future development of the List of Wetlands of International Importance of the Convention on Wetlands (Ramsar, Iran, 1971) – 2012 revision* (2012).

[115] 'Criterion 1: A wetland should be considered internationally important if it contains a representative, rare, or unique example of a natural or near-natural wetland type found within the appropriate biogeographic region.'

[116] Resolution XIII.12, Appendix E2 of the Strategic Framework and guidelines for the future development of the List of Wetlands of International Importance of the Convention on Wetlands [10].

In designating peatlands as Ramsar Sites, 'special attention' should be paid to:

- 'peatland areas with specific attributes' – including, among others, large carbon storage and active carbon sequestration – and peatlands with high potential as 'nature-based solutions', to reduce the risks of impacts related to climate change;[117] and
- 'vulnerable peatlands (for example, where minor impacts could lead to major degradation), ... degraded peatlands with high potential for restoration and ... peatlands that reduce the vulnerability of nearby human populations in the face of climate change.'[118]

Furthermore, the revised Annex highlights the application of Criterion 1 with respect to carbon storage. Peatlands:

> for which the relevance of climate-change adaptation and mitigation is considered in the process of their designation as demonstration sites with respect to Criterion 1 would feature (some of) the following attributes: a. Large peat volume that can be preserved, always in proportion to the area of the territory of the contracting party, which makes the request/proposal; b. Information on the area's history, land use, hydrology, and peat volume, to enable assessment of the effects of restoration, as appropriate, on carbon store capacity and GHG fluxes to be used for communication and awareness raising ...[119]

Similar provisions can be found on the designation criteria for mangroves as Ramsar Sites (Appendix E4): in determining the appropriate boundaries for site designation, consideration should be given, among other things, to 'provision for the effects of sea-level rise and human-induced climate changes that may otherwise lead to loss of habitat and genetic processes; and consideration of the possible landward migration of mangroves in response to sea level rise'.[120]

Although the criteria allow for the designation of areas and ecosystems that are important for climate change mitigation and adaptation, there is no absolute obligation to give protected status to important habitats such as peatlands. The Ramsar Convention does not contain quantitative obligations (with the exception of the obligation in the Convention to designate at least one site in the international Ramsar list).[121] The Aichi Targets under the Biodiversity Convention set quantitative obligations for protected areas: under Target

---

[117] Ibid, [11].
[118] Ibid, [12].
[119] Ibid, [13].
[120] Ramsar Convention, *Resolution XI.8 Annex 2 (Rev COP13)*, note 114, [67].
[121] Article 2.4: 'Each Contracting Party shall designate at least one wetland to be included in the List when signing this Convention or when depositing its

11, at least 17% of terrestrial and 10% of marine areas should be conserved through effectively and equitably managed, ecologically representative and well-connected systems of protected areas and other effective area-based conservation measures by 2020.[122] According to *Global Biodiversity Outlook 5*, this target has been partially achieved: the quantitative targets are likely to be reached, but progress on the other commitments of Target 11 (ensuring that protected areas safeguard the most important areas for biodiversity and are ecologically representative, well connected, and equitably and effectively managed) has been modest.[123]

The ongoing biodiversity and climate crisis requires the protection of a larger part of this planet.[124] As mentioned above, the Global Biodiversity Framework under the CBD seeks protected designations of 30% of global terrestrial and marine areas by 2030.[125] At the EU level, this target has also been adopted in the EU Biodiversity Strategy to 2030.[126] For Ramsar Sites, the *Global Wetland Outlook* points to opportunities for the designation of many more Ramsar Sites.[127] The Ramsar Convention can be considered as the lead partner for wetlands for the CBD[128] and can help to realise the Global Biodiversity Framework.

Although these quantitative targets for the expansion of protected areas are important, designation is only a first step; protected areas should also be effectively managed. The *Global Biodiversity Outlook 5* revealed that only a small proportion of protected areas is assessed for management effectiveness.[129] The indicator under the Aichi Target concerns the number of assessments made

---

instrument of ratification amongst others or accession, as provided in Article 9.' Convention on Wetlands, note 4.

[122] Target 11: 'By 2020, at least 17% of terrestrial and inland water, and 10% of coastal and marine areas, especially areas of particular importance for biodiversity and ecosystem services, are conserved through effectively and equitably managed, ecologically representative and well connected systems of protected areas and other effective area-based conservation measures, and integrated into the wider landscapes and seascapes.'

[123] Secretariat of the Convention on Biological Diversity, note 50, 82.

[124] See, eg, E Dinerstein et al, 'An Ecoregion-Based Approach to Protecting Half the Terrestrial Realm' (2017) 6 *BioScience* 534; EO Wilson, *Half-Earth. Our Planet's Fight for Life* (Norton & Co, New York, 2016).

[125] Target 3, Global Biodiversity Framework.

[126] EU Biodiversity Strategy, note 75, Section 2.1.

[127] Gardner and Finlayson, note 10, 58.

[128] See A Tolentino, 'The Ramsar, Biodiversity and Climate Change Conventions – Inter-conventions Synergies' (2015) 45 *Environmental Policy and Law* 209.

[129] Secretariat of the Convention on Biological Diversity, note 50, 83.

by states, rather than the actual effectiveness of management in protected areas. Under the Ramsar Convention, Ramsar Site management authorities have been encouraged to evaluate the effectiveness of the management of each Ramsar Site.[130] Also, a Ramsar Site Management Effectiveness Tracking Tool (R-METT) has been approved as a voluntary self-assessment tool for evaluating the management effectiveness of Ramsar Sites and other wetlands.[131] Effective management should start with a management plan, but around half of all Ramsar Sites have no management plan.[132]

The management and restoration of protected sites can be challenging in the face of climate change and other human pressures. The Ramsar Convention requires states to notify the Ramsar Secretariat if the ecological character of a Ramsar Site in their territory has changed, is changing, or is likely to change.[133] Climate change can affect the ecological character of wetlands directly – for example, due to the effects of warming – or indirectly through interactions with other stressors;[134] although there has been discussion as to whether climate change leading to change in ecological character should give rise to a notification obligation under Article 3.2.[135]

Ramsar Handbook 18 on managing wetlands clarifies that the management objectives for Ramsar Sites should identify the factors that influence site features – essentially, 'those activities that are causing, or are likely to cause, change in ecological character'. Both negative and positive factors should be considered, as they both have implications for management. Uncontrollable factors such as climate change must also be taken into account.[136]

---

[130] Ramsar Convention, *Resolution XII.15: Evaluation of the management and conservation effectiveness of Ramsar Sites* (2015) [23].

[131] Ibid, [22].

[132] RC Gardner et al, *Global Wetland Outlook: Technical Note on Responses* (Ramsar Convention Secretariat, Gland, 2018), at: https://static1.squarespace.com/static/5b256c78e17ba335ea89fe1f/t/5bcaebf2104c7bc58460d91b/1540025353605/Ramsar+Technical+Note-Responses+Sept+2018a.pdf.

[133] Convention on Wetlands, note 4, Art 3.2; see further Chapter 3 in this volume.

[134] CM Finlayson et al, 'Policy considerations for managing wetlands under a changing climate' (2017) 68 *Marine and Freshwater Research* 1803.

[135] D Pritchard, *Change in ecological character of wetland sites – a review of Ramsar guidance and mechanisms* (2014) [4.29]–[4.35].

[136] Ramsar Convention Secretariat, *Handbook 18: Managing wetlands: Frameworks for managing Wetlands of International Importance and other wetland sites* (Ramsar Handbooks, 4th edition, Ramsar Convention Secretariat, Gland, 2010) 47.

Climate change might render it difficult, and in some circumstances impossible, to achieve a good conservation status for some habitats and the species that depend on them. Is this a reason to remove sites from the Ramsar List? Partial loss or deterioration of a designated wetland's ecological character is likely to be the most common scenario that could lead to the consideration of a boundary restriction.[137] Resolution VIII.22, paragraph 6(b)[138] concerns sites which 'unavoidably' lose their importance. According to the *Ramsar Handbook*, it is important to distinguish between what is 'avoidable' and what is 'unavoidable'; but it may be hard to make such a distinction, especially when the loss of ecological character is caused by indirect changes such as climate change.[139]

Because of the difficulties in determining unavoidable and avoidable losses of value and functions and attributes for which a site was designated, it is better to consider whether the damage or change is or is not reversible. If it is possible that the situation may reverse itself or could be reversed through appropriate management measures such as restoration, there is no need for a decision on delisting or restriction.[140] Restoration of wetlands is thus preferred over delisting, in line with the mitigation hierarchy (avoid-mitigate-compensate) that has also been accepted in the Ramsar Convention framework.[141] However, restoration of wetlands, as a preferred alternative to delisting, can still present challenges given climate change impacts and can give rise to uncertainties – for instance, regarding the reference system to be used for setting restoration objectives.[142]

---

[137] Ramsar Convention Secretariat, *Handbook 19: Addressing change in wetland ecological character: Addressing change in the ecological character of Ramsar Sites and other wetlands* (Ramsar Handbooks, 4th edition, Ramsar Convention Secretariat, Gland, 2010) 55. See further Chapter 5 in this volume.

[138] Ramsar Convention, *Resolution VIII.22: Issues concerning Ramsar sites that cease to fulfil or never fulfilled the Criteria for designation as Wetlands of International Importance* (2002) [6b].

[139] Ramsar Convention Secretariat, note 136, 55.

[140] Ibid, 56–57.

[141] Ramsar Convention, *Resolution XI.9: An Integrated Framework and guidelines for avoiding, mitigating and compensating for wetland losses* (2012).

[142] See A Cliquet, 'Ecological restoration as a legal duty in the Anthropocene' in M Lim (ed), *Charting Environmental Law Futures in the Anthropocene* (Springer, 2019) 59–70; G Gann et al, *International principles and standards for the practice of ecological restoration* (Society for Ecological Restoration, 2nd edition, Washington DC, 2019), in particular Principle 3: Ecological restoration practice is informed by native-referenced ecosystems, while considering environmental change.

4 CONCLUSION

The relationship between wetlands and climate change is a double-edged sword: on the one hand, wetlands are important for climate mitigation and adaptation; but on the other hand, they face additional pressure due to climate change. International climate law, as well as the Ramsar Convention framework, has increasingly recognised the role of wetlands in climate change mitigation and adaptation. While climate mitigation efforts initially focused predominantly on forests, greater attention is now being paid to other ecosystems – notably wetlands and peatlands. Initiatives at the international and national level have also been launched, such as the restoration of peatlands. However, it may nonetheless be asked whether the current legal framework under the Ramsar Convention is sufficient to protect and restore remaining wetlands and peatlands. Are the soft law resolutions under both the Ramsar Convention and the CBD strong enough to protect and restore wetlands and peatlands?

The *Global Wetland Outlook* and the *Global Biodiversity Outlook* reports confirm that we are losing wetlands at an alarming rate. The international climate change regime could provide a more binding framework; in order to achieve the goals of the Paris Agreement, states must strengthen their commitments under their NDCs, which could include the protection and restoration of wetlands. There is also a call for a 'Paris-style' agreement for biodiversity that includes binding targets for biodiversity – not only to promote mitigation and adaptation to climate change, but also for the sake of biodiversity itself. The new EU nature restoration law, which includes binding restoration targets and non-deterioration clauses, could set an important example in this regard. As well as binding conservation and restoration targets for wetlands and peatlands, strong non-deterioration obligations are required. The most recent IPCC report from 2022[143] demonstrates that it is more urgent than ever to immediately protect and restore remaining wetlands. Strategic climate and biodiversity litigation can play an important role in pushing governments to take the necessary measures, demanding greater action from states for the protection and restoration of wetlands.[144] Arguments in future litigation can be drawn from both international climate law and biodiversity law. These arguments can refer, among other things, to the international recognition of the role of

---

[143] IPCC, note 16.

[144] A Cliquet, 'Ecologisch herstel als juridische win-win bij klimaatswijziging: een positief verhaal?' in H Schoukens and C Billiet (eds), *Klimaatrechtspraak. Waarom rechters het klimaat (niet) zullen redden* (Die Keure, Brugge, 2021) 241–71.

wetlands in climate change mitigation and compensation, as is explicitly and increasingly recognised by the Ramsar Convention, as well as the contribution of well-functioning wetlands to achieving the goals of the Paris Agreement.

12. 'Within and without': EU nature conservation law and the protection of wetlands

Richard Caddell

1 INTRODUCTION

As in all regions of the planet, wetlands constitute a vital component of the collective natural environment of the European Union (EU). While wetland coverage across the EU has fluctuated with its changing membership, the most recent official statistics indicate that wetlands constitute approximately 73 000 square kilometres of the bloc's territory, or some 1.7% of its total landmass.[1] Many individual Member States feature higher aggregated concentrations within their national boundaries; and across the EU, wetlands continue to encompass key habitats for a vast array of species, help to improve water quality, play an increasingly significant role in mitigating climate change, and provide a plethora of tangible and intangible benefits for humans through their myriad contributions to ecosystem services.

Despite this extensive footprint, modern wetland coverage pales into near insignificance in comparison to historical baselines; and in a bleakly familiar manner to other jurisdictions, much of this rich natural heritage has been squandered over the previous century. In 1995, in what strikingly remains the sole occasion on which the EU has undertaken concerted reflection on the plight of wetlands, these habitats were considered to be 'among the most threatened ecosystems and landscapes', with an estimated 'two thirds of Europe's wetlands ... [disappearing] since the beginning of our century,

---

[1] See Eurostat, 'Wetlands Cover 2% of the EU's Land', 1 February 2018, at: https://ec.europa.eu/eurostat/web/products-eurostat-news/-/edn-20180201-1. Nevertheless, these statistics – which were compiled in 2015 – incorporate a significant contribution by the United Kingdom to the aggregated total (at the time, the Member State with the fifth greatest volume of wetlands within its territory), which has subsequently withdrawn from the European Union.

mainly lost through development processes which did not take their functions and values adequately into account'.[2] In individual Member States, the picture is even starker: since 1700, Ireland has lost over 90% of its fabled wetlands; while more than 80% of national coverage has disappeared across the Czech Republic, Hungary, Lithuania, and Germany.[3]

The EU has an intriguing regulatory relationship with its remaining wetlands. Although an active party to many international biodiversity treaties, it remains precluded from formal membership of the Ramsar Convention;[4] yet its constituent Member States host and maintain a vast network of Ramsar Sites, including an ever-growing number of those designated under the Convention as being Wetlands of International Importance.[5] As an entity composed of Member States that have ceded elements of national sovereignty to supranational oversight, the EU itself has no territory and thus no wetlands of its own, but it prescribes and polices significant obligations towards wetland conservation across this combined landmass. Each of its 27 Member States is simultaneously a contracting party to the Ramsar Convention and also subject to overarching requirements of EU environmental law that considerably expand the ambit of wetland commitments prescribed by international law. The EU further combines concerted vigilance over potential non-compliance, alongside powerful political and judicial enforcement capabilities, with extensive structural funding mechanisms to promote conservation strategies, thus providing clear scope to facilitate a resurgence in the conservation fortunes of wetlands across its Member States. However, even with this protective architecture, the conservation status of European wetlands remains a cause for collective anxiety.[6]

The EU has provided for the protection of waterbirds and their habitats on a near-contemporaneous basis to that of the Ramsar Convention. Since the first tentative environmental policies were adopted in the 1970s, the array of EU provisions that are prospectively applicable to wetlands has expanded mark-

---

[2] European Commission, *Communication from the Commission to the Council and the European Parliament: Wise Use and Conservation of Wetlands*, COM(95) 189 (final).

[3] E Fluet-Chouinard et al, 'Extensive Global Wetland Loss Over the Past Three Centuries' (2023) 614 *Nature* 281. This most recent and authoritative global review indicates that virtually all 27 current EU Member States, plus the United Kingdom, have suffered extensive wetland losses relative to historical coverage.

[4] Convention on Wetlands of International Importance especially as Waterfowl Habitat, adopted 2 February 1971, entered into force 21 December 1975, 996 UNTS 245 (amended 1982 & 1987).

[5] Ibid, Art 2; see further Chapter 3 in this volume.

[6] European Commission, note 2; Fluet-Chouinard et al, note 3.

edly, ranging from nature conservation laws to legal instruments concerning fresh, ground, and seawater quality and flood defences, environmental assessment obligations, infrastructure and development rules, chemical and pollution control regulations, and even environmental information requirements. An exhaustive accounting of this extensive mosaic of legal and technical standards would inevitably fall beyond the spatial confines of this Chapter. Accordingly, this contribution focuses expressly on the application of the key provisions of EU nature conservation law to wetland species and ecosystems.

EU nature conservation law is primarily addressed through two core Directives, which must be transposed into national law by the Member States and represent the primary supranational process through which many of the aspirations of the Ramsar Convention are promoted in these jurisdictions. The first such instrument was adopted in 1979 to promote the conservation of birds and their habitats,[7] setting out the inaugural provisions of EU law applicable to wetlands. Subsequently, the pioneering Birds Directive was complemented in 1992 by a broader instrument for the holistic protection of species and habitats across the EU.[8] Together, the Birds and Habitats Directives prescribe obligations concerning the protection of listed species and collectively preside over an extensive network of protected areas – christened Natura 2000[9] – which includes a significant number of wetland ecosystems and species. However, notwithstanding the voluminous literature on the European Union's cornerstone biodiversity legislation, its application to wetlands has barely been explored. This Chapter therefore evaluates the extent to which the current legislative framework applies to the protection of wetland ecosystems within the EU and how this regulatory architecture dovetails with the approaches and priorities advanced under the Ramsar Convention. It further seeks to account for why this corpus of law, policy, and jurisprudence has largely failed to stem wetland losses across the territory of the EU; while also considering how future developments might further enhance wetland conservation across and within the Member States.

To this end, Section 2 of this Chapter examines the role of the longstanding Birds Directive in the conservation of wetlands and waterbirds, outlining the influence of the Ramsar Convention in its conception and design, and

---

[7] Directive 2009/147/EC of the European Parliament and of the Council of 30 November 2009 on the Conservation of Wild Birds [2010] *Official Journal* L 20/7 (consolidated version) (hereinafter, 'Birds Directive').

[8] Council Directive 92/43/EEC of 21 May 1992 on the Conservation of Natural Habitats and of Wild Fauna and Flora [1992] *Official Journal* L 206/7 (hereinafter, 'Habitats Directive').

[9] Reproduced fully at https://natura2000.eea.europa.eu/.

considering its subsequent implementation. Section 2 further evaluates recent contentious developments towards the eradication of lead shot in wetlands and the legal challenges that have been brought against this hard-fought legislation. Section 3 then addresses the overarching role of the Habitats Directive and the consolidated Natura 2000 network of protected areas in the conservation and management of wetland ecosystems, with the ethos of this regime having moved recently from emphasising site-based protections towards promoting the restoration of degraded ecosystems. Section 4 considers the coexistence between the Ramsar Convention and the European Union, and Section 5 concludes.

2 WETLAND PROTECTION AND THE BIRDS DIRECTIVE

2.1 The Ramsar Convention, Wetlands, and the Genesis of the Birds Directive

On the morning of 2 February 1971, when the Ramsar Convention was formally concluded, there was little obvious indication that the European Economic Community (EEC) might have any meaningful part to play in implementing the aspirations and commitments of this nascent instrument – or indeed in advancing the cause of wetland conservation more generally. The Ramsar Convention does not provide for accession by entities other than states; hence, participation by the EEC was never envisaged as an active possibility. Equally importantly, while the EEC did exercise limited and tightly defined competences over the activities of its constituent Member States to facilitate the effective functioning of the Common Market, these powers did not at the material time expressly extend to environmental regulation. Strikingly, however, by the end of the decade, the EEC would not only have fashioned an expanding remit over environmental affairs, but also have enacted far-sighted legislation to advance the protection of birds and their habitats across the Community, with the implementation of the Ramsar Convention having motivated the elaboration of obligations towards the protection of wetlands and waterbirds as a key component of this process.

That wetland conservation would eventually form a key part of the EEC's flagship environmental legislation (which is now binding on the 27 Member States of the modern-day European Union)[10] is attributable to a chain of overlapping and complementary approaches throughout the 1970s in two separate

---

[10] In 1993 the EEC, which had previously become an umbrella term for three separate integrative Communities, was renamed the European Community (EC),

European integration organisations. In elaborating its Birds Directive, the EEC was heavily influenced by parallel developments within the Council of Europe – a separate, older, and far larger intergovernmental body concerned with an extensive array of political and policy initiatives across Europe, with both organisations sharing a progressively increasing number of mutual Member States. A specific remit for nature conservation within the Council of Europe was formalised in 1962 with the creation of the European Committee for the Conservation of Nature and Natural Resources (CDSN).[11] Shortly afterwards, the European Information Centre for Nature Conservation – latterly known as the Naturopa Centre – was established in 1967 under the auspices of the CDSN.[12] Dubbed 'the mouthpiece of the Council of Europe in everything concerning the natural environment',[13] the Naturopa Centre fulfilled a vital role as a conduit for coordination between the wider framework of the Council of Europe, the CDSN, and the national conservation agencies of the respective Member States, while also spearheading a series of prominent outreach programmes to raise public awareness of ecological problems across Europe through the widespread distribution of its free publications.[14] The CDSN would ultimately play a key role in entrenching wetland conservation as an operative legislative priority throughout Europe over the course of the subsequent decade, including the emerging EEC policies towards birds.

---

to identify that its mandate for integration had moved beyond purely economic affairs, before becoming fully absorbed into the European Union in 2009.

[11] Resolution 62 (31): Nature Conservation, at: https://rm.coe.int/09000016804 beffe. It has been contended by one of the institution's former Environmental Directors that the Council of Europe had indirectly considered numerous environmental problems since its inception in 1949, although an explicit remit towards nature conservation was established by the Consultative Assembly in 1961, which called for a permanent system of cooperation for nature conservation that was duly established the following year. S Renborg, 'Environmental Protection Work in the Council of Europe' (1971) XIX *European Yearbook* 42, 42.

[12] See JC Robertson, 'European Information Centre for Nature Conservation' (1968) 1 *Biological Conservation* 87.

[13] C Meyer, 'The Council of Europe, Conservation and Environmental Concern' (1990) 17 *Environmental Conservation* 74, 75.

[14] The Naturopa Centre was an often outspoken advocate for ecological awareness, and its glossy flagship publication *Naturopa* did not sugarcoat its public messages concerning environmental problems, memorably assessing the biological quality of Dutch wetlands as having 'reached the level of a rotting corpse'. A van Wijngaarden, 'Every Sort and Kind' (1980) 34–35 *Naturopa* 8, 8.

Initially – and perhaps most visibly, as the then sole regional body with an express remit for nature conservation[15] – the Council of Europe was identified as a key facilitator of intergovernmental activity during the early negotiations towards the Ramsar Convention.[16] Subsequently, the early work of the CDSN prioritised the conservation threats facing waterbirds and wetlands. In 1967, the Council of Europe adopted its first targeted resolutions on species and habitats, exhorting Member States to accelerate national protections for habitats 'bearing in mind existing recommendations and projects (particularly the MAR project on wetlands)'[17] and to take measures to protect 28 named species of birds and their habitats, among which waterbirds were amply represented.[18] In 1973, following an extensive survey by the CDSN, the Council of Europe's list of protected bird species expanded considerably and Member States were urged to ratify the Ramsar Convention as soon as possible.[19] Then, in 1975, significant practical impetus for the prioritisation of European waterbirds and wetland conservation policies was generated through the establishment by the CDSN of an Ad-hoc Group on Wetlands, charged with making 'proposals for

---

[15] In 1970, the Council of Europe identified itself as 'the most appropriate framework for the successful conduct of European action for the protection of man's natural milieu', while presciently observing that its nature conservation activities across the next decade were likely to be duplicated by international and European bodies. Recommendation 586 (1970): Council of Europe activities on nature conservation and the protection of amenities, at: https://pace.coe.int/en/files/14621/html.

[16] DA Stroud et al, 'Development of the Text of the Ramsar Convention: 1965–1971' (2022) 73 *Marine and Freshwater Research* 1107, 1108. However, by 1969, in the light of the Soviet Union's drafting initiatives, the Council of Europe become increasingly sidelined as a potential advisory body to the emerging Convention in favour of less overtly political ornithological organisations. GVT Matthews, *The Ramsar Convention on Wetlands: Its History and Development* (Ramsar Convention Bureau, Gland, 1993) 20.

[17] Resolution 67 (25): *Various Causes of the Disappearance of Wild Life*, at: https://rm.coe.int/CoERMPublicCommonSearchServices/DisplayDCTMContent?documentId=09000016804bc50a.

[18] Resolution 67 (24): *Birds in Need of Special Protection in Europe*, at: https://rm.coe.int/CoERMPublicCommonSearchServices/DisplayDCTMContent?documentId=090000168050bb32.

[19] Resolution 73 (31): *Birds in Need of Special Protection in Europe*, at: https://rm.coe.int/CoERMPublicCommonSearchServices/DisplayDCTMContent?documentId=09000016804ee39a. Echoing parallel developments within the EEC at the time, the Resolution called for 'special attention to the migratory species'.

practical and operational measures providing adequate protection of wetlands in the Council of Europe member states'.[20]

Meanwhile, momentum towards the promotion of a greater environmental focus in general, and the active pursuit of bird conservation in particular, was simultaneously building within the EEC. With the Community poised to expand to nine participants in 1973, the acceding states – Denmark, Ireland, and the United Kingdom – met with the six founding Member States in 1972 to discuss proposals by the European Commission to tentatively commence the enactment of environmental measures,[21] despite ecological concerns being expressly absent from the original Treaty of Rome. Although the European Commission acknowledged that this constituent treaty would require an expansive interpretation in order to support a tangible environmental remit,[22] in November 1973 the EEC adopted its first Environmental Action Programme, mandating a series of priorities for collective activity over the next two years.[23] This included consideration of the conservation status of birds – albeit exclusively from the specific perspective of unsustainable hunting practices – pledging to promote 'joint action by the Member States in the Council of Europe and other international organizations' and to study the prospective harmonisation of national provisions 'on the protection of animal species and migratory birds in particular'.[24]

This mandated study was duly completed in 1974,[25] whereafter the European Commission addressed a formal Recommendation to the Member

---

[20] European Committee for the Conservation of Nature and Natural Resources, *Ad hoc Group – Wetlands, Report of Actions Taken*, Document CE/Nat/VS (75) 12, 13 June 1975, at: https://rm.coe.int/090000168067bce4, 2.

[21] L Krämer, 'The Interdependency of Community and Member State Activity on Nature Protection Within the European Community' (1993) 20 *Ecology Law Quarterly* 25, 28–29.

[22] European Commission, *First Communication of the Commission about the Community's Policy on the Environment*, SEC(71) 2616 final, 8–10.

[23] Declaration of the Council of the European Communities and of the representatives of the Governments of the Member States meeting in the Council of 22 November 1973 on the programme of action of the European Communities on the environment [1973] *Official Journal* C112/1.

[24] Ibid, 40.

[25] The study itself, however, received a lukewarm reception and almost blunted initial legislative momentum. While its lead author, Bernhard Grzimek, was a veterinarian of significant repute, he had little ornithological expertise and much of the work was ultimately undertaken by two doctoral students. Notwithstanding their assiduous efforts, the European Commission considered that the final document lacked the requisite prestige, was focused unduly on conditions in par-

States concerning the protection of birds and their habitats.[26] In this regard, the Commission identified the key mortality factors afflicting European populations of birds as being intensive capture; habitat loss – notably through drainage schemes; the poisoning of food sources; and environmental pollution. Pointedly, from a wetland perspective, the Commission observed that bird protection could be 'significantly improved' if all Member States acceded 'as quickly as possible' to the newly inaugurated Ramsar Convention, which was endorsed as 'being of vital importance for the protection of the ecological balance and an irreplaceable natural heritage; in scope it goes far beyond the mere protection of waterfowl habitats'.[27] Concurrently, public pressure was also building for legislative activity by the EEC, with a prominent campaign prompting the European Parliament in 1975 to demand additional Community-wide protections for birds.[28] While these concerns were still primarily framed around hunting, the Parliament nevertheless also observed the need to create protected areas, to improve environmental conditions, and to externally promote further international protections for birds.

Although this political advocacy was primarily driven by prominent ornithological organisations, strong public support for bird conservation was nevertheless readily and organically apparent across Europe – and attributable largely to the pioneering and extensive outreach work undertaken by the CDSN and the Naturopa Centre. In 1970, the Council of Europe launched its first major environmental campaign – the European Conservation Year – compris-

---

ticular Member States, and took an unpalatably strong position against hunting activities to be the catalyst for a new departure into supranational nature conservation. See J-H Mayer, 'Saving Migrants: A Transnational Network Supporting Supranational Bird Protection Policy' in W Kaiser, B Leucht, and M Gehler (eds), *Transnational Networks in Regional Integration: Governing Europe 1945–83* (Palgrave McMillan, Basingstoke, 2010) 176, 185–87. Later surveys in 1976 and 1977 by more established avian experts would further galvanise the Commission's commitment to developing bird legislation.

[26] Commission Recommendation of 20 December 1974 to the Member States concerning the protection of birds and their habitats [1974] *Official Journal* L 21/24.

[27] Ibid, [3]–[6]. The Commission also recommended participation in the International Convention for the Protection of Birds 1950, adopted 18 October 1950, entered into force 17 January 1963, 638 UNTS 186 (amended 1973 & 2016). The 1950 Convention was largely concerned with unsustainable hunting and trading of birds and eggs, but Articles 10 and 11 also prescribed obligations pertinent to the conservation of wetland habitats.

[28] European Parliament, Resolution on Petition No 8/74: 'Save the migratory birds' [1975] *Official Journal* C60/51.

ing some 200 000 separate events across Europe to promote an unprecedented degree of public environmental education, as well as facilitating significant improvements to national nature conservation, protected area coverage, and environmental research.[29] Building on these achievements, in 1972 the CDSN launched the first in a series of biennial campaigns on individual ecological issues. In 1976, as part of the remit of the CDSN's Ad-hoc Group on Wetlands, the Naturopa Centre launched the European Campaign for the Protection and Management of Wetlands. Unlike its previous two campaigns on soil and water, respectively – whose themes had been pre-determined by the Council of Europe's overarching policy priorities at the time – the 1976–77 wetlands campaign was popularly selected by mutual consent among the various national conservation agencies, resulting in a far more collaborative, cohesive, and successful initiative.[30] The campaign had a significant impact in focusing sustained regulatory attention on wetlands, leading to the collation of a number of national inventories and prompting additional European participants to sign the Ramsar Convention, enact implementing legislation and institute further protective designations.[31] More significantly, these developments engendered a shift in regulatory mindset away from the simple imposition of temporal or numerical constraints on hunting entitlements or restrictions on the deployment of the more indiscriminate and wasteful forms of capture towards the need to establish clear obligations on habitat conservation as a wider holistic strategy to halt and reverse losses of birds, with wetland protection increasingly recognised as a crucial part of this strategy.

Back in the EEC, this shift was reflected in an initial proposal for a Directive on bird conservation, published by the European Commission in 1976, in which, *inter alia*, habitat protection was identified as having a 'decisive impact' on population levels of birds – an element considered 'particularly important in the case of species dependent on specific habitats such as wetlands or woodlands'.[32] In 1977, the initial Environmental Action Programme of 1973 was updated and fresh priority objectives were established for the

---

[29] See further European Committee for the Conservation of Nature and Natural Resources, *Results of ECY*, Document CE/Nat/Centre (71) Bil 1 Revised, at: https://rm.coe.int/0900001680670f28.

[30] M Stegers, 'Arousing Public Interest' (1980) 34–35 *Naturopa* 51.

[31] See European Committee for the Conservation of Nature and Natural Resources, *Report on National Activities and the Impact of the 1976 Campaign on Wetlands*, Document E/Nat/Centre (76) 8, at: https://rm.coe.int/090000168099a5ed; MF Broggi, 'Vote for the Environment: The Centre's Campaigns' (1977) 27 *Naturopa* 7, 10.

[32] European Commission, *Proposal for a Council Directive on Bird Conservation*, COM(76) 676 final [5].

years 1977–81.[33] Significantly, this second iteration of the EEC's environmental aspirations featured explicit recognition of the value of the Ramsar Convention and the need to protect wetlands of international importance, while also declaring for the first time that wetland conservation could – and arguably should – also become a matter for Community regulation.

To this end, reiterating its previous appeal in 1974 for Member States to accede to the Ramsar Convention, the EEC nevertheless considered this to be merely an 'initial step towards the protection of wetlands' that 'must be backed up by other initiatives at national and, if necessary, Community level'.[34] While taking 'due account' of the parallel work undertaken under the auspices of the Council of Europe, the European Commission was charged with collating a:

> coherent inventory of Community wetlands which it thinks need protection either because their intrinsic characteristics make them important to the Community internationally or because their geographical location is such that they provide vital staging areas for certain species of migratory birds.[35]

Upon submission of this inventory, the Commission would then, following consultations with national experts and taking into consideration developments within the Council of Europe, 'if necessary … make appropriate proposals to the Council, particularly as regards the protection and managements of certain wetlands and contiguous zones'.[36] These commitments presented considerable scope for wetland conservation to be prioritised as a core element of the emergent Birds Directive. As outlined further below, wetland conservation was expressly incorporated into the final text of this instrument, which the Commission has interpreted with emphasis upon the designation of broad spatial protections and the control of deleterious impacts upon them.

Meanwhile, the Council of Europe was moving apace with an even more ambitious programme of work, having mandated the drafting of a holistic regional treaty for nature conservation in 1976[37] – an initiative that also featured the participation of the EEC. While birds and wetland ecosystems had

---

[33] Resolution of the Council of the European Communities and of the Representatives of the Governments of the Member States meeting within the Council of 17 May 1977 on the continuation and implementation of a European Community policy and action programme on the environment [1977] *Official Journal* C 139/1.

[34] Ibid, [157].

[35] Ibid, [158].

[36] Ibid, [159].

[37] Recommendation 783 (1976): Protection of Birds and Their Habitats in Europe, at: https://pace.coe.int/en/files/14817/html.

long formed a central component of its regulatory intentions, the scope of what would shortly become the Council of Europe's Bern Convention[38] ultimately took a markedly more expansive approach to biodiversity protection. During the drafting process, and in the light of the CDSN's ongoing work on wetlands, particular attention was given to these habitats, which were identified as 'the most threatened natural milieu'.[39] However, in identifying candidate wetland sites for both the nascent Bern Convention and the UNESCO Man and Biosphere Programme, launched earlier in 1971, the CDSN advised that these initiatives should extend beyond merely seeking to facilitate or support the work of the Ramsar Convention, since this regime:

> deals only with wetlands of *international* importance and essentially from the waterfowl angle, whereas the CDSN considers that general safeguarding action must be taken and although birds may be the primary inhabitants of these areas, they are not the only creatures worthy of interest.[40]

The final text of the Bern Convention thus makes no mention of wetlands *per se*, or indeed any other named category of ecosystem, adopting instead a deliberately broad approach to the habitats and species under its purview.[41] Likewise, it applies to a more holistic cohort of wetland-dependent species,

---

[38] Convention for the Conservation of European Wildlife and Natural Habitats, adopted 19 September 1979, entered into force 1 June 1982, 1982 UKTS 56.

[39] European Committee for the Conservation of Nature and Natural Resources, *Report of the Third Meeting*, Document No 2950, 27 November 1978, at: https://rm.coe.int/0900001680915972, 1.

[40] Ibid, 2; emphasis in the original documentation. This perspective exemplifies the modern regulatory dilemma confronting the Ramsar Convention. Had Schrödinger contemplated waterfowl, rather than cats, a different sense of quantum superposition may have become apparent, as commentators have both lamented and celebrated the Convention's focus on birds. See MJ Bowman, 'The Ramsar Convention Comes of Age' (1995) 42 *Netherlands International Law Review* 1, 7 (considering this thematic emphasis disadvantageous); P Bridgewater and RE Kim, '50 Years on, W(h)ither the Ramsar Convention? A Case of Institutional Drift' (2021) 30 *Biodiversity and Conservation* 3919, 3921 (criticising the Convention's expansion beyond its titular remit). Both the Council of Europe and eventually the European Union preferred a more expansionist legal approach to wetlands.

[41] Indeed, the drafters considered that the wording 'should not be too explicit in order to keep it open for developing co-operation between the Contracting Parties, *inter alia* in respect of the creation of a network of biogenetic reserves, the protection of wetlands, etc'. Council of Europe, *Explanatory Report to the Convention on the Conservation of European Wildlife and Natural Habitats*, at: https://rm.coe.int/16800ca431 [23].

listed in its Appendices, duly extending beyond the Ramsar Convention's initial, narrower focus on waterbirds to further include reptiles, amphibians, and plants. Moreover, wetland ecosystems were firmly within the contemplation of the Council of Europe to be prioritised for protection by its Member States at this time, having noted with concern that 'wetlands are one of the most threatened ecosystems in Europe, are in constant regression and are in danger of disappearing'.[42] Applying both a demonstrably European and subtly different tinge to the central commitments of the Ramsar Convention and paraphrasing its definition of a 'wetland',[43] the Council of Europe further urged Member States to deposit inventories of 'wetlands of *European interest*' with the CDSN, to protect and manage 'wetlands of *particular* interest', and to incorporate 'the *most valuable* of these protected wetlands' within the European network of biogenetic reserves.[44]

These Recommendations, however, lacked coherence in relation to their equivalent Ramsar Convention commitments. For instance, it is suggestive – but not necessarily axiomatic, given waterfowl requirements and other specific criteria – that a wetland of 'European interest' would also be one of 'international importance' for the purposes of the Ramsar Convention.[45] Arguably, the other exhortations within the Council of Europe's Recommendation to identify and manage wetlands of 'particular' interest and 'most' value – which remain fundamentally undefined – would fall under the Ramsar Convention's obligations towards the 'wise use' of wetlands, which applied to all wetlands within the territory of a contracting party that was also a Member State of the Council of Europe at the material time.[46] Yet while these terminological divergences may been unhelpful to national actors charged with implementing the youthful Ramsar Convention, they nevertheless provided further regional steering towards prioritising wetland conservation as a region-wide concern – a development that would be further buttressed for EEC Member States with the concurrent adoption in 1979 of the Birds Directive, which would also advance specific obligations regarding wetlands and waterbirds.

---

[42] Recommendation No R (79) 11 of the Committee of Ministers to Member States Concerning the Identification and Evaluation Card for the Protection of Wetlands, at: https://rm.coe.int/0900001680504564.
[43] Convention on Wetlands, note 4, Art 1.1.
[44] Recommendation No R (79) 11, note 42 (emphasis added).
[45] Convention on Wetlands, note 4, Art 2.
[46] Ibid, Art 3.1; see further Chapter 4 in this volume. The application of the latter part of Art 3.1 was seemingly overlooked by the CDSN in its consideration of candidate sites (note 39) – albeit that its broader approach facilitated the designation of sites not classed on the Ramsar List and those that were ecologically significant wetlands from the perspective of species other than waterfowl.

2.2 Wetland Protection and the Implementation of the Birds Directive

Following hard-fought negotiations, the final text of the original Birds Directive[47] was adopted in April 1979 as a watershed for the EEC, constituting its first legislative provisions expressly developed for the purposes of nature conservation. While the legal basis for the adoption of the Directive was arguably less than watertight at the material time, and despite some Member States requiring a greater degree of persuasion to the merits of supranational biodiversity law,[48] this was never formally challenged before the (then) European Court of Justice ('Court').[49] By 1985, a landmark judgment of the Court pre-dating the first cases brought regarding the Birds Directive concluded that environmental protection was 'one of the Community's essential obligations'[50] and this instrument has accordingly formed a cornerstone of EU environmental law since its inception.

As with the Bern Convention, wetland conservation was evidently within the contemplation of the EEC institutions as an operative priority for early initiatives towards the implementation of the Birds Directive. Concurrent with the adoption of the Directive, the Council directed an accompanying Resolution to Member States calling upon them, among other things, to notify the Commission within two years of 'the areas which they have or intend to have designated as wetlands of international importance'.[51] Moreover, 'account shall be taken of the need to protect biotopes and flora and fauna without, however, delaying the action of primary importance for bird conservation,

---

[47] Council Directive of 2 April 1979 on the Conservation of Wild Birds (79/409/EEC) [1979] *Official Journal* L 103/1. In 2009, the Directive was consolidated in the light of multiple additions to its Annexes, a significant influx of new Member States, textual adjustments applied by the subsequent Habitats Directive, and the withdrawal of Greenland from the European Communities. See note 7.

[48] ALR Jackson, *Conserving Europe's Wildlife: Law and Policy of the Natura 2000 Network of Protected Areas* (Routledge, Abingdon, 2018) 72.

[49] WPJ Wils, 'The Birds Directive 15 Years Later: A Survey of the Case Law and a Comparison with the Habitats Directive' (1994) 6 *Journal of Environmental Law* 219, 223–24 (noting an initial challenge in France, but by the time this litigation came before the *Conseil d'Etat*, the issue had become moot).

[50] Judgment of the European Court of Justice (Court) in Case 240/83 *Procureur de la République v Association de Défense des Brûleurs d'Huiles Usagées* (1985) ECR 531.

[51] Council Resolution of 2 April 1979 concerning Directive 79/409/EEC on the Conservation of Wild Birds [1979] *Official Journal* C 103/6 [I].

particularly in wetlands',[52] as advanced under the prevailing Environmental Action Programme.[53] Specific obligations on wetland protection were further set out in the Birds Directive itself and are central to the EU's regulatory requirements towards these habitats.

The Birds Directive applies to 'all species of naturally occurring birds in the wild state in the European territory of the Member States',[54] including their nests, eggs, and habitats.[55] The jurisdictional extent of the EU's nature conservation laws – both the Birds Directive and the Habitats Directive refer respectively to 'territory' – had initially raised interpretive complications, especially in the marine context, where extended national entitlements were being negotiated under what would become the UN Convention on the Law of the Sea 1982[56] and a number of national agencies considered this to be analogous with the territorial sea.[57] This had considerable practical implications: unlike the Ramsar Convention, which applies only to waters of up to six metres at low tide,[58] a broad interpretation of 'territory' would engender conservation obligations throughout the entirety of the maritime jurisdiction of the Member States. An expansive approach was long advocated to ensure greater ecological coherence,[59] and was subsequently confirmed by the Court in the context of the Habitats Directive.[60] Geographically, waterbirds are thus protected to a far more extensive degree under the Birds Directive than the Ramsar Convention. Another important implication of Article 1 is that it applies with reference to the collective territories of the EU, requiring Member States to contemplate protected species holistically, as opposed to merely considering those that are ordinarily present within their individual territories – which again goes further than the Ramsar Convention. This more communal view of protected species was emphasised by the Court at an early stage in the operation of the

---

[52] Ibid.

[53] Council of the European Communities, note 33.

[54] Birds Directive, note 7, Art 1(1).

[55] Ibid, Art 1(2).

[56] United Nations Convention on the Law of the Sea, adopted 10 December 1982, entered into force 16 November 1994, 1833 UNTS 397.

[57] R Caddell, 'The Maritime Dimensions of the Habitats Directive: Past Challenges and Future Opportunities' in G Jones QC (ed), *The Habitats Directive: A Developer's Obstacle Course?* (Hart, Oxford, 2012) 183, 187–88.

[58] Convention on Wetlands, note 4, Art 1.1.

[59] D Owen, 'The Application of the Wild Birds Directive beyond the Territorial Sea of European Community Member States' (2001) 12 *Journal of Environmental Law* 38.

[60] Judgment of the European Court of Justice (Court) in Case C-6/04 *Commission v UK* (2005) ECR I-9017; see Caddell, note 57.

Directive, with bird conservation described as 'a transfrontier environment problem entailing common responsibilities', necessitating the widest possible transposition of its requirements on a national level.[61]

The Directive requires Member States to take the 'requisite measures' to maintain populations 'at a level which corresponds in particular to ecological, scientific and cultural requirements, while taking into account economic and recreational requirements'.[62] As discussed below, this suggests a degree of flexibility in the designation of spatial protections, but one that has been interpreted narrowly by the Court. In keeping with the broad concerns raised in the initial Environmental Action Programmes of the EEC, the Birds Directive adopts a two-pronged approach to the conservation of the birds under its purview, with discrete clusters of provisions addressing species and habitat protections, respectively. In the case of the former, Articles 5–9 impose restrictions and qualifications on the hunting, taking, sale, and trade of protected species, identified within the Annexes to the Directive, with birds that may be hunted pursuant to national legislation listed on Annex II[63] and those species subject to trade assigned to Annex III.[64] In addressing the hunting of birds, the Directive calls upon Member States to comply with 'principles of wise use and ecologically balanced control'[65] – the language of this provision considered to advance this central tenet of the Ramsar Convention in this context.[66]

Of express significance to the conservation of wetlands are the Directive's habitat-related provisions, elaborated through Articles 3 and 4. Under Article 3, Member States are required to 'preserve, maintain or re-establish a sufficient diversity and area of habitats' for all birds addressed by the Directive through the designation of protected areas, the upkeep and management of the ecological needs of habitats both within and beyond the confines of protected areas,

---

[61] Judgment of the European Court of Justice (Court) in Case 247/85 *Commission v Belgium* (1987) ECR 3057. Here Belgium had sought to limit its transposing legislation on trade only to species found locally, rather than those that might be prospectively imported from within the entirety of the Community.

[62] Birds Directive, note 7, Art 2.

[63] Ibid, Art 7. Nevertheless, under Article 8, Annex IV identifies a series of hunting gear and techniques that are to be prohibited due to their large-scale or non-selective nature that has the propensity to cause the local disappearance of particular species.

[64] Ibid, Art 8.

[65] Ibid, Art 7(4).

[66] G Tucker et al, 'Nature Conservation Policy, Legislation and Funding in the EU' in G Tucker (ed), *Nature Conservation in Europe: Approaches and Lessons* (Cambridge University Press, Cambridge, 2023) 59, 75.

the re-establishment of destroyed biotopes, and the creation of biotopes.[67] This provision evidently includes wetlands and ultimately encompasses far-reaching conservation obligations upon the respective Member States.

Protected area management is advanced under Article 4, whereby species listed on Annex I to the Directive are subject to 'special conservation measures' to ensure their survival and reproduction within their area of distribution.[68] This list, which has expanded considerably since the inception of the Directive, includes substantial numbers of waterbirds. Annex I incorporates those species that have been assessed to be in danger of extinction, vulnerable to specific changes in habitat, or are considered rare by virtue of their small populations or restricted local habitat; or that require particular attention due to the specific nature of their habitats.[69] Where a species is listed on Annex I, Member States are required 'in particular' to designate spatial protections – termed 'Special Protection Areas' (SPAs)[70] – encompassing 'in particular the most suitable territories in numbers and size', which in practice will necessarily include significant wetland habitats.

Representing the most explicit obligation concerning wetlands across the EU's nature conservation Directives, Article 4(2) mandates that 'similar measures' be taken for 'regularly occurring migratory species' that are not listed on Annex I, bearing in mind the need to protect their breeding, moulting, and wintering areas, and migratory staging posts. Article 4(2) further states that '[t]o this end, Member States shall pay particular attention to the protection of wetlands and particularly to wetlands of international importance'. Cumulatively, Articles 4(1) and (2) provide scope for extensive individual designations of spatial protections for wetlands by Member States – either because they have been identified as suitable habitats for Annex I species or for migratory species that are regularly present across the EU.

The final wording of Article 4(2) is less elegant than initially intended and arguably less ostensibly focused on the objectives of the Ramsar Convention, despite its use of familiar terminology. Intriguingly, the textual history of the Birds Directive reveals that many of its provisions were drafted by members of the EEC's legal cadre who coincidentally were also keen ornithologists – notably John Temple Lang, a redoubtable Dubliner who played a highly influential role in composing its far-reaching Articles 3 and 4. As originally

---

[67] Birds Directive, note 7, Art 3.
[68] Ibid, Art 4(1).
[69] Ibid.
[70] Ibid. The phrase 'in particular' means that this obligation does 'not exclusively' consist of designating SPAs. Opinion of Advocate General van Gerven in Case C-355/90 *Commission v Spain* (1993) ECR I-4241 [22].

articulated, the obligation towards migratory species in draft Article 4(2) simply instructed Member States to 'take the requisite measures to preserve recognised wetlands of international importance',[71] with Temple Lang's use of this express Ramsar vocabulary allegedly calculated 'to transpose the convention through the back door'.[72] Indeed, speaking in retirement, Temple Lang considered that this wording sought to extend the identification of areas of international importance to all migratory species and to facilitate a continental list of all such locations for subsequent protection.[73]

The express focus on the Ramsar Convention within Article 4(2) was diluted somewhat in 1977, when the Economic and Social Committee of the EEC indicated that it was 'concerned' by this initial formulation, which 'should be qualified so as to make it clear that "recognized wetlands" referred to wetlands meeting the criteria agreed at the International Conference on the Conservation of Wetlands and Waterfowl held in Heiligenhafen' for identifying such wetlands for the purposes of the Ramsar Convention.[74] This provision was duly modified,[75] but ultimately disregarded the suggestion to limit its scope purely to those wetlands designated on the Ramsar List.

Meanwhile, political pressures during the drafting process also impacted on the application of Article 4(2), with hunting organisations keen to limit both the numbers of individual species protected under the instrument and the scope of any attendant hunting restrictions. Accordingly, some Member States 'put emphasis more on the principles of habitat protection than on the politically hazardous route of regulations curtailing hunting'.[76] Indeed, in the drafting meetings convened in 1977, the French representatives suggested that protected areas for the purposes of the burgeoning Article 4 should be listed on a separate Annex to the Directive, with the Ramsar List identified as a basis for this – a proposal based less on an unequivocal endorsement of the merits of the 1971 Convention and one that instead 'appears to have been aimed at limiting Article 4 to wetlands'.[77] From the opposite direction, particular Member States were also concerned about relinquishing sovereignty over protective designa-

---

[71] European Commission, *Proposal for a Council Directive on Bird Conservation*, COM(76) 676 (final).

[72] Mayer, note 25, 187.

[73] Jackson, note 48, 42.

[74] Opinion on the Proposal for a Council Directive on Bird Conservation [1977] *Official Journal* C 152/3.

[75] Reproduced in Jackson, note 48, 63–64.

[76] R Boardman, *The International Politics of Bird Conservation: Biodiversity, Regionalism and Global Governance* (Edward Elgar Publishing, Cheltenham, 2006) 111.

[77] Jackson, note 48, 61.

tions, with Denmark, for instance, keen to curtail SPAs to those areas already established as national Ramsar Sites.[78]

The final version of Article 4(2) thus requires Member States to promote wetland conservation as a priority for non-Annex I migratory species, with particular emphasis on Wetlands of International Importance. This mirrors the approach under Article 4.1 of the Ramsar Convention – in the words of Temple Lang himself, under Article 4(2) of the Birds Directive:

> [t]he fact that an area does not qualify under any of the criteria normally used for determining 'international importance' does not, of course, mean that it need not be protected, only that it is the responsibility of the national authorities to protect it, and that it is not of high enough priority to necessitate international action.[79]

Although the Ramsar Convention's terminology is deployed without direct attribution – despite the earlier wishes of the Economic and Social Committee – subsequent judicial consideration indicates that Article 4(2) 'refers without doubt' to the Ramsar List in the context of 'wetlands of international importance'.[80]

Designation is a vital component of the Birds Directive, but one further buttressed by the obligations in Article 4(4) in respect of SPAs to take 'appropriate steps to avoid pollution or deterioration of habitats or any disturbances affecting the birds, in so far as these would be significant having regard to the objectives of this Article'. Furthermore, outside these areas, Member States must also strive to avoid pollution or deterioration of habitats. However, Article 4(4) has since been modified by the application of the subsequent Habitats Directive, which – as observed further below – has broadened the degree to which Member States may justify proceeding with development activities in protected areas for imperative reasons of overriding public importance.

Compliance with the requirements of EU law is subject to scrutiny and oversight by the European Commission, which may initiate remedial steps culminating in litigation against Member States considered to have fallen short of expectations. Prospective infringements are identified by the Commission either through its own investigations or in response to external complaints. Hence, in the context of wetlands, vigilant non-governmental organisations

---

[78] Ibid, 72.

[79] J Temple Lang, 'The European Community Directive on Bird Conservation' (1982) 22 *Biological Conservation* 11, 16.

[80] Opinion of Advocate General van Gerven in Case C-57/89 *Commission v Germany* (1991) ECR I-903 [47]. The European Commission has also considered this provision 'an implicit reference at Community level to the Ramsar Convention'. European Commission, note 2, 9.

have proved pivotal in alerting the EU authorities to potential breaches of these provisions.[81] The pursuit of infringement proceedings first triggers a collaborative dialogue between the Member State and the Commission, whereby an initial letter of notice is communicated with a deadline for a response. Thereafter, the Commission may address a Reasoned Opinion to the Member State, outlining the basis for the alleged infraction and requesting compliance. Should this fail to achieve the desired outcome, the Commission may ultimately refer the matter to the Court, although corrective action is often forthcoming without recourse to judicial intervention. For instance, the *Eilean na Muice Duibhe*/Duich Moss wetland in Scotland was subject to a complaint to the Commission by the Royal Society for the Protection of Birds (RSPB) over government approval for peat extraction in a critical wetland habitat for the Greenland white-fronted goose. Developmental approval was revoked following the commencement of infringement proceedings and the United Kingdom duly designated this area a Site of Special Scientific Interest under domestic planning law, an SPA under the Birds Directive, and a Wetland of International Importance under the Ramsar Convention.[82]

Other complaints have nevertheless necessitated judicial interpretation of the Directive and a formal assessment of whether protective designations have been made to the requisite ecological extent, which has clarified the scope of the wetland requirements under Article 4. One especially contentious element of the Directive concerns the coexistence between conservation obligations and socio-economic aspirations in these 'most suitable territories', for which the Court has consistently adopted a restrictive approach to development activities. In one of the Directive's earliest disputes, the *Leybucht Dykes* case,[83] the Commission brought infringement proceedings against Germany for modifying a significant wetland SPA. Following the designation, urgent reinforcements to flood defences became necessary and the authorities adjusted the line of the dyke to favour harbour access and fishing interests, thereby reducing the size of the SPA.

Although Article 2 refers to economic and recreational activities, this was deemed not to be an autonomous derogation from the general system of

---

[81] As indeed have the administrative mechanisms of the Bern Convention itself, with some leading wetland actions brought under the Birds and Habitats Directives having started as complaints filed under its case files process, including concerns over the Duich Peat Moss (Case File 1985/1), Santoña Marshes (Case File 1987/3), and Doñana National Park (Case File 1998/5).

[82] N Haigh, The European Community and International Environmental Policy' (1991) 3 *International Environmental Affairs* 163.

[83] Judgment of the European Court of Justice (Court) in Case C-57/89 *Commission v Germany* (1991) ECR I-883.

protection established by the Directive; and since these interests are expressly omitted from the considerations to be assessed by Member States in implementing Article 4, they cannot be taken into account in designation practices.[84] Thus, while Articles 4(1) and 4(2) provide 'a certain margin of discretion' over site selections, no such freedoms apply to the obligation under Article 4(4) to prevent pollution, deterioration, or disturbances in SPAs once they are designated, thereby allowing little scope to revise existing SPA boundaries; otherwise 'a Member State could unilaterally escape from the obligations imposed on them'.[85] Nevertheless, the Court ruled that a general interest superior to the ecological objective of the Directive could be applied, provided that the construction activities deemed vital to the protection of human life were no more than minimally detrimental to the SPA[86] and the ecological impacts were fully offset.[87] *Leybucht Dykes* remains a leading evaluation of SPA requirements and one that has had a significant bearing on subsequent cases. While the judgment is rather concise, as in many EU cases, important interpretive information may be gleaned from the preceding Opinion delivered by the Advocate General. Significantly, Advocate General Van Gerven's conclusion that the SPA boundary could be adjusted was based on consideration of the approach of the Ramsar Convention,[88] which permits a contracting party to revise or delete the boundaries of a Ramsar-listed wetland 'because of its urgent national interests'[89] – the Convention thus playing a key, yet unheralded, role in litigation that would frame the subsequent application of Article 4 of the Birds Directive.

As originally formulated, the requirements of Article 4(4) represent a powerful check on the economic aspirations of Member States for wetlands that would qualify for protective designations under Articles 4(1) and 4(2). Indeed,

---

[84] See Judgment of the European Court of Justice (Court) in Case C-247/85 *Commission v Belgium* (1987) ECR 3029 [8] and Judgment of the European Court of Justice (Court) in Case C-262/85 *Commission v Italy* (1987) ECR 3073 [8]. Both cases involved insufficient protections against hunting, but *Leybucht Dykes* further applied this approach to protected area considerations. Ibid, [22].

[85] Ibid, [20].

[86] Ibid, [8].

[87] Ibid, [22].

[88] Opinion of Advocate General Van Gerven, note 80, [30].

[89] Convention on Wetlands, note 4, Art 2.5; see Chapter 5 in this volume. The Advocate General considered that the obligation in Article 4(4) of the Birds Directive 'is not expressly dealt with in the Ramsar Convention or any other international convention', ibid, [36], further indicating that the EU provisions move considerably beyond the wetland protections contemplated under international law.

efforts to expand the position in *Leybucht Dykes* have been unequivocally rejected by the Court, which has emphasised that decisions on the classification and delimitation of SPAs shall be based solely on the ornithological criteria prescribed in the Directive. Chiefly, in the *Santoña Marshes* case,[90] Spain had instituted a series of national protections over a substantial wetland ecosystem but had stopped short of classifying it as an SPA, while also approving extensive infrastructure projects within and in close vicinity to the marshes. The European Commission considered this to have breached Articles 4(1), 4(2), and 4(4) of the Directive; while Spain contended that protective designations ought to be gradual and progressive and, moreover, that economic interests could override the ecological importance of the marshes. This was tersely rejected by the Court, which considered *Leybucht Dykes* exceptional; and once the ornithological criteria had been met, the obligation arose to institute an SPA, for which economic and recreational requirements 'do not enter into consideration'.[91] This was swiftly endorsed in the *Lappel Bank* case,[92] where the RSPB challenged a decision to exclude a small but ornithologically significant area of inter-tidal mudflats from the Medway Estuary and Marshes SPA – a wider site that the UK had listed under the Ramsar Convention as a Wetland of International Importance – to prospectively expand an important port. Endorsing the judgment in *Santoña Marshes*, the Court ruled that a Member State is not permitted to consider economic requirements when establishing an SPA and determining its boundaries.[93]

Consequently, there is considerable scrutiny of Member States' application of ornithological criteria, especially where this results in more limited SPA designations than had been expected by the European Commission. To this end, the Court has consistently reiterated that Member States are required to classify as SPAs the territories that meet the ornithological criteria in Articles 4(1) and 4(2)[94] – especially those that, in light of these criteria, appear to be the most suitable territories for the conservation of the species in question.[95]

---

[90] Judgment of the European Court of Justice (Court) in Case C-355/90 *Commission v Spain* (1993) ECR I-4221.

[91] Ibid, [19].

[92] Judgment of the European Court of Justice (Court) in Case C-44/95 *Regina v Secretary of State for the Environment*, ex parte *RSPB* (1996) ECR I-3851.

[93] Ibid, [27]. This was however a pyrrhic victory since construction proceeded before the Court's ruling was delivered. In 2004, a larger area of agricultural land was rewilded as a compensatory measure for the development.

[94] Case C-355/90, note 90, [26–27], [32]; Judgment of the European Court of Justice (Court) in Case C-378/01 *Commission v Italy* (2003) ECR I-2857 [14].

[95] Judgment of the European Court of Justice (Court) in Case C-3/96 *Commission v Netherlands* (1998) ECR I-3031 [62].

A Member State cannot avoid its obligations to protect SPAs under the first sentence of Article 4(4) by simply declining to establish such an area in the first place – these obligations will instead apply to an area that *should* have been designated an SPA[96] – or by applying a lesser protective designation under national law.[97] Nor can the ultimate designations of SPAs be 'manifestly less' than the number and total area of the sites considered most suitable.[98]

For its part, in assessing whether a Member State fulfilled its designation obligations under Article 4, the European Commission has relied on the inventories of the Important Bird Area (IBA) programme[99] as *prima facie* evidence that an SPA designation is necessary in a particular location. The IBA concept is intimately connected to the Birds Directive and was initiated in 1979 as a means of aiding the implementation of Article 4 by identifying the most important known areas for birds to prioritise the establishment of SPAs.[100] Waterbirds have been a central element of IBAs since their inception, with the initial document commissioned from the International Council for Bird Protection and the International Waterfowl and Wetlands Research Bureau,[101] following an earlier study specifically on Wetlands of International Importance.[102] While the expansion of the criteria to species other than waterfowl raised some initial interpretive questions,[103] its probative value to the assessment of SPA coverage has subsequently become firmly entrenched. Indeed, although it has been categorically established that IBAs are not legally binding,[104] unless a Member State can adduce evidence from a more persuasive document, the Court is entitled to use the IBA as a basis for assessing whether sufficient territories have been classified under Article 4.[105] Overturning this presumption is a high threshold for a Member State to cross in practice – IBAs

---

[96] Case 355/90, note 90, [22].

[97] Case C-3/96, note 95, [55].

[98] Ibid, [63].

[99] The IBA programme was subsequently expanded as 'Important Bird and Biodiversity Areas', given its importance to other species.

[100] See further PF Donald et al, 'Important Bird and Biodiversity Areas (IBAs): The Development and Characteristics of a Global Inventory of Key Sites for Biodiversity' (2019) 29 *Bird Conservation International* 177.

[101] Ibid, 178.

[102] Written Question No 1858/82 (10 January 1983); Answer of 15 April 1983 [1983] *Official Journal* C 150/4.

[103] Temple Lang, note 79, 14–16.

[104] Judgment of the European Court of Justice (Court) in Case C-374/98 *Commission v France* (2000) ECR I-10799 [25]; Case C-3/96, note 95, [68–70].

[105] Judgment of the European Court of Justice (Court) in Case C-378/01 *Commission v Italy* (2003) ECR I-02857 [18].

have been considered 'a basis for reference' for assessing compliance and 'substantial evidence' of deficient classification of SPAs;[106] and while these 'can be invalidated by better scientific knowledge',[107] attempts to challenge the scientific legitimacy of national inventories in versions subsequent to the authoritative IBA of 1989 have received short shrift.[108] Moreover, the 'mere possibility' that the application of the ornithological criteria of an IBA could have led different actors to produce markedly different classifications of SPAs 'cannot as such be taken into consideration in order to undermine the probative value' of these documents.[109]

Nevertheless, the Ramsar Convention has not been without interpretive utility in overturning this presumption in individual cases. In the *Seine Estuary* case,[110] the construction of a titanogypsum plant was considered compatible with Article 4(2) because it had been built on land that had dried out several decades previously and no longer met the Ramsar Convention's definition of a 'wetland'[111] – a point not contested by the European Commission. Likewise, in *Commission v Greece*, smaller designations for particular wetlands were justified because Greece had applied the Ramsar criteria in selecting the specific sites for protection and successfully rebutted the Commission's erroneous assertion that the pertinent IBA indicated that certain areas should be considered potential Ramsar Sites.[112]

Ultimately, the Birds Directive remains a landmark provision of EU law and one that has provided demonstrable conservation improvements for birds

---

[106] Opinion of Advocate General Kokott in Case C-334/04 *Commission v Greece* (2006) ECR I-9218 [34].

[107] Ibid, [41].

[108] Ibid, [36–40]; Opinion of Advocate General Kokott in Case C-235/04 *Commission v Spain* (2006) ECR I-5418 [47–56]; Judgment of the European Court of Justice (Court) in Case C-418/04 *Commission v Ireland* (2007) ECR I-10997 [47–52]; Judgment of the Court of Justice of the European Union (CJEU) in Case C-97/17 *Commission v Bulgaria* ECLI:EU:C:2018:285 [82–84].

[109] Case C-3/96, note 95, [71].

[110] Judgment of the European Court of Justice (Court) in Case C-166/97 *Commission v France* (1999) ECR I-1729.

[111] Opinion of Advocate General Fennelly in Case C-166/97 *Commission v France* (1998) ECR I-1721 [21]. France was nevertheless considered to have breached the Directive by failing to designate a sufficiently large SPA at an earlier point.

[112] Opinion of Advocate General Kokott, note 106, [55]. Greece was held to be in breach of the Directive in respect of other non-designations.

across the continent,[113] as well as many key areas of wetland habitats. As Krämer observes, the requirements for protective designations, combined with meaningful compliance and oversight mechanisms, have advanced the rehabilitation of habitats and species to an extent that eludes international treaties such as the Ramsar Convention.[114] Indeed, enforcement processes have often engendered significant long-term benefits for the wetland sites in question: the *Poitevin Marsh* case,[115] for instance – in which a series of breaches of Article 4 was upheld – led to significant regulatory improvements to the management of the site.[116] Moreover, beyond the legal obligations of the Directive regarding hunting and habitats, a series of European Bird Species Action Plans have been formulated, from which individual species of waterfowl have been particular beneficiaries.[117] Nevertheless, Member States have consistently chafed at the restrictions of Article 4 and the lack of flexibility for economic considerations, thus generating a lengthy chain of litigation. Meanwhile, many of the vital documentary requirements of the Directive – notably the collation and review of implementation reports – have been 'poor',[118] thereby inhibiting a more panoramic view of individual and collective approaches to the core tenets of this pioneering legislation.

2.3 Lead Shot Restrictions

Another issue affecting the conservation of waterfowl and their wetland habitats – but one addressed outside the Birds Directive – concerns the use of lead shot by hunters. Vast numbers of birds regularly ingest lead from spent ammunition, either as gizzard stones to aid digestion or by consuming fragments that remain within scavenged animals, with significant scope for adverse health consequences.[119] Phasing out the use of lead in ammunition has thus been considered an important long-term conservation strategy to reduce

---

[113] PF Donald et al, 'International Conservation Policy Delivers Benefits for Birds in Europe' (2007) 307 *Science* 810.

[114] L Krämer, *EU Environmental Law* (Sweet and Maxwell, London, 2012) 198.

[115] Judgment of the European Court of Justice (Court) in Case C-96/98 *Commission v France* (1999) ECR I-8548.

[116] P Commenville, 'France' in Tucker, note 66, 311, 317.

[117] See https://circabc.europa.eu/ui/group/3f466d71-92a7-49eb-9c63-6cb0fadf29dc/library/882eeeb3-86e9-4944-adbe-edf7001c5eb1?p=1.

[118] Krämer, note 114, 189. This is scathing criticism, given the author was previously a leading official within the European Commission's environmental infrastructure.

[119] See DJ Pain, R Mateo, and RE Green, 'Effects of Lead Ammunition on Birds and Other Wildlife: A Review and Update' (2019) 48 *Ambio* 935.

anthropogenic stressors on waterbirds and their habitats.[120] This has, however, proved to be a challenging regulatory journey for the EU. Although the Birds Directive facilitates restrictions on hunting activities, it does not regulate the composition of ammunition, which is instead the preserve of the complex EU regime on chemical regulation: the Registration, Evaluation, Authorisation and Restriction of Chemicals (REACH).[121] Furthermore, the passage of any such legislation must navigate the political complexities of powerful hunting lobbies within particular Member States and the European Parliament.

While the Bern Convention's Standing Committee advocated the use of non-lead ammunition since 1991,[122] and the European Commission identified phasing in 'non-toxic shot' as part of its approach to the wise use of wetlands in 1995,[123] with individual Member States imposing variable restrictions on the use of lead shot under national law[124] and key treaties seeking the discontinuation of these products,[125] specific EU legislation has nevertheless been slow to emerge. This was undesirable from the standpoints of ensuring regulatory consistency and promoting coherent policies towards migratory birds, which

---

[120] N Kanstrup et al, 'Hunting with Lead Ammunition is Not Sustainable: European Perspectives' (2018) 47 *Ambio* 846.

[121] Regulation (EC) No 1907/2006 of the European Parliament and of the Council of 18 December 2006 concerning the Registration, Evaluation, Authorisation and Restriction of Chemicals (REACH), establishing a European Chemicals Agency, amending Directive 1999/45/EC and repealing Council Regulation (EEC) No 793/93 and Commission Regulation (EC) No 1488/94 as well as Council Directive 76/769/EEC and Commission Directives 91/155/EEC, 93/67/EEC, 93/105/EC and 2000/21/EC [2006] *Official Journal* L 396/1.

[122] Recommendation No. 28 (1991) on the Use of Non-toxic Shot in Wetlands, at: https://rm.coe.int/1680746b41.

[123] European Commission, note 2, 15.

[124] See VG Thomas and R Guitart, 'Limitations of European Union Policy and Law for Regulating Use of Lead Shot and Sinkers: Comparisons with North American Regulation' (2010) 20 *Environmental Policy and Governance* 57; R Mateo and N Kanstrup, 'Regulations on Lead Ammunition Adopted in Europe and Evidence of Compliance' (2019) 48 *Ambio* 989.

[125] Agreement on the Conservation of African-Eurasian Migratory Waterbirds, Resolution 1.14: Phasing Out Lead Shot in Wetlands (1999), Resolution 2.2: Phasing Out Lead Shot for Hunting in Wetlands (2002), Resolution 4.1: Phasing Out Lead Shot for Hunting in Wetlands (2008); see further Chapter 7 in this volume. A number of parties had entered reservations to this objective within the AEWA Action Plan. See M Lewis, 'AEWA at Twenty: An Appraisal of the African-Eurasian Waterbird Agreement and its Unique Place in International Environmental Law' (2016) 19 *Journal of International Wildlife Law and Policy* 22.

have long constituted central objectives of the Birds Directive. However, in 2015, the Commission requested the European Chemicals Agency (ECHA) to collate information on the uses of lead ammunition with a view towards amending REACH. In 2017, the ECHA recommended further restrictions, whereafter in 2018 its Committee for Risk Assessment advised that a transition period would be necessary to phase in these restrictions.[126] In January 2021, a Regulation was duly adopted,[127] amending Annex XVII of REACH to introduce prohibitions on the use of lead shot in and within the vicinity of wetlands, effective from 15 February 2023 onwards.[128]

Regulation 2021/57 modifies REACH by prohibiting the discharge of gunshot containing a concentration of lead equal to or greater than 1% by weight in or within 100 metres of a 'wetland', and the carrying of any such gunshot where this occurs while out wetland shooting or as part of going wetland shooting.[129] These restrictions have been hard fought and definitionally contentious. Notably, the Regulation expressly applies the definition of a 'wetland' advanced under Article 1.1 of the Ramsar Convention to REACH,[130] 'since that definition is comprehensive, covering all types of wetlands (including peatlands, where many waterbirds are also found), and since the Ramsar Convention has also developed a classification system for wetland types to help in the identification of wetlands'.[131] A specific buffer zone of 100 metres was considered necessary to ensure that spent ammunition that may not have been fired within a wetland does not ultimately land within a wetland.[132] Furthermore, it was deemed necessary to apply the prohibition to the act of 'carrying' shot, due to the practical difficulties of proving the *mens rea* of the offence. Accordingly, the legislative amendments carry the presumption that

---

[126] European Chemicals Agency, *Committee for Risk Assessment (RAC) Committee for Socio-economic Analysis (SEAC) Opinion on an Annex XV dossier proposing restrictions on Lead in Gunshot*, at: https://echa.europa.eu/documents/10162/07e05943-ee0a-20e1-2946-9c656499c8f8.

[127] Commission Regulation (EU) 2021/57 of 25 January 2021 amending Annex XVII to Regulation (EC) No 1907/2006 of the European Parliament and of the Council concerning the Registration, Evaluation, Authorisation and Restriction of Chemicals (REACH) as regards lead in gunshot in or around wetlands [2021] *Official Journal* L 24/19.

[128] Member States whose territory comprises 20% wetlands (excluding territorial waters) have an additional one-year phase-in period. REACH, note 121, Annex XVII, [12].

[129] Ibid, [11].

[130] Ibid, [13(a)].

[131] Regulation 2021/57, note 127, Recital 24.

[132] Ibid, Recital 20.

any person in possession of lead shot within these areas was intending to use this ammunition, unless they can demonstrate that there were merely traversing the wetland to discharge the shot in a non-wetland area.[133]

Strong objections have nevertheless been made by both landowners and hunting advocates, who have contested both the application of the Ramsar definition and the practicalities of enforcement, leading to questions being raised within the EU's political institutions and a formal challenge to the legislation brought before the Court. Particular unease has been generated over the types of landscapes that might conceptually be caught by the definition and its attendant obligations to create buffer zones around such sites – especially given the notion of a 'temporary' wetland considered under Article 1.1 of the Ramsar Convention. Peatlands also proved to be a strong source of anxiety during public consultations – especially those that are dry in nature and which may not obviously be considered a wetland by hunters, for which a derogation was unsuccessfully sought.[134]

Subsequently, a series of scenarios presented to the European Parliament indicates that the European Commission is keen to take an expansive approach to the concept of a 'wetland', with limited tolerance for a failure to create appropriate buffer zones. Testing the *de minimis* concept of a 'wetland', the Commission has been asked whether a 1 square metre area of temporary water which appeared in an otherwise dry arable field after a shower of rain equates to a 'wetland' for the purposes of the Ramsar Convention, and whether it would accordingly be appropriate to expect a buffer zone to be established. In response, and to the chagrin of hunting organisations, the Commission indicated that 'any area temporarily covered by water is a wetland, independently of its size', and considered the use of the Ramsar Convention's definition to be 'appropriate as it is possible for hunters to apply and for the enforcement officers to enforce it', for which buffer zones could be readily established.[135] Similar questions have sought to determine whether a snow-covered area meets the Ramsar definition of a 'temporary' wetland – which, in the view of the Commission, would do so only if the area beneath the snow would so

---

[133] Ibid, Recital 19.

[134] ECHA, note 126, 32. 'Peated areas' are not formally defined by the EU, but the Court has interpreted these sites 'according to the ordinary meaning of that word' rather prosaically as 'a wetland characterised by the presence of peat'. Judgment of the Court of Justice of the European Union (CJEU) in Case C-234/20 *Request for a Preliminary Ruling from the Augstākā tiesa (Senāts) 'Sātiņi-S' SIA* ECLI:EU:C:2022:56 [28].

[135] Written Question E-002271/2020 (15 April 2020); Answer of 1 July 2020, at: https://www.europarl.europa.eu/doceo/document/E-9-2020-002271-ASW_EN.html#def1

qualify.[136] Such interventions are not merely exercises in interpretation or hypothetical analysis, but serve to ensure the political visibility of objections to these restrictions – especially in a forum in which the hunting lobby has traditionally exerted concerted influence.

In September 2021, a collective of Polish hunting interests brought an action to annul Regulation 2021/57, arguing, *inter alia*, that the European Commission's conclusions concerning risks to human health and the environment were incorrect; alternatives to lead shot caused unnecessary suffering to animals; the ECHA report exhibited bias; the definition of 'wetlands' was so imprecise as to contravene the principle of legal certainty; the measure was disproportionate; and the Regulation violated the presumption of innocence. In December 2022, the Court rejected these claims and dismissed the action,[137] in the process bringing a further degree of clarity to the purpose of the legislation and the definitional parameters of a 'wetland' – especially those that are more temporary and ephemeral in nature. Perhaps mindful of the Commission's response to earlier questions, the applicants argued that the Ramsar definition in this context was sufficiently broad as to capture 'a simple puddle of water that appears after rain in a depression of land and thus almost the entire territory of the European Union'.[138] While the Court accepted that the definition 'does not describe in figures the minimum size of a wetland or the minimum duration of the existence of a wetland in order to respond to the temporary nature of such an area'[139] and this temporality does not preclude classification as a 'wetland', it was considered equally true that the words 'presuppose a certain stability, which excludes phenomena such as puddles that appear and disappear from one day to the next'.[140] Moreover, and in a manner reflective of the Court's jurisprudence concerning obligations to establish SPAs, it was

---

[136] Parliamentary Question P-004575/2020(ASW) (18 August 2020); Answer of 28 October 2020, at: https://www.europarl.europa.eu/doceo/document/P-9-2020-004575_EN.html.

[137] Judgment of the Court of Justice of the European Union (CJEU) in Case T-187/21 *Firearms United Network and Others v Commission* ECLI:EU:T:2022:848.

[138] Ibid, [113].

[139] Ibid, [106].

[140] Ibid, [129]. The contracting parties have also observed that the Ramsar Convention does not define a 'small' wetland – but have considered this in terms of 'springs, ponds and headwater streams' as opposed to puddles or temporary inundations. Ramsar Convention, *Resolution XIII.21: Conservation and Management of Small Wetlands* (2018) [9]. The term 'small wetland' is also used in the EU's new Nature Restoration Law (discussed further below), also without an express definition.

determined that the purpose of the Regulation was to protect waterbirds and their dependent predators; thus, the concept of a 'wetland' 'clearly does not include waters which, by their ephemeral nature, are not habitats for waterbirds'[141] and the provision was deemed sufficiently precise. Nevertheless, this does illustrate a note of caution for the Ramsar Convention where its terms are co-opted into legislation likely to be subject to judicial scrutiny, whereby national (or supranational) courts may interpret these terms in ways and contexts that may not necessarily have been envisaged by the Convention's institutions.

In February 2023, the applicants filed an appeal against this ruling,[142] which had yet to be heard at the time of writing. In the meantime, the current formulation indicates that the Ramsar definition – at least as applied to REACH – will be interpreted with reference to the capacity of temporary areas to constitute waterfowl habitat. While less extensive than the European Commission's initial interpretation, there remains a degree of ambiguity regarding when protective designations and buffer zones will be required. The Commission has indicated that this is a matter for Member States to determine – albeit that additional training and steering by the ECHA may be necessary.[143] Nevertheless, the application of these restrictions is fiercely contested[144] and subsequent practice within the individual Member States will likely test the parameters of implementing legislation, as well as the enforcement capability and resolve of national authorities.

---

[141] Case T-187/21, note 137, [130]. In keeping with the Court's assessment that a degree of habitat functionality is required, the contracting parties also view a small wetland as capable of supporting vulnerable populations of threatened species and important for the conservation of biodiversity. Ramsar Convention, *Resolution XIV.15: Enhancing the Conservation and Management of Small Wetlands* (2022) [8].

[142] Appeal brought on 21 February 2023 by Firearms United Network, Tomasz Walter Stępień, Michał Budzyński and Andrzej Marcjanik against the judgment of the General Court delivered on 21 December 2022 in Case T-187/21, *Firearms United Network and Others v Commission* (Case C-105/23 P) [2023] *Official Journal* C 134/5.

[143] Parliamentary question – P-004575/2020(ASW), note 136.

[144] The Court considered that the Regulation did not infringe the presumption of innocence, since it provided a basis for rebuttal – that is, that the hunter was merely crossing a wetland to demonstrably shoot elsewhere – but the obligation to apply safety zones remains highly contentious and likely to be subject to additional litigation.

3 WETLANDS AND THE HABITATS DIRECTIVE

As outlined above, the Birds Directive was developed by the EEC contemporaneously to the elaboration of the Bern Convention by the Council of Europe; and while bird conservation was a significant factor in negotiating the latter instrument, it was ultimately a more holistic regime that would apply to a much wider array of species and habitats. In 1981, the EEC became a contracting party to the Bern Convention[145] and, given this broader focus, the creation of additional nature conservation legislation to implement these commitments beyond the avian context became inevitable.[146] By 1987, and in the light of the Single European Act reconstituting the Treaties of Rome and incorporating a clearer series of environmental competences for the Community,[147] the fourth iteration of the Environmental Action Programme was launched,[148] in which the now European Community (EC) noted the thematic limitations of its existing bird-focused legislation and advocated 'a major new thrust in the field of nature conservation'.[149] Rueing the 'poor' implementation of the Bern Convention by the EC Member States, it was considered that a 'comprehensive framework of nature protection measures, at Community level, would undoubtedly help to improve the situation of endangered species of plants and animals within the Community'.[150]

The Bern Convention requires contracting parties to 'take appropriate and necessary legislative and administrative measures' to protect endangered natural

---

[145] Council Decision 82/72/EEC of 3 December 1981 concerning the conclusion of the Convention on the conservation of European wildlife and natural habitats [1982] *Official Journal* L 38/1.

[146] See further G Jones QC, 'The Bern Convention and the Origins of the Habitats Directive' in Jones, note 57, 1.

[147] Single European Act [1987] *Official Journal* L 169/1. These powers were subsequently finessed by the Maastricht Treaty in 1992, which created the European Union. Treaty on European Union [1992] *Official Journal* C 191/1; see further D Wilkinson, 'Maastricht and the Environment: Implications for the EC's Environmental Policy of the Treaty on European Union' (1992) 4 *Journal of Environmental Law* 221.

[148] Resolution of the Council of the European Communities and of the representatives of the Governments of the Member States, meeting Within the Council of 19 October 1987 on the continuation and implementation of a European Community policy and action programme on the environment (1987–1992) [1987] *Official Journal* C 328/1.

[149] Ibid, [5.1.2]–[5.1.4].

[150] Ibid, [5.1.6].

habitats and the habitats of species of flora and fauna listed on its Appendices[151] – a commitment buttressed in 1989 by its Standing Committee's aspiration to establish a region-wide network of Areas of Special Conservation Interest comprising protective designations by the contracting parties advanced under Article 4, termed the 'Emerald Network'.[152] The resultant Habitats Directive, concluded in 1992, therefore represents the core provision by which the EU implements its commitments under the Bern Convention – notably through the aggregation of an EU-wide network of protected areas termed 'Natura 2000'. However, the initially symbiotic relationship between the two instruments and their respective protected area networks has subsequently become dominated by EU provisions in the light of subsequent EU enlargements.[153]

Given the intention to create a more holistic instrument, wetlands were significantly less prominent in the elaboration of the Habitats Directive in comparison to the earlier Birds Directive. Nevertheless, the prior experience of Member States regarding the enforcement of the Birds Directive, including its wetland protections, also stiffened their resolve in negotiating key elements of the Habitats Directive. As initially proposed, the European Commission envisaged that the Natura 2000 network would encompass SPAs, as well as new protective designations under the Habitats Directive and, unless contrary notice was given within two years of the entry into force of the Directive, 'areas classified by the Member States under the Ramsar Convention'.[154] This express incorporation of Ramsar Sites failed to make it into the final version of the Directive, with Member States keen to wrest control from the Commission over future protective designations.[155] More significantly, by this stage a number of Member States had become increasingly perturbed by the environmental enforcement zeal of the Commission and the limited room

---

[151] Bern Convention, note 38, Art 4.

[152] Recommendation No. 16 (1989) of the standing committee on areas of special conservation interest, at: https://search.coe.int/bern-convention/Pages/result_details.aspx?ObjectId=0900001680746c25.

[153] Y Epstein, 'The Habitats Directive and the Bern Convention: Synergy and Dysfunction in Public International Law' (2016) 26 *Georgetown International Environmental Law Review* 139. The Bern Convention is not expressly referred to within the text of the Habitats Directive. Jackson observes that the launch of the Emerald Network was motivated in part by some mutual participants to usurp the need for a nascent Habitats Directive: Jackson, note 48, 140–42. If so, these particular negotiators would be sorely disappointed.

[154] European Commission, *Proposal for a Council Directive on the Protection of Natural and Semi-natural Habitats and of Wild Fauna and Flora* COM(88) 381 (final) [1988] *Official Journal* C 247/3, draft Art 6(1).

[155] Jackson, note 48, 145.

for manoeuvre under Article 4 of the Birds Directive. Notably, the *Eilean na Muice Duibhe*/Duich Moss episode galvanised British opposition to the more ecologically ambitious elements of the Commission's proposal, while the steadily mounting judgments of Article 4 infringements – embodied most clearly by the key wetland cases of *Leybucht Dykes* and *Santoña Marshes* – prompted other Member States to view the drafting balance of competing interests more clearly in favour of socio-economic concerns.[156]

Consequently, while the core obligations of the Directive are to maintain or restore at a favourable conservation status the natural habitats and species of wild fauna and flora of Community interest,[157] such measures will take account of 'economic, social and cultural requirements and regional and local characteristics'.[158] Member States are to create Special Areas of Conservation (SACs)[159] comprising sites hosting natural types listed in Annex I to the Directive and the habitats of species designated in Annex II. Annex I incorporates a considerable number of habitat types falling within the Ramsar Convention's definition of a 'wetland'; while Annex II includes many species of plants and animals for which wetlands constitute important habitats. Accordingly, as under the Birds Directive, there is considerable scope for widespread protective designations for wetland areas under the Habitats Directive. However – and in a manner that sought to dilute aspects of *Leybucht Dykes* and subsequent Article 4 case law – the Habitats Directive expressly promotes additional flexibility in designation practice. Member States must take appropriate steps to avoid habitat deterioration and the disturbance of protected species,[160] and must subject 'any plan or project' likely to have a 'significant impact' on an SAC to an 'appropriate assessment'.[161] If this assessment reveals negative implications and if there is 'an absence of alternative solutions', the plan or project may nevertheless proceed if there are 'imperative reasons of overriding public interest, including those of a social or economic nature', albeit – and echoing Article 4.2 of the Ramsar Convention – that the Member State must take all compensatory measures necessary to ensure that the overall coherence of the Natura 2000 network is protected.[162] Significantly, Articles 6(2)–(4) subsequently replace the obligation incumbent in the first sentence of Article 4(4) of the Birds Directive.[163]

---

[156] Ibid, 268–70.
[157] Habitats Directive, note 8, Art 2(2).
[158] Ibid, Art 2(3).
[159] Ibid, Art 3(3).
[160] Ibid, Art 6(2).
[161] Ibid, Art 6(3).
[162] Ibid, Art 6(4); see further Chapter 5 in this volume.
[163] Habitats Directive, note 8, Art 7.

As with the Birds Directive, infringement proceedings have been brought over perceived deficiencies in designation practice. One particularly prominent dispute concerned the alarming levels of damage inflicted on the wetlands of the Doñana National Park in Spain through the abstraction of groundwater, which has concurrently troubled the Ramsar Convention's institutions since 1990.[164] Groundwater is not protected as such under the Habitats Directive and is instead addressed through the pertinent provisions of EU water law.[165] Nevertheless, the Court found breaches of both the Water Framework Directive and the Habitats Directive.[166] On the latter provision, as parts of the wetland encompassed Mediterranean temporary ponds – a habitat type protected under the Directive – it was considered that the abstraction of the groundwater had rendered the ponds ecologically reliant upon rainwater and that they were therefore unduly affected during dry periods. This was ruled to have breached the obligation in Article 6(2) to prevent the deterioration of habitats, which Spain was considered to have failed to halt.[167]

While the obligations towards protected areas are highly significant, the value of the Habitats Directive to wetland conservation extends beyond this requirement. Notably, Article 18 encourages Member States to undertake research and scientific work, with particular emphasis on transboundary cooperation.[168] This provides further scope to buttress activities in respect of transboundary Ramsar Sites, the vast majority of which are presently designated in Europe, including many that are between EU Member States and have been listed as SACs.[169] More significantly, however, the Habitats Directive was accompanied by a root-and-branch restructuring of EC funding mechanisms, reformulated in 1992 as the *L'Instrument Financier pour l'Environnement* (LIFE) funding programme,[170] which has played a vital role in underwriting the creation of protected areas and promoting conservation and restoration

---

[164] Ramsar Convention, *Recommendation 4.9.1: Doñana National Park, Spain* (1990).

[165] Notably the Water Framework Directive: Directive 2000/60/EC of the European Parliament and of the Council of 23 October 2000 establishing a framework for Community action in the field of water policy [2000] *Official Journal* L 327/1.

[166] Judgment of the Court of Justice of the European Union (CJEU) in Case C-559 *Commission v Spain* ECLI:EU:C:2021:512.

[167] Ibid, [169].

[168] Habitats Directive, note 8, Art 18(2).

[169] See further https://www.ramsar.org/sites/default/files/documents/library/list_of_transboundary_sites.pdf.

[170] Council Regulation (EEC) No 1973/92 of 21 May 1992 establishing a financial instrument for the environment (LIFE) [1992] *Official Journal* L 206/1.

projects in the Member States and candidate countries. The LIFE programme replaced a less coordinated series of funding programmes which, from a wetland perspective, were considered to have had a limited impact and generally involved small wetland sites.[171] LIFE funding has subsequently proved crucial to wetland conservation – especially in the more recent EU Member States, where financial support for conservation and restoration projects would otherwise have proved considerably more challenging to sustain.[172]

Beyond the Directive itself, in 1995 the European Commission elaborated its first – and to date only – strategy document for the conservation and wise use of wetlands[173] as part of the 'Towards Sustainability' programme of activities pursued under the fifth iteration of the Environmental Action Programme.[174] To this end, the Commission considered that the wise use and conservation of wetlands correspond to the concept of 'sustainability' that lay at the heart of the (then) new Environmental Action Programme,[175] while further lamenting that although wetlands facilitate vital ecosystem functions, they have been chronically underappreciated from an ecological, economic, and regulatory

---

[171] European Commission, note 2, 27.

[172] See V Šefferová Stanová and R Rybanič, 'Slovakia' in Tucker, note 66, 555, 567; R Baškytė and Ž Obelevičius, 'Lithuania' in Tucker, note 66, 451, 465: A Lotman and S Lotman, 'Estonia' in Tucker, note 66, 273, 288; B Barov, 'Bulgaria' in Tucker, note 66, 181, 195; K Sipos, 'Hungary' in Tucker, note 66, 374, 386; C Papazoglou and A Demetropoulos, 'Cyprus' in Tucker, note 66, 219, 231. LIFE funding has also proved to be exceptionally valuable to wetland conservation efforts in some of the more longstanding Member States. See E Martins and J Ventocilla, 'Belgium' in Tucker, note 66, 160, 174; C Olmeda and JC Blanco, 'Spain' in Tucker, note 66, 593, 606; MOG Eriksen and M Pantzar, 'Sweden' in Tucker, note 66, 612, 618. To put this into further perspective, the European Commission's estimated annual financing needs at the EU level for 2021–27 for wetland management and restoration was approximately €565 million. Parliamentary Question E-001147/2022(ASW) (21 March 2022); Answer of 20 May 2022, at: https://www.europarl.europa.eu/doceo/document/E-9-2022-0 01147-ASW_EN.html.

[173] European Commission, note 2.

[174] *Towards Sustainability: A European Community Programme of Policy and Action in Relation to the Environment and Development* [1993] *Official Journal* C 138/5.

[175] European Commission, note 2, 42. The Commission has also previously declared that the Community 'agrees with the definition of wise use of wetlands given in the Ramsar Convention and, accordingly, considers it important to promote this philosophy'. Written Question No 2543/91 (8 November 1991); Answer of 14 May 1992 [1992] *Official Journal* C 235/7.

perspective.[176] The Commission further observed that although numerous provisions of EU law were applicable to wetland protection, 'a continuing regression of wetlands in the European Union is noted', while 'a coherent Community wetland policy still does not exist'.[177] With this in mind, the Commission elaborated a series of objectives for a Community wetland policy – namely no further wetland loss or degradation, wise use of remaining wetlands, wetland improvements and restoration, and international cooperation and action favouring wise use and conservation of wetlands.[178]

To implement these objectives, the Commission identified five operational priorities for wetland conservation. First – as had been advocated since 1979 – a definitive European network of wetlands should be established, incorporating 'all wetlands of Union importance' identified under the nature conservation Directives, for which Ramsar Sites 'represent the global link'.[179] Reflecting the Commission's original intentions for the Habitats Directive, the Member States were encouraged to integrate their respective Ramsar Sites into the Natura 2000 network, 'even if, exceptionally, they are not Natura sites'.[180] Practice in this respect remains mixed, although a number of Member States have directly incorporated all,[181] or a significant majority,[182] of their Ramsar designations into the Natura 2000 network. Allied to this, improving knowledge of wetlands and their values was a further operational priority and one intended to be promoted primarily by the Commission, although there is relatively limited evidence to suggest that the EU institutions have presided over any great advancement in this respect.

The remaining operational priorities largely addressed the integration of wetland conservation into alternative policies and provisions – an issue that had long been a distinct weakness of EU environmental policies to that point. In particular, land use planning and other sectoral policies presented considerable scope for conflict and siloed approaches. The Birds Directive in particular had to negotiate a challenging – and sometimes incoherent – coexistence with other sectoral policies, especially regarding agriculture. Indeed, Temple Lang initially warned against the absurdity of seeking to protect wetland species

---

[176] European Commission, note 2, 42.
[177] Ibid, 43.
[178] Ibid, 43–44.
[179] Ibid, 44.
[180] Ibid, 44.
[181] See E Gerritsen, 'The Netherlands' in Tucker, note 66, 468, 474; H Toivonen and O Ojala, 'Finland' in Tucker, note 66, 291, 300; I Christopoulou, 'Greece' in Tucker, note 66, 353, 363.
[182] Eriksen and Pantzar, note 172, 622; G Tucker et al, 'United Kingdom' in Tucker, note 66, 634, 651.

while simultaneously granting generous subsidies for drainage projects;[183] yet the early application of the Directive reveals this was an enduring problem,[184] and one that required attention in subsequent reforms of agricultural policy.

While the European Commission expressed confidence that its legislative framework provided a strong platform to promote the conservation and wise use of wetlands, the degree to which the loose collection of objectives that constitutes the 'Community wetland policy' has been advanced and achieved remains questionable. Indeed, progress has not been objectively, coherently, or demonstrably measured, and wetland conservation has not been as expressly considered by the EU institutions in subsequent years; accordingly, it is a cause for disappointment that three decades later, the 'policy' has not been substantively reviewed or revisited. Tellingly, perhaps, in May 2022 the European Commission declared that it 'does not have a specific action plan for all the wetlands in the EU'.[185] This suggests that, beyond the letter of these core provisions, significant emphasis has been placed on national implementation of the Ramsar Convention as opposed to revitalising the 1995 strategy.

This concern notwithstanding, more recent legislative developments have reoriented the EU's nature conservation objectives away from a predominantly site-based approach towards a more holistic and ecosystem-centred vision, with considerable scope to advance the conservation of wetlands. Drawing upon the objectives of the EU Biodiversity Strategy to 2030,[186] in 2022 the Commission proposed a Regulation on nature restoration, establishing a series of targets for the elaboration of effective and area-based restoration measures for degraded ecosystems.[187] This included objectives for the restoration of wetlands – a policy that had previously been identified by the Commission as

---

[183] Temple Lang, note 79, 18.

[184] See for example Written Question No 844/86 (10 July 1986), lamenting the 'blind policy of draining our marshes and wetlands', which the European Commission deemed 'not usually compatible with obligations stemming from the Directive': [1986] *Official Journal* C 271/4. This was also criticised within individual RAMs; see below.

[185] Parliamentary question – E-001147/2022(ASW), note 172.

[186] European Commission, *Communication from the Commission to the European Parliament, the Council, the European Economic and Social Committee and the Committee of the Regions. EU Biodiversity Strategy for 2030: Bringing Nature Back Into Our Lives* COM (2020) 380 final.

[187] Proposal for a Regulation of the European Parliament and of the Council on nature restoration COM/2022/304 final.

a significant element of future climate strategy, given the prospective role of wetlands as climate sinks.[188]

While the initial proposal was warmly received by the scientific and ecological community, its subsequent political reception proved to be significantly more volatile, and one evocatively characterised as 'an unprecedented rollercoaster in the context of EU environmental governance'.[189] In order to proceed further, the proposed legislation required approval by both the European Council and the European Parliament – a hurdle that was intensified in a number of Member States by a hostile media response, concerted disinformation campaigns, and intense lobbying by particular sectoral interests.[190] This resulted in a series of compromises that diluted important elements of the original proposal, including the contentious addition of an 'emergency brake' to suspend the application of the legislation, and the future of the EU's Nature Restoration Law hung precipitously in the balance until a final version of the legislation was approved by the Council in June 2024. The Nature Restoration Law duly entered into force on 18 August 2024[191] and retains a number of provisions with the potential to substantively advance the cause of wetland conservation.

Wetlands are considered prominently within the Nature Restoration Law, with a particular emphasis on their prospective value to future EU climate policy, observing that it is important that 'ecosystems in all land categories, including forests, grasslands, croplands and wetlands, are in good condition in order to be able to capture and store carbon effectively'.[192] The primary objective of the Nature Restoration Law is for Member States to put in place the necessary restoration measures to ensure that particular habitat types that

---

[188] European Commission, *Communication from the European Commission to the European Parliament, the Council, the European Economic and Social Committee and the Committee of the Regions. Forging a climate-resilient Europe – the new EU Strategy on Adaptation to Climate Change* (COM/2021/82 final); see further Chapter 11 in this volume.

[189] A Cliquet et al, 'The Negotiation Process of the EU Nature Restoration Law Proposal: Bringing Nature Back in Europe against the Backdrop of Political Turmoil?' (2024) 32 *Restoration Ecology* e14158, 2.

[190] Ibid, 306; K DeCleer and A Cliquet, 'Nature Restoration: Proposed EU Law under Threat' (2023) 619 *Nature* 252.

[191] Regulation (EU) 2024/1991 of the European Parliament and of the Council of 24 June 2024 on nature restoration and amending Regulation (EU) 2022/869 [2024] *Official Journal* L 2024/1991.

[192] Ibid, Recital 19.

are not currently in a 'good condition' are improved to meet this standard.[193] The habitats for which such measures are to be taken are listed on Annex I of the legislation and specifically include coastal and inland wetlands. The Nature Restoration Law aspires to a graduated recovery of these habitat types, measured against specific targets, with immediate priority given to restoration in areas located within Natura 2000 sites.[194] Given that the restoration targets apply to distinct ecosystems, with variable representative coverage across different Member States, each individual Member State is required to elaborate a national restoration plan to quantify restoration priorities on a domestic level.[195] The scale of wetland restoration undertaken through the Nature Restoration Law will therefore also vary between individual Member States, although the legislation elaborates a clearer series of obligations to promote restorative activity than had previously applied under the more species-oriented nature conservation legislation of the EU.

Given its recent provenance, the Nature Restoration Law remains fundamentally untested. On one level, it marks an important new direction in EU environmental law and represents the largest multilateral scale on which such an approach has been applied, and has thus been received with a degree of cautious optimism.[196] Nevertheless considerable challenges remain – not least the entrenchment of overriding public interest where restoration conflicts with renewable energy projects[197] and national security,[198] and the ability of Member States to apply the emergency brake to temporarily suspend activities in the case of an 'unforeseeable, exceptional and unprovoked event',[199] which constitute compromises familiar to pre-existing EU legislation but were vital in brokering an uneasy agreement to advance the Nature Restoration Law into existence. Similarly, the success of the Nature Restoration Law – including

---

[193] Ibid, Art 4(1). The condition of a habitat is considered to be good if 'the key characteristics of the habitat type, in particular its structure, functions and typical species or typical species composition reflect the high level of ecological integrity, stability and resilience necessary to ensure its long-term maintenance and thus contribute to reaching or maintaining favourable conservation status for a habitat, where the habitat type concerned is listed in Annex I to Directive 92/43/EEC, and, in marine ecosystems, contribute to achieving or maintaining good environmental status': Art 3(4).
[194] Ibid.
[195] Ibid, Art 14.
[196] D Hering, 'Securing Success for the Nature Restoration Law' (2023) 382 *Science* 1248.
[197] Regulation (EU) 2024/1991, note 191, Art 6.
[198] Ibid, Art 7.
[199] Ibid, Art 27.

and beyond the context of wetland restoration – will hinge upon a ready supply of significant revenue streams from public and private sources, as well as its full and assiduous implementation by the Member States – a number of which have exhibited scepticism at best, and open hostility at worst, to the central tenets of the legislation.

4 THE EUROPEAN UNION AND THE RAMSAR CONVENTION

Unlike subsequent biodiversity treaties, which have welcomed prospective accession by 'Regional Economic Integration Organizations'[200] (REIOs), participation in the Ramsar Convention has always remained strictly confined to states.[201] This membership qualification had a clear practical logic given that a core obligation incumbent upon the contracting parties is the identification of Wetlands of International Importance under national sovereignty,[202] which lies outside the purview of a supranational body with no sovereign territory to designate – although each of the Member States is a contracting party to the Ramsar Convention in its individual capacity. Nevertheless, the EU retains a spectral presence in these proceedings, with Member States required to coordinate on issues falling within EU competences,[203] given that they are acting

---

[200] See, for example, Convention on the Conservation of Migratory Species of Wild Animals (CMS), adopted 23 June 1979, entered into force 1 November 1983, 1651 UNTS 67, Art I(1)(k); Agreement on the Conservation of African-Eurasian Migratory Waterbirds (AEWA), adopted 16 June 1995, entered into force 1 November 1999, 2365 UNTS 203, Art XIII(1); Bern Convention, note 38, Art 19. A similar position to the Ramsar Convention was initially taken by the Convention on International Trade in Endangered Species of Wild Fauna and Flora, adopted 3 March 1973, entered into force 1 July 1975, 993 UNTS 243 (CITES). In 1983 CITES was amended to allow the participation of REIOs, although this was a contentious development that would take until 2013 to formally enter into effect.

[201] Convention on Wetlands, note 4, Art 9.2.

[202] Ibid, Art 2.1.

[203] On these modalities, see EJ Goodwin, 'State Delegations and the Influence of COP Decisions' (2019) 31 *Journal of Environmental Law* 235, 254. EU coordination practices are not always welcomed in international forums, from the standpoints of both transparency and the collective bloc voting power of this aggregation of parties. See R Caddell, 'Biodiversity Loss and the Prospects for International Cooperation: EU Law and the Conservation of Migratory Species of Wild Animals' (2008) 8 *Yearbook of European Environmental Law* 218.

as trustees in a forum in which the EU is itself disbarred from exercising these competences directly.[204]

While keen to ensure a truly global cadre of participants in the Ramsar Convention,[205] the contracting parties through the Conference of the Parties (COP) have sought to cultivate a cooperative relationship with the EU institutions. In 1990, the COP strongly supported:

> the establishment of closer links with the Commission of the European Communities with a view to facilitating combined action for conservation and wise use of wetlands in Community Member States, and provision of technical assistance for wetland conservation and wise use of wetlands in developing countries.[206]

Iterations of the Ramsar Convention's Work Plan have also noted the value of the Habitats Directive in promoting the objectives of the Convention.[207] The most tangible point of cooperation between the two bodies remains the Mediterranean Wetlands Initiative – a joint programme between the Ramsar Convention, the European Commission, and key scientific and advocacy partners[208] – as one of the most longstanding and prominent Ramsar Regional Initiatives that has been heavily underwritten by the EU since its inception.[209] Indeed, more recent expressions of cooperation with the EU have been expressed in exclusively financial terms, noting valuable funding possibilities as opposed to more conceptual partnership advantages;[210] and the EU has been barely identified in more recent iterations of the Ramsar Convention's

---

[204] M Cremona, 'Member States as Trustees of the European Union: Participating in International Agreements on Behalf of the European Union' in A Arnull et al, *A Constitutional Order of States?* (Hart, Oxford, 2011) 435.

[205] Ramsar Convention, *Recommendation 1.1: Expanding the Convention's membership* (1980) (noting the initial contracting parties were largely drawn from Europe).

[206] Ramsar Convention, *Recommendation 4.11: Cooperation with international organizations* (1990).

[207] Ramsar Convention, *Resolution VII.27: The Convention Work Plan 2000–2002* (1999), General Objective 7.2.8.

[208] See further NE Farantouris, 'The International and EU Legal Framework for the Protection of Wetlands with Particular Reference to the Mediterranean Basin' (2009) 6 *Macquarie Journal of International and Comparative Environmental Law* 31.

[209] European Commission, note 2, 25.

[210] See, for example, Ramsar Convention, *Resolution X.11: Partnerships and synergies with Multilateral Environmental Agreements and other institutions* (2008); Ramsar Convention, *Resolution XI.6: Partnerships and synergies with Multilateral Environmental Agreements and other institutions* (2012).

Strategic Plan. A Cooperative Agreement was in place between the Ramsar Convention and the European Environment Agency in 2006–11,[211] although this lapsed with little indication of prospective renewal.

Beyond formal linkages, a series of Ramsar Advisory Missions (RAMs) have been conducted with respect to an array of wetlands within the EU, including sites familiar to the case law of the Court. For instance, the contentious development at Leybucht was subject to a prior RAM in which a series of compensatory measures was suggested – with the Court subsequently ruling in favour of Germany, having been satisfied by the efforts made by the authorities to mitigate the impacts of the development.[212] Other RAMs have simultaneously highlighted the value of EU nature conservation law for wetland conservation, while also criticising the schizophrenic approach to funding, with the EU simultaneously demanding wetland protections while also underwriting extensive drainage and development schemes.[213] Meanwhile, the multiple RAMs in respect of the Doñana National Park[214] appear to have ultimately precipitated the Commission's infringement proceedings against Spain.

5 CONCLUSIONS

The European Union is, to borrow F Scott Fitzgerald's celebrated literary flourish, 'within and without' the Ramsar regime: simultaneously exercising extensive supranational authority over aspects of wetland management while possessing no wetlands of its own and precluded from formal participation in the Convention itself, which remains the preserve of its constituent Member States. Nevertheless, the EU can be seen to have provided a vital regulatory spur for the conservation of wetlands, exercised through its sprawling mosaic of environmental legislation, case law, and sectoral policies.

EU nature conservation law continues to represent the primary legal vehicle for the protection of wetlands within the Member States. Since 1979, the pioneering Birds Directive has engendered a widespread series of spatial protections for waterfowl habitats, with these obligations catalysed in no small part by the Ramsar Convention. These provisions have been subject to asser-

---

[211] Ramsar Convention, *Document SC54-16: Review of current and proposed cooperative agreements* (2018).

[212] RAM 19: Ostfriesisches Wattenmeer and Dollart (1990); see also Ramsar Convention, *Recommendation 4.9.4: Conservation of the Leybucht, Federal Republic of Germany* (1990).

[213] RAM 6: Ramsar Sites in Greece (1988); RAM 11: Greek Ramsar Sites (1989).

[214] RAM 51: Doñana (2002); RAM 70: Doñana (2011); RAM 95: Doñana (2020).

tive oversight by the EU institutions, including the application of meaningful enforcement mechanisms, which has ensured that protective designations for wetlands have been established to a far fuller extent and less subordinated to industrial interests than would likely have otherwise been the case. Since 1992, the scope for wetland conservation has both broadened and narrowed: broadened through the development of more holistic habitat protections under the Habitats Directive, and crucially underpinned by consistent and generous sources of funding for rehabilitation projects; yet also narrowed through adjustments to the margin of appreciation for economic development.

Ultimately, the strengths and frailties of the EU's wetland protections are representative of the vagaries of the Directives themselves: the utility of the enforcement processes is counterbalanced by the multi-annual duration of legal proceedings; funding opportunities are vital but also finite and competitive; protective designations are ubiquitous but often ecologically fragmented; and environmental conservation remains an essential endeavour of the EU, but one that co-exists uneasily with competing sectoral interests in a dynamic, multifaceted, and multi-state context. More specifically, it remains a strong cause for disappointment that the EU's sole strategic review of the ecological plight of wetlands occurred three decades ago; while attempts to reorient the central tenets of EU biodiversity law towards environmental restoration, with its myriad potential benefits for wetland protection, weathered considerable turbulence within a testing political climate.

These challenges notwithstanding, the multifaceted array of EU laws and policies pertinent to wetlands ensures that Member States are arguably in the strongest position among the contracting parties to promote the core principles and objectives of the Ramsar Convention. Indeed, the rich array of regulatory advantages applicable to wetlands within the EU has prompted commentators to conclude that the true value of the Ramsar Convention lies more in its capacity to strengthen conservation approaches in regions in which this legal machinery is largely absent.[215] By consistently pushing beyond the comparatively narrow objectives of the Ramsar Convention, the European Union has thus posited an essential and complementary contribution to the cause of wetland conservation.

---

[215] D Stroud et al, 'The International Drivers of Nature Conservation, Their Objectives and Impacts on Nature Conservation Policies and Actions in Europe' in Tucker, note 66, 41, 48.

13. Rights of wetlands and the Ramsar Convention
Erin Okuno[1]

1 INTRODUCTION

The 'Rights of Nature' movement has recently emerged as another legal and policy tool to help address the rapid degradation and destruction of the Earth's biodiversity and ecosystems, including wetlands.[2] At international, national, and sub-national levels, the rights of Nature have been formally recognised for Nature as a whole, for certain types of ecosystems or specific natural features, and for particular species through constitutions, statutes, Indigenous and tribal laws, ordinances, and other local laws, as well as in non-binding declarations, resolutions, and other policy statements.[3] From a philosophical perspective,

---

[1] Many thanks to my co-editors, Dr Richard Caddell and Professor Royal C Gardner, for their insightful feedback on this Chapter. I am also grateful to the members of the Society of Wetland Scientists Rights of Wetlands Initiative for the opportunity to collaborate with and learn from them over the last several years – their work has informed much of this Chapter.

[2] The movement evolved in part from the Earth jurisprudence theoretical framework, which has been described as 'the fastest growing legal movement of the twenty-first century' and takes a biocentric (or ecocentric) – rather than an anthropocentric – approach to law and human-Nature relationships. D Takacs, 'We Are the River' (2021) 2021 *University of Illinois Law Review* 545, 551 (quoting United Nations Secretary General, *Harmony with Nature: Report of the Secretary General*, UN Doc A/74/236 [129] (2019)); M Maloney, 'Building an Alternative Jurisprudence for the Earth: The International Rights of Nature Tribunal' (2016) 41 *Vermont Law Review* 129, 131–32; S Borràs, 'New Transitions from Human Rights to the Environment to the Rights of Nature' (2016) 5 *Transnational Environmental Law* 113, 114–15; Cormac Cullinan, 'The Legal Case for the Universal Declaration of the Rights of Mother Earth' (2010) 1, at: https://www.garn.org/wp-content/uploads/2021/09/Legal-Case-for-Universal-Declaration-Cormac-Cullinan.pdf.

[3] See *infra* Section 2 for specific examples.

the Rights of Nature movement reframes rights and values through a biocentric lens,[4] and from a pragmatic perspective, the movement provides additional means to protect the environment that can be an alternative to or can complement existing legal frameworks,[5] including the Ramsar Convention on Wetlands of International Importance.[6]

This Chapter describes the history and status of the Rights of Nature movement, highlighting successful examples and discussing the types of rights holders and rights, the different forms of recognition, and the various approaches to enforcement. It also identifies some of the challenges faced by the Rights of Nature movement. The Chapter then explains the history and goals of the proposed Universal Declaration of the Rights of Wetlands and closes by evaluating how the rights of wetlands might be incorporated into Ramsar Convention processes.

2   THE RIGHTS OF NATURE MOVEMENT

Although many of the concepts and principles behind the Rights of Nature movement have existed for millennia,[7] the explicit incorporation of the rights of Nature into legal and policy frameworks has gained much momentum since the turn of the century. This section describes the growth of the Rights of Nature movement, provides an overview of some fundamental concepts, and offers examples of how governments, courts, organisations, communities, individuals, and others have acknowledged, supported, and advocated for the rights of Nature throughout the world.

2.1   History

The notion that Nature has rights is not new. Indeed, for hundreds and even thousands of years, many Indigenous peoples[8] and cultures have seen humans as an interconnected part of Nature and have viewed Nature and natural

---

[4]   E Ryan, H Curry, and H Rule, 'Environmental Rights for the 21st Century: A Comprehensive Analysis of the Public Trust Doctrine and Rights of Nature Movement' (2021) 42 *Cardozo Law Review* 2447, 2500–01.

[5]   Ibid, 2501.

[6]   Convention on Wetlands of International Importance especially as Waterfowl Habitat, adopted 2 February 1971, entered into force 21 December 1975, 996 UNTS 245 (amended 1982 & 1987).

[7]   Takacs, note 2, 552–53; Ryan, Curry, and Rule, note 4, 2502–03; GT Davies et al, 'Towards a Universal Declaration of the Rights of Wetlands' (2021) 72 *Marine and Freshwater Research* 593, 597.

[8]   See further Chapter 6 in this volume.

elements as living beings or 'persons' with inherent rights.[9] Over the last several decades, however, the concept of recognising and protecting Nature's rights has moved into mainstream Western discourse through the works of Christopher Stone, Thomas Berry, Cormac Cullinan, and others,[10] and the Rights of Nature movement has grown significantly during the twenty-first century. Examples at the national and especially sub-national level have proliferated: several countries now legally recognise the rights of Nature at the national level, and dozens of municipalities and other sub-national governments have passed laws as well.[11]

Perhaps the first modern legal recognition of the rights of Nature in an ordinance occurred in 2006, when Tamaqua Borough, Pennsylvania in the United States passed the Tamaqua Borough Sewage Sludge Ordinance.[12] The ordinance made it 'unlawful for any corporation ... to interfere with the existence and flourishing of natural communities or ecosystems', and explained that 'Borough residents, natural communities, and ecosystems ... [are] "persons" for purposes of the enforcement of the civil rights of those residents, natural communities, and ecosystems'.[13] Shortly after, in 2008, Ecuador became the first country to recognise the rights of Nature through an amendment to its national constitution, which stated that 'Nature, or Pacha Mama ... has the right to integral respect for its existence and for the maintenance and regeneration of its life cycles, structure, functions and evolutionary processes'.[14] In 2010,

---

[9] See Davies et al, note 7, 596, 597; CM Kauffman and PL Martin, *The Politics of Rights of Nature: Strategies for Building a More Sustainable Future* (MIT Press, Cambridge, Massachusetts, 2021) 9.

[10] Takacs, note 2, 554–57. Stone famously argued that 'the natural environment as a whole' should have rights and that guardians could be appointed to defend those rights in legal proceedings. See CD Stone, 'Should Trees Have Standing? Toward Legal Rights for Natural Objects' (1972) 45 *Southern California Law Review* 450, 456, 464–73. In 2001, Berry suggested a set of principles about Nature's rights that informed the Earth jurisprudence (or 'Wild Law') movement spearheaded by Cullinan and others. Takacs, note 2, 556–57.

[11] See Harmony with Nature United Nations, 'Rights of Nature Law and Policy', *Harmony with Nature UN* at: http://harmonywithnatureun.org/rightsOfNature; DR Boyd, *The Rights of Nature: A Legal Revolution That Could Save the World* (ECW Press, Toronto, 2017) 127.

[12] LC Pecharroman, 'Rights of Nature: Rivers That Can Stand in Court' (2018) 7 *Resources* 1, 4; Boyd, note 11, 112–13.

[13] Tamaqua Borough, Schuylkill County, Pennsylvania, Ordinance No 612 of 2006 § 7.6, at: http://files.harmonywithnatureun.org/uploads/upload666.pdf.

[14] Ryan, Curry, and Rule, note 4, 2514; Republic of Ecuador, Constitution of the Republic of Ecuador, 20 October 2008, Art 71, at: https://pdba.georgetown.edu/Constitutions/Ecuador/english08.html.

Bolivia became the first country to pass a rights of Nature statute,[15] which enumerates seven specific rights of Mother Earth and the attendant obligations and duties of the state and society.[16] Throughout the next decade, numerous other examples of rights of Nature laws and policies also materialised.[17]

2.2 Types of Rights Holders and Rights

Laws and policy instruments recognise different constituents or elements of Nature as rights holders. Some identify or acknowledge the rights of Mother Earth or Nature as a whole[18] – Ecuador's Constitution and Bolivia's statute, noted above, offer examples of this approach.[19] In some instances, however, the instruments focus on a certain type of ecosystem, such as all rivers, which Bangladesh and the Universal Declaration of the Rights of Rivers recognise.[20] Or the rights might be those of an individual natural feature, such as a specific river or lake – for example, Río Atrato, Lake Tota, and the Colombian Amazon.[21] Still other approaches identify a particular species as the rights holder: the White Earth Band of Ojibwe in Minnesota in the United States, for example, has a Rights of Manoomin (wild rice) law.[22]

---

[15] Ryan, Curry, and Rule, note 4, 2511; Boyd, note 11, 191–92.

[16] Estado Plurinacional de Bolivia, Ley No 071, 21 December 2010, Art 1, 7–9, at: https://static1.squarespace.com/static/5e3f36df772e5208fa96513c/t/5fbd1 2014b66d17947f72dce/1606226434545/Bolivia+Ley-Derechos-Madre-Tierra+20 10+Spanish.pdf; Boyd, note 11, 191–94. Bolivia subsequently enacted a more detailed and complementary law, the Framework Law on Mother Earth and Holistic Development for Well Being. Boyd, note 11, 194.

[17] For helpful timelines, see Earth Law Center, 'Timeline: Developments in Earth law and the Rights of Nature', at: https://www.earthlawcenter.org/timeline; Community Environmental Legal Defense Fund, 'Rights of Nature: Timeline', at: https://celdf.org/rights-of-nature/timeline. Additional examples of laws and policies also are provided throughout this Section of the Chapter.

[18] N Pain and R Pepper, 'Can Personhood Protect the Environment? Affording Legal Rights to Nature' (2021) 45 *Fordham International Law Journal* 315, 334.

[19] Ibid, 335–38; Ryan, Curry, and Rule, note 4, 2507.

[20] E O'Donnell et al, 'Stop Burying the Lede: The Essential Role of Indigenous Law(s) in Creating Rights of Nature' (2020) 9 *Transnational Environmental Law* 403, 413.

[21] P Wesche, 'Rights of Nature in Practice: A Case Study on the Impacts of the Colombian Atrato River Decision' (2021) 33 *Journal of Environmental Law* 531, 533.

[22] United Nations Secretary General, note 2, [36].

The types of rights also vary.[23] Ecuador's Constitution provides that Nature has 'the right to integral respect for its existence and for the maintenance and regeneration of its life cycles, structure, functions and evolutionary processes' and 'the right to be restored'.[24] The 2019 National Environment Act in Uganda likewise states that 'Nature has the right to exist, persist, maintain and regenerate its vital cycles, structure, functions and its processes in evolution'.[25] Some rights are more specific, such as the seven listed rights in Bolivia's Rights of Mother Earth Law: life, diversity of life, water, clean air, balance, restoration, and freedom from pollution.[26] Similarly, Panama's 2022 statute identifies six minimum rights of Nature: the right to exist, persist, and regenerate; the right to biodiversity; the right to the preservation of the functions of its water cycles; the right to the preservation of air quality; the right to restoration; and the right to be free from contamination and toxic and radioactive waste.[27] In the ecosystem-specific context, the Universal Declaration of the Rights of Rivers lists some of the fundamental rights of rivers, including 'the right to flow, … to perform essential functions within its ecosystem, … to be free from pollution, … to feed and be fed by sustainable aquifers, … to native biodiversity, and … to regeneration and restoration'.[28]

Some instruments provide for Nature's right to thrive or flourish. For instance, Principle 2 of the 2016 International Union for Conservation of Nature (IUCN) World Declaration on the Environmental Rule of Law notes that 'Nature has the inherent right to exist, thrive, and evolve'.[29] The Lake Erie Bill of Rights (which was quickly invalidated) similarly stated that the lake and

---

[23] See Ryan, Curry, and Rule, note 4, 2512 (noting the 'great diversity in the type of rights granted by different rights of nature initiatives').

[24] Republic of Ecuador, note 14, Arts 71, 72.

[25] The National Environment Act, 2019, Acts Supplement to The Uganda Gazette No 10, Volume CXII, 7 March 2019, part I(4)(1), at: https://static1.squarespace.com/static/5e3f36df772e5208fa965 13c/t/5fbd12907dd1bd0a7624e251/16062265 77880/UGANDA+National-Environment-Act-2019.pdf.

[26] Estado Plurinacional de Bolivia, note 16, Art 7.

[27] República de Panamá, Ley No 287 De 24 de febrero de 2022, Gaceta Oficial Digital No 29484-A, 24 February 2022, Art 10, at: https://www.gacetaoficial.gob.pa/pdfTemp/29484_A/GacetaNo_29484a_20220224.pdf; K Surma, 'Panama Enacts a Rights of Nature Law, Guaranteeing the Natural World's "Right to Exist, Persist and Regenerate"', 25 February 2022, *Inside Climate News* at: https://insideclimatenews.org/news/25022022/panama-rights-of-nature/.

[28] *Universal Declaration of the Rights of Rivers* § 3, at: https://www.rightsofrivers.org/#declaration.

[29] IUCN and World Commission on Environmental Law, *IUCN World Declaration on the Environmental Rule of Law* (2016) at: https://www.iucn.org/

its watershed 'possess the right to exist, flourish, and naturally evolve'.[30] The Universal Declaration of the Rights of Mother Earth includes other rights, such as 'the right to be respected', 'the right to maintain its identity and integrity as a distinct, self-regulating and interrelated being', and 'the right to not have its genetic structure modified or disrupted in a manner that threatens its integrity or vital and healthy functioning'.[31] The Lafayette Climate Bill of Rights (passed in Lafayette, Colorado in the United States in 2017) even identifies the right of ecosystems to live in a healthy climate,[32] and the city's Community Rights Act further provides that all residents and ecosystems in the city have the right to clean water and clean air, among other rights.[33]

In some instances, laws acknowledge Nature as a legal person with the same rights as a person. In New Zealand, for example, legislation has recognised the legal personhood of Te Awa Tupua (the Whanganui River watershed) and Te Urewera (formerly a national park).[34] New Zealand has also passed legislation to recognise Mount Taranaki as a legal person.[35]

Rather than identifying specific rights, a less common approach in some instruments is to set out management requirements for the protection of certain natural features.[36] For example, the state of Victoria in Australia passed a law acknowledging that the Yarra River is a 'living and integrated natural entity' and requiring development and implementation of a strategic management plan to protect the river and certain nearby land.[37] The act also sets out management

---

sites/default/files/2022-10/world_declaration_on_the_environmental_rule_of_law_final_2017-3-17.pdf.

[30] City of Toledo, Ohio, Toledo Municipal Code, Chapter XVII, § 254, at: https://codelibrary.amlegal.com/codes/toledo/latest/toledo_oh/0-0-0-158818. Recognition of the right to flourish is common among ordinances passed by local governments in the United States. Kauffman and Martin, note 9, 63.

[31] *Universal Declaration of the Rights of Mother Earth*, Art 2, at: https://www.garn.org/universal-declaration/.

[32] Colorado Community Rights Network, 'Lafayette Climate Bill of Rights', 14 October 2017, *COCRN* at: https://cocrn.org/lafayette-climate-bill-rights/; Boyd, note 11, 127.

[33] Colorado Community Rights Network, note 32.

[34] Pain and Pepper, note 18, 320; Ryan, Curry, and Rule, note 4, 2517–18.

[35] Ryan, Curry, and Rule, note 4, 2518.

[36] Ibid, 2512–13.

[37] Ibid, 2513; Yarra River Protection (Wilip-gin Birrarung murron) Act 2017, No 49 of 2017, § 1, at: https://content.legislation.vic.gov.au/sites/default/files/2020-04/17-49aa005%20authorised.pdf.

principles and 'Yarra protection principles' to inform decision-making and development in the area.[38]

2.3 Forms of Recognition

Governments, courts, and international organisations have employed different forms of recognition for the rights of Nature. Some governments have identified rights of Nature through constitutional amendments. Ecuador amended its Constitution to include rights of Nature and, as of 2023, is the only country to do so.[39] The state of Guerrero in Mexico amended its Constitution to indicate that the state must protect the rights of Nature in its legislation.[40] Mexico City also amended its Constitution to recognise the rights of Nature.[41]

Some governments have followed what appears to be the more common legislative route by passing national or sub-national laws.[42] As noted above, Bolivia's statute identifies specific rights of Mother Earth.[43] Similarly, Uganda identified rights of Nature in its 2019 statute, the National Environment Act.[44] Panama has also enacted national legislation to recognise the rights of Nature and the state's and people's obligations to respect and protect those rights.[45]

At the sub-national level, the City Council of Pittsburgh in Pennsylvania passed an ordinance in 2010 banning commercial natural gas extraction and creating a bill of rights for the 'residents, natural communities and ecosystems'

---

[38] Yarra River Protection (Wilip-gin Birrarung murron) Act 2017, note 37, §§ 8, 13.

[39] See Ryan, Curry, and Rule, note 4, 2514. Although other national constitutions do not yet recognise the rights of Nature, a country's constitution can help provide the basis for subsequent national statutes that recognise the rights of Nature, as was the case in Bolivia. See P Villavicencio Calzadilla and LJ Kotzé, 'Living in Harmony with Nature? A Critical Appraisal of the Rights of Mother Earth in Bolivia' (2018) 7:3 *Transnational Environmental Law* 397, 399, 407, 411.

[40] Constitución Política del Estado Libre y Soberano de Guerrero, Art 2, at: http://files.harmonywithnatureun.org/uploads/upload665.pdf; Harmony with Nature United Nations, note 11.

[41] Constitución Política de la Ciudad de México, Art 18, at: http://files.harmonywithnatureun.org/uploads/upload687.pdf; Harmony with Nature United Nations, note 11.

[42] See Ryan, Curry, and Rule, note 4, 2510.

[43] Estado Plurinacional de Bolivia, Ley No 071, note 16.

[44] The National Environment Act, 2019, note 25.

[45] República de Panamá, Ley No 287, note 27, Art 1.

in the city.[46] Grant Township in Pennsylvania also enacted, by ordinance, a community bill of rights that recognised the rights of natural communities,[47] but a federal district court struck down several sections of the community bill of rights because the township did not have the necessary legal authority.[48] In response, the township voted to become a home rule municipality and adopted a home rule charter in 2015 that also referred to the rights of Nature.[49] In 2013, the City of Santa Monica in California passed a Sustainability Rights Ordinance that included a section on the rights of natural communities and ecosystems in the city.[50] Other examples include the Municipality of Paudalho in Brazil, which amended its law in 2018 to recognise Nature's right to exist, thrive, and evolve,[51] and the municipality of Curridabat in Costa Rica, which recognised pollinators, native plants, and trees as citizens of the city.[52]

New Zealand has employed a unique approach to the rights of Nature. A 2011 treaty claims settlement agreement recognised the rights of Te Awa Tupua as a legal person, and the settlement agreement was formalised through national legislation in 2017.[53] Through another legislative act in 2014, New Zealand recognised the Te Urewera ecosystem as a legal entity with 'all the rights, powers, duties, and liabilities of a legal person'.[54] Te Awa Tupua even

---

[46] City of Pittsburgh, Pennsylvania, Code of Ordinances, Chapter 618: Marcellus Shale Natural Gas Drilling, at: https://library.municode.com/pa/pittsburgh/codes/code_of_ordinances?nodeId=COOR_TITSIXCO_ARTIRERIAC_CH618MASHNAGADR.

[47] Boyd, note 11, 114–16.

[48] Ibid, 119–20. In 2018, the Pennsylvania Department of Environmental Protection revoked an injection well permit because it would have violated Grant Township's Home Rule Charter. Letter from Pennsylvania Department of Environmental Protection to Pennsylvania General Energy Co., LLC, 19 March 2018, at: https://celdf.org/wp-content/uploads/2020/03/Yanity-Letter-2.pdf.

[49] Boyd, note 11, 120.

[50] City of Santa Monica, California, Santa Monica Municipal Code, Chapter 12.02: Sustainability Rights, at: https://ecode360.com/42743266#42743266. As a result of the amendment, the San Severino Ramos Natural Water Spring's rights were recognised. United Nations Secretary General, note 2, [25].

[51] Câmara Municipal de Paudalho, Emenda à Lei Orgânica No 03, 5 January 2018, at: http://files.harmonywithnatureun.org/uploads/upload720.pdf.

[52] P Greenfield, '"Sweet City": the Costa Rica suburb that gave citizenship to bees, plants and trees', 29 April 2020, *The Guardian* at: https://www.theguardian.com/environment/2020/apr/29/sweet-city-the-costa-rica-suburb-that-gave-citizenship-to-bees-plants-and-trees-aoe.

[53] Boyd, note 11, 134.

[54] CM Kauffman and PL Martin, 'Constructing Rights of Nature Norms in the US, Ecuador, and New Zealand' (2018) 18 *Global Environmental Politics* 43, 50

owns itself,[55] as does Te Urewera[56] – a remarkable and distinctive acknowledgement among rights of Nature instruments, which is likely due, at least in part, to the historical and legal context in which the legislation arose.[57]

Indigenous and tribal laws and decisions also have acknowledged the rights of Nature. Several tribes have laws recognising the rights of Nature.[58] In 2017, the Ponca Tribe in Oklahoma, United States passed a rights of Nature statute to ban fracking, and in 2022, the tribe recognised the rights of the Arkansas and Salt Fork Rivers through the adoption of a statute.[59] As noted earlier, the White Earth Band of Ojibwe also passed a law in 2018 to make it illegal for government or businesses to infringe the rights of Manoomin.[60] A pipeline permit was challenged in tribal court based on the law, with Manoomin as the lead plaintiff in the case,[61] but the court dismissed the case in 2022 for lack of subject-matter jurisdiction.[62] In 2019, the Yurok Tribal Council in California, United States

---

(citing Te Awa Tupua Act 2017, clause 14; Te Urewera Act 2014, Art 11) (internal quotation marks omitted).

[55] S Biggs, 'When Rivers Hold Legal Rights', 17 April 2017, *Earth Island Journal* at: https://www.earthisland.org/journal/index.php/articles/entry/when_rivers_hold_legal_rights/.

[56] Boyd, note 11, 134, 150.

[57] For more on the history and context that led to the legal recognition of Te Awa Tupua and Te Urewera, see, eg, Takacs, note 2 and Kauffman and Martin, note 9.

[58] See Bioneers, *Guide to Rights of Nature in Indian Country* (2024) 5–6, at: https://bioneers.org/wp-content/uploads/2023/09/Bioneers-Rights-of-Nature-Guide-2023.pdf.

[59] A Brown, 'Cities, Tribes Try a New Environmental Approach: Give Nature Rights', 30 October 2019, *Pew Charitable Trusts* at: https://www.pewtrusts.org/en/research-and-analysis/blogs/stateline/2019/10/30/cities-tribes-try-a-new-environmental-approach-give-nature-rights#:~:text=In%202017%2C%20the%20Ponca%20Nation,rising%20cancer%20and%20asthma%20rates; 'Ponca Nation Resolution Recognizing the Rights of Nature', *Eco Jurisprudence Monitor* at: https://ecojurisprudence.org/initiatives/ponca-nation-statute/; 'Ponca Nation Statute Recognizing Rights Of Rivers', *Eco Jurisprudence Monitor* at: https://ecojurisprudence.org/initiatives/ponca-nation-statute-recognizing-rights-of-rivers/.

[60] United Nations Secretary General, note 2, [36].

[61] K Surma, 'To Stop Line 3 Across Minnesota, an Indigenous Tribe Is Asserting the Legal Rights of Wild Rice', 22 October 2021, *Inside Climate News* at: https://insideclimatenews.org/news/22102021/line-3-rights-of-nature-minnesota-white-earth-nation/.

[62] *Minnesota Department of Natural Resources v. Manoomin*, Opinion, File No AP21-0516, 10 March 2022, at: https://whiteearth.com/assets/files/programs/judicial/cases/Manoomin%20opinion%20AP21-0516.pdf.

approved a resolution on the rights of the Klamath River.[63] Additionally, in Australia, First Law recognises the inherent right to life of Martuwarra (the Fitzroy River).[64] An interesting example of governments working together can be found in Quebec, Canada, where the Innu Council of Ekuanitshit and the Minganie Regional County Municipality (in collaboration with other organisations, including the Muteshekau-shipu Alliance) passed parallel resolutions in 2021 to recognise the Magpie River as a legal person and provide for nine rights of the river.[65]

In some jurisdictions, rights of Nature have manifested through judicial decisions. In India, the Uttarakhand High Court determined that the Ganges and Yamuna Rivers are legal persons with rights (although subsequently on appeal, the Supreme Court of India stayed the decision pending completion of the appeal).[66] The High Court in Bangladesh in 2019 recognised the legal rights of *all* rivers in the country as 'legal persons' – a decision that the Appellate Division of the Supreme Court later upheld.[67] Multiple courts in

---

[63] United Nations Secretary General, note 2, [34]; Ryan, Curry, and Rule, note 4, 2538.

[64] O'Donnell et al, note 20, 422–24; Martuwarra RiverOfLife et al, 'Recognizing the Martuwarra's First Law Right to Life as a Living Ancestral Being' (2020) 9 *Transnational Environmental Law* 541, 541, 544, of which Martuwarra is listed as the lead author.

[65] Muteshekau-shipu Alliance, 'For the first time, a river is granted official rights and legal personhood in Canada', 23 February 2021, at: http://files.harmonywithnatureun.org/uploads/upload1070.pdf; Conseil des Innu de Ekuanitshit, No consecutif 919-082, 18 January 2021, 11–12, at: http://files.harmonywithnatureun.org/uploads/upload1072.pdf; Province de Québec Municipalité Régionale de Comté de Minganie, Résolution no 025-21: Reconnaissance de la personnalité juridique et des droits de la rivière Magpie – Mutehekau Shipu, 16 February 2021, at: http://files.harmonywithnatureun.org/uploads/upload1069.pdf; Ryan, Curry, and Rule, note 4, 2538.

[66] S Jolly and KSR Menon, 'Of Ebbs and Flows: Understanding the Legal Consequences of Granting Personhood to Natural Entities in India' (2021) 10 *Transnational Environmental Law* 467, 467–69; E O'Donnell and I Jahan, 'Rivers Are Still People In South Asia Despite Court Showdown', 1 February 2023, *CodeBlue* at: https://codeblue.galencentre.org/2023/02/01/rivers-are-still-people-in-south-asia-despite-court-showdown/. Interestingly, the main rationale for the Uttarakhand High Court's decision was the protection of faith and religion. Jolly and Menon 475–76.

[67] Mari Margil, 'Bangladesh Supreme Court Upholds Rights of Rivers', 24 August 2020, *Center for Democratic and Environmental Rights* at: https://www.centerforenvironmentalrights.org/news/bangladesh-supreme-court-upholds-rights-of-rivers; Ryan, Curry, and Rule, note 4, 2508.

Colombia have issued rulings regarding the rights of Nature: the Colombian Constitutional Court ruled that Río Atrato is a subject of legal rights, and the Superior Court of Medellín determined that the Cauca River is a subject of rights as well.[68] In Ecuador, the Provincial Court of Loja upheld the constitutional rights of the Vilcabamba River in an 'action for protection' against a highway project.[69]

In addition, governments and groups have used non-binding policies and other methods to advance the rights of Nature. One example is the Blue Mountains City Council in New South Wales, Australia, which decided in April 2021 that it will consider the rights of Nature in its municipal planning.[70] According to the City Council's website, rights of Nature 'principles will be progressively added to all current and future strategic documents, planning and decision making processes and the operational delivery of Council's functions'.[71] Building on the Blue Mountains City Council's efforts, the Donegal County Council in Ireland adopted a motion in late 2021 signalling its intention to recognise the rights of Nature as a core value there.[72] Political parties and other groups have also begun to embrace the Rights of Nature movement. The Green Party of England and Wales adopted a policy in which the party 'advocates the legal recognition of rights of nature as a legal concept to protect ecosystems'.[73] After working with the Community Environmental Legal Defense Fund (CELDF), the Florida Democratic Party incorporated rights of Nature into its platform in 2019.[74] CELDF is also working with

---

[68] Ryan, Curry, and Rule, note 4, 2512; United Nations Secretary General, note 2, [25]. Other courts in Colombia have also issued rulings regarding the rights of rivers and other natural features. See United Nations Secretary General, note 2, [27–29].

[69] Boyd, note 11, 160–64.

[70] Blue Mountains City Council, 'Environment: Rights of Nature', at: https://www.bmcc.nsw.gov.au/rights-of-nature.

[71] Ibid.

[72] CELDF, 'Donegal County Council Votes Unanimously to Adopt Rights of Nature', 7 January 2022, *CELDF* at: https://celdf.org/2022/01/donegal-county-council/.

[73] CELDF, 'Green Party of England and Wales' Rights of Nature Policy', 28 February 2016, *CELDF* at: https://celdf.org/2016/02/green-party-of-england-and-wales-rights-of-nature-policy/. The Scottish Green Party previously adopted a similar policy in 2015. The Ecologist, 'Greens commit to Rights of Nature law', 29 February 2016, at: https://theecologist.org/2016/feb/29/greens-commit-rights-nature-law.

[74] S Powers, 'Florida Democratic Party adopts "rights of nature" into platform', 16 October 2019, *Florida Politics* at: https://floridapolitics.com/archives/

a newly formed network of French groups to promote rights of Nature efforts in Europe.[75]

Non-binding international declarations, organisations, and tribunals also provide support for the rights of Nature. In 2009, the United Nations General Assembly adopted its first Harmony with Nature Resolution,[76] and the next year, the World People's Conference on Climate Change and the Rights of Mother Earth, held in Bolivia, produced the Universal Declaration of the Rights of Mother Earth.[77] The Global Alliance for the Rights of Nature (GARN) was also established in 2010 as 'a global network of organizations and individuals committed to the universal adoption and implementation of legal systems that recognize, respect and enforce "Rights of Nature"'.[78] In 2014, GARN created and held the first International Rights of Nature Tribunal as 'a forum for people from all around the world to speak on behalf of nature, to protest the destruction of the Earth ... and to make recommendations about Earth's protection and restoration'.[79] The Tribunal and its Regional Tribunals investigate, hold non-binding hearings, and make recommendations about alleged violations of the rights of Nature.[80] The judges on the panel are lawyers and ethics experts.[81] The International and Regional Tribunals[82] are not without their critics, but they arguably serve an important educational purpose

---

308603-florida-democratic-party-adopt-rights-of-nature-into-platform/; Harmony with Nature United Nations, note 11; Ryan, Curry, and Rule, note 4, 2529.

[75] CELDF, 'International Rights of Nature Solidarity', 3 March 2021, *CELDF* at: https://celdf.org/2021/03/international-rights-of-nature-solidarity. For examples of other policy efforts, see United Nations Secretary General, note 2, [46–65].

[76] See United Nations General Assembly, *Harmony with Nature*, UN Doc A/RES/64/196 (2009); Harmony with Nature United Nations, 'Programme', at: http://www.harmonywithnatureun.org.

[77] See *Universal Declaration of the Rights of Mother Earth*, note 31. Some of the listed inherent rights of Mother Earth include the rights to live and exist, to regenerate, to clean air, and to integral health. Ibid, Art 2.

[78] GARN, 'Who we are', at: https://www.garn.org/about-garn.

[79] International Rights of Nature Tribunal, 'Our Tribunals', at: https://www.rightsofnaturetribunal.org (emphasis removed); International Rights of Nature Tribunal, 'History of the Rights of Nature Tribunal', at: https://www.rightsofnaturetribunal.org/about-us; Maloney, note 2, 129, 135.

[80] Maloney, note 2, 130; CM Kauffman and PL Martin, 'Can Rights of Nature Make Development More Sustainable? Why Some Ecuadorian Lawsuits Succeed and Others Fail' (2017) 92 *World Development* 130, 131.

[81] Maloney, note 2, 136.

[82] The Regional Tribunals serve the same functions as the International Tribunal. Ibid, 136–37.

for society and governments and provide rights of Nature advocates with an opportunity to gain practical experience.[83]

The IUCN included the rights of Nature in its 2017–2020 Programme, which 'aim[ed] to secure the rights of nature', and as part of Target 6 ('The implementation of commitments under biodiversity-related conventions and international agreements is accelerated'), IUCN was to 'advance rights regimes related to the rights of nature'.[84] IUCN also called for the incorporation of the rights of Nature in its decision-making processes at the IUCN World Conservation Congress in 2012,[85] and rights of Nature issues, including a Rights of Rivers session,[86] featured in the IUCN World Conservation Congress in September 2021.[87] Also in 2021, the Earth Law Center and the GARN Youth Hub led a youth declaration regarding the rights of Nature.[88] The youth declaration 'support[s] the recognition of Rights of Nature within the IUCN and other institutions ... and Conventions' and '[pledges] to take

---

[83] Boyd, note 11, 217–18.

[84] IUCN, *IUCN Programme 2017–2020*, Approved by the IUCN World Conservation Congress, September 2016, 16, 26, at: https://portals.iucn.org/library/sites/library/files/documents/WCC-6th-001.pdf; Pecharroman, note 12, 5. IUCN's Programme 2021–24 also notes that '[a] just and fair legal system that protects the rights of nature and people is particularly essential in the face of the climate and biodiversity crisis'. IUCN, *Nature 2030: One Nature, One Future: A Programme for the Union 2021–2024*, Adopted by IUCN members by electronic vote on 10 February 2021, 11, at: https://portals.iucn.org/library/sites/library/files/documents/WCC-7th-001-En.pdf.

[85] WCC-2012-Res-100-EN, *Incorporation of the Rights of Nature as the organizational focal point in IUCN's decision making*, September 2012, at: https://portals.iucn.org/library/sites/library/files/resrecfiles/WCC_2012_RES_100_EN.pdf.

[86] 'Rights of Rivers at the IUCN World Conservation Congress 2021', 10 September 2021, *The Freshwater Blog* at: https://freshwaterblog.net/2021/09/10/rights-of-rivers-at-the-iucn-world-conservation-congress-2021/; 'The Universal Declaration of the Rights of Rivers: Updates on a Global Movement', *IUCN World Conservation Congress Marseille* at: https://www.iucncongress2020.org/programme/official-programme/session-52640; 'Energizing the rights of nature', *IUCN World Conservation Congress Marseille* at: https://www.iucncongress2020.org/programme/official-programme/session-43272.

[87] Earth Law Center, 'Rights of Nature a focus at the IUCN Congress', 10 September 2021, at: https://www.earthlawcenter.org/elc-in-the-news/2021/9/rights-of-nature-a-focus-at-the-iucn-congress.

[88] Earth Law Center and GARN Youth Hub, 'Youth and the Rights of Nature Movement: shifting the paradigm for all future generations' (2021) at: https://www.garnyouth.org/publications.

further action in deepening and spreading the awareness of the fundamental and inalienable rights of Nature and of future generations'.[89]

Another example of recent – and arguably significant – recognition of the rights of Nature at the international level occurred under the auspices of the Convention on Biological Diversity (CBD). At the Fifteenth Meeting of the Conference of the Parties to the CBD (Part Two) in December 2022, the parties adopted the Kunming-Montreal Global Biodiversity Framework through Decision 15/4,[90] which included a reference to the rights of Nature. Specifically, with respect to the Framework's implementation, it states: 'The Framework recognizes and considers these diverse value systems and concepts, including, for those countries that recognize them, rights of nature and rights of Mother Earth, as being an integral part of its successful implementation.'[91] This high-level international acknowledgement of the rights of Nature could open the door for other international treaties and organisations to recognise the rights of Nature in the future.

2.4 Enforcement Approaches

Rights of Nature legal instruments take a variety of approaches to enforcement. In some instances, the means of enforcement are clearly provided for by law, decision, or other instrument; in others, however, the enforcement mechanism or process is not mentioned. The rights of Nature may be enforced by different entities or people, depending on the instrument. Sometimes, the ecosystem itself has enforcement rights,[92] or the instrument may appoint a specific group or commission as a guardian that is responsible for protecting the rights and interests of the natural feature.[93] In other instances, enforcement power is

---

[89] Ibid.

[90] Convention on Biological Diversity, *Decision 15/4: Kunming-Montreal Global Biodiversity Framework* (2022). The Framework 'sets out an ambitious plan to implement broad-based action to bring about a transformation in our societies' relationship with biodiversity by 2030, in line with the 2030 Agenda for Sustainable Development and its Sustainable Development Goals'. Ibid, 4.

[91] Ibid, 5.

[92] As an example, the Crystal Spring ecosystem filed a motion to intervene to defend its own rights in a lawsuit. CELDF, 'Press Release: Ecosystem, Community Group, and Municipal Authority File for Intervention in Lawsuit to Defend Community from Injection Well', 11 August 2015, *CELDF* at: https://celdf.org/2015/08/press-release-ecosystem-community-group-and-municipal-authority-file-for-intervention-in-lawsuit-to-defend-community-from-injection-well.

[93] See Kauffman and Martin, note 54, 51. It has been suggested that the guardianship model (which is used for Te Awa Tupua and Te Urewera in New Zealand)

restricted to local residents or citizens of a municipality,[94] and occasionally, *any* person may enforce the rights of Nature.[95]

3 CHALLENGES TO THE RIGHTS OF NATURE

As this Chapter has outlined, the rights of Nature have been recognised and upheld in many instances over the last decade or so. But the successes have not been easy, and the Rights of Nature movement has faced and will continue to face many actual and perceived challenges. Political opportunity, the types of organisations and coalitions involved, and cultural context are among the factors that have impacted how and whether rights of Nature laws have been passed.[96] Challenges in terms of implementation and enforcement have included lack of precedent, lack of a clear enforcement mechanism, lack of capacity (including funding and time), conflicting rights, lack of specificity, and lack of political will or other political issues.[97] For example, Ecuador is arguably a leader and an overall success story in terms of advancing the rights of Nature, as it enshrined the rights of Nature in its national Constitution; however, the government in Ecuador was initially slow to enact the necessary laws to effectuate the rights enshrined in the Constitution.[98] Furthermore, under Ecuador's Constitution, other rights (eg, socio-economic rights) may conflict with the rights of Nature.[99] Bolivia's implementation also has been inconsistent and slow in some respects.[100]

Moreover, although the rights of Nature have been recognised through constitutional, legislative, or judicial actions, they often have been contested in court, and courts have not always upheld these laws due to lack of authority on

---

may be a stronger approach because guardians are appointed and required to protect the natural feature's rights. Ibid.

[94] Ibid.

[95] This is the case in Ecuador. Ibid; Ryan, Curry, and Rule, note 4, 2509.

[96] See Kauffman and Martin, note 9, 59–60 (describing in detail how these factors have influenced rights of Nature laws in Ecuador, the United States, and New Zealand).

[97] For more information on enforcement and other challenges, see, eg, E O'Donnell and J Talbot-Jones, 'Creating legal rights for rivers: lessons from Australia, New Zealand, and India' (2018) 23 *Ecology and Society* Article 7.

[98] Kauffman and Martin, note 9, 83; Boyd, note 11, 175–84.

[99] Kauffman and Martin, note 9, 64, 67, 128; see also L Schimmöller, 'Paving the Way for Rights of Nature in Germany: Lessons Learnt from Legal Reform in New Zealand and Ecuador' (2020) 9 *Transnational Environmental Law* 569, 579.

[100] See Boyd, note 11, 197–201.

the part of the local government, pre-emption by federal and/or state law,[101] or overly vague laws, among other reasons.[102] In the United States, for instance, a federal court struck down Grant Township's initial community bill of rights ordinance as being beyond the township's authority.[103] A court also struck down Toledo's Lake Erie Bill of Rights[104] for being beyond the local government's authority and for being unconstitutionally vague; the city appealed, but budgetary constraints caused the city to dismiss the appeal.[105] Some of the early rights of Nature lawsuits in Ecuador may have become politicised, and it is also possible that judges in some of the initial lawsuits did not fully understand the rights of Nature concepts.[106] And even when rights of Nature advocates prevail in court, companies and governments may not comply, or may be slow to comply, with court orders.[107]

The hierarchy of governmental authority in a country may also impact success, as state and/or federal legislative action may pre-empt local governments' efforts.[108] This challenge has been borne out in the United States. The Ohio legislature, for example, pre-empted Toledo's Lake Erie Bill of Rights by prohibiting people from filing lawsuits on behalf of ecosystems.[109] Similarly, the Florida legislature passed a state law to pre-empt local rights of Nature laws, including the Wekiva River and Econlockhatchee River Bill of Rights in

---

[101] Kauffman and Martin, note 9, 64–65; Boyd, note 11, 129.

[102] D Zartner, 'Watching Whanganui & the Lessons of Lake Erie: Effective Realization of Rights of Nature Laws' (2021) 22 *Vermont Journal of Environmental Law* 1, 37; E Macpherson, 'The (Human) Rights of Nature: A Comparative Study of Emerging Legal Rights for Rivers and Lakes in the United States of America and Mexico' (2021) 31 *Duke Environmental Law and Policy Forum* 327, 376; MJ Moutrie, 'The Rights of Nature Movement in the United States: Community Organizing, Local Legislation, Court Challenges, Possible Lessons and Pathways' (2020) 10 *Barry University Environmental and Earth Law Journal* 5, 57.

[103] Kauffman and Martin, note 9, 166–67.

[104] Ryan, Curry, and Rule, note 4, 2526–27. The Lake Erie Bill of Rights declared the rights of the ecosystem and restricted corporate rights. Ibid, 2527.

[105] Ibid, 2525–27.

[106] Kauffman and Martin, note 9, 83, 89, 114–15; Kauffman and Martin, note 80, 134–35.

[107] See, eg, Boyd, note 11, 163–64; Kauffman and Martin, note 9, 94.

[108] Kauffman and Martin, note 9, 64.

[109] Ryan, Curry, and Rule, note 4, 2527; CELDF, 'Media Release: Ohio Legislature Bans Rights of Nature Enforcement', 18 July 2019, at: https://celdf.org/2019/07/rights-of-nature-ban.

Orange County, Florida.[110] Bills were also introduced in the Missouri legislature in 2021 and 2022 in an attempt to block rights of Nature laws.[111]

But even where implementation or enforcement efforts are unsuccessful, these actions often still serve to increase support for and raise awareness about the rights of Nature – they can 'mobili[se] society and plac[e] [rights of Nature] on the national agenda, making it a salient issue'.[112] Lawsuits also may help to improve judges' (and others') understanding and knowledge of the issues.[113] Furthermore, additional court rulings can help to establish precedent and strengthen the Rights of Nature movement over time.[114]

4 RIGHTS OF WETLANDS AND THE RAMSAR CONVENTION

Humans have had significant negative effects on the world's wetlands, through drainage, conversion, pollution, and other direct and indirect actions.[115] Wetland degradation and losses continue worldwide, suggesting that efforts

---

[110] CELDF, 'CELDF Statement on Orange County, FL "Rights of Nature" Law', 4 November 2020, at: https://celdf.org/2020/11/celdf-statement-on-orange-county-fl-rights-of-nature-law/; Ryan, Curry, and Rule, note 4, 2533–34. Section 403.412(9)(a) of the Florida Statutes provides that '[a] local government regulation, ordinance, code, rule, comprehensive plan, charter, or any other provision of law may not recognize or grant any legal rights to a plant, an animal, a body of water, or any other part of the natural environment that is not a person or political subdivision as defined in s. 1.01(8) or grant such person or political subdivision any specific rights relating to the natural environment not otherwise authorized in general law or specifically granted in the State Constitution'. Florida Statutes, Environmental Protection Act, § 403.412, at: http://www.leg.state.fl.us/Statutes/index.cfm?App_mode=Display_Statute&Search_String=&URL=0400-0499/0403/Sections/0403.412.html.

[111] Missouri House of Representatives, HB 54, 101st General Assembly, 1st Regular Session, at: https://house.mo.gov/Bill.aspx?bill=HB54&year=2021&code=R; Missouri House of Representatives, HB 1463, 101st General Assembly, 2nd Regular Session, at https://house.mo.gov/Bill.aspx?bill=HB1463&year=2022&code=R. See also CELDF, 'Newsletter: The Rights of Nature Are Blooming', 9 May 2021, at: https://celdf.org/2021/05/newsletter-the-rights-of-nature-are-blooming/.

[112] Kauffman and Martin, note 9, 115; Kauffman and Martin, note 80, 139.

[113] See Kauffman and Martin, note 9, 113–16; Kauffman and Martin, note 80, 139.

[114] See Kauffman and Martin, note 9, 80–84, 113–16; Kauffman and Martin, note 80, 138, 139.

[115] Davies et al, note 7, 595.

to protect and conserve wetlands have been insufficient.[116] And the rate at which the world is losing wetlands is staggering.[117] Recognising the threats to – and the critical importance of – wetlands, a group of scientists and attorneys proposed a Universal Declaration of the Rights of Wetlands in 2020.[118] This section discusses the proposed declaration and the potential role for the Ramsar Convention in supporting the rights of wetlands.[119]

4.1 Universal Declaration of the Rights of Wetlands

The proposed Universal Declaration of the Rights of Wetlands grew, in part, out of meetings of the Society of Wetland Scientists (SWS), as well as the SWS Climate Change and Wetlands Initiative.[120] The proposed Declaration seeks 'to promote a paradigm shift and step-change in the human-wetlands relationship that could lead to a fundamental change in the trajectory for global wetland ecosystems'.[121] As Davies et al explain:

> Because of the central role of wetlands in supporting life on Earth and for combatting the ongoing destabilisation of the earth's climate, support for the Universal Declaration of the Rights of Wetlands will provide a timely basis for a step-change in effective and morally robust re-visioning of the human relationship with wetlands and Nature to include people as members of the Earth community, bearing responsibilities and duties, as well as rights, rather than being primarily resource extractors or managers of natural resources who currently so often are identified as the sole possessors of personhood and rights.[122]

---

[116] Ibid, 593, 595.

[117] For recent estimates, see, eg, M Courouble et al, *Global Wetland Outlook: Special Edition 2021* (Ramsar Convention Secretariat, Gland, 2021) 6; E Fluet-Chouinard et al, 'Extensive Global Wetland Loss Over the Past Three Centuries' (2023) 614 *Nature* 281, 283.

[118] Davies et al, note 7, 593. To learn more about the Universal Declaration of the Rights of Wetlands, see https://www.rightsofwetlands.org. Two of the editors and authors in this book, Royal Gardner and Erin Okuno, were among the authors of the article that proposed the Universal Declaration of the Rights of Wetlands.

[119] Perhaps unsurprisingly, the drafters of the Ramsar Convention did not incorporate rights of Nature concepts in the text of the Convention, despite their somewhat concurrent development in time.

[120] Ibid.

[121] 'Universal Declaration of the Rights of Wetlands', at: https://wwwrights.ofwetlands.org.

[122] Davies et al, note 7, 597.

The proposed Declaration is consistent with and builds on the Rights of Nature movement and previous declarations, such as the 1982 World Charter for Nature.[123] It aligns with the 2010 Universal Declaration of Rights of Mother Earth (which provides that 'all other beings also have rights which are specific to their species or kind and appropriate for their role and function within the communities within which they exist'),[124] as well as the 2016 IUCN World Declaration on the Environmental Rule of Law (which emphasises that 'humanity exists within nature and that all life depends on the integrity of the biosphere and the interdependence of ecological systems').[125] The proposed Declaration also complements and follows a similar path as other declarations that focus on promoting the rights of particular types of ecosystems or natural features, such as the Universal Declaration of the Rights of Rivers[126] and the Kawsak Sacha (Living Forest) declaration of the Indigenous Kichwa People of Sarayaku in Ecuador.[127]

Finlayson et al identified seven relevant considerations for wetland users and decisionmakers with respect to the rights of wetlands and the proposed Declaration, including the benefits provided by wetlands to humans and the natural world, existing implicit support for the rights of wetlands, and the growth of the Rights of Nature movement.[128] The proposed Declaration and the recognition and support of specific rights of wetlands also fit quite naturally within the larger Rights of Nature movement. Wetlands (particularly rivers and lakes) are already 'leaders' in terms of which ecosystems or natural features have recognised rights in legal and policy instruments.[129] As noted

---

[123] Ibid.

[124] See *Universal Declaration of the Rights of Mother Earth*, note 31. The Universal Declaration of Rights of Mother Earth was inspired in part by the Universal Declaration of Human Rights. P Burdon, 'Earth Rights: The Theory' (2011) *IUCN Academy of Environmental Law eJournal* 1.

[125] See IUCN and World Commission on Environmental Law, note 29, § II, Principle 2.

[126] See 'Universal Declaration of the Rights of Rivers', at: https://www.rightsofrivers.org/#declaration. To date, over 160 organisations have endorsed the Universal Declaration of the Rights of Rivers. See ibid.

[127] See Pueblo Originario Kichwa De Sarayaku, 'Declaration: *Kawsak Sacha* – The Living Forest' (June 2018) at: https://ecojurisprudence.org/wp-content/uploads/2022/08/KAWSAK_SACHA_DECLARATION-2018.pdf.

[128] CM Finlayson et al, 'Reframing the human–wetlands relationship through a Universal Declaration of the Rights of Wetlands' (2022) 73 *Marine and Freshwater Research* 1278, 1279.

[129] See Ryan, Curry, and Rule, note 4, 2559; Mihnea Tănăsescu, 'Rights of Nature, Legal Personality, and Indigenous Philosophies' (2020) 9 *Transnational Environmental Law* 429, 431.

throughout this Chapter, examples of wetlands with recognised rights include the Whanganui River, Río Atrato, Lake Tota, the Colombian Amazon, the Ganga and Yamuna Rivers, Lake Erie, the Klamath River, Martuwarra (the Fitzroy River), the Vilcabamba River, and all rivers in Bangladesh. Given the importance of wetlands and their prominent role in the Rights of Nature movement thus far, it seems logical to identify and promote specific rights of wetlands, and the proposed Universal Declaration of the Rights of Wetlands is one tool to this end.

BOX 13.1 UNIVERSAL DECLARATION OF THE RIGHTS OF WETLANDS

The following excerpts are from the proposed text of the declaration:

- ACKNOWLEDGING that wetlands are essential to the healthy functioning of Earth processes and provision of essential ecosystem services ...
- ACKNOWLEDGING that wetlands have significance for the spiritual or sacred inspirations and belief systems of many people worldwide, but particularly for Indigenous peoples and local communities living in close relationship to wetlands ...
- FURTHER ACKNOWLEDGING that humans and the natural world with all of its biodiversity depend upon the healthy functioning of wetlands and the benefits that they provide ...
- ALARMED that existing wetland conservation and management approaches have failed to stem the loss and degradation of wetlands of all types around the globe ...
- CONVINCED that recognising the enduring rights and the legal and living personhood of all wetlands around the world will enable a paradigm shift in the human–Nature relationship towards greater understanding, reciprocity and respect leading to a more sustainable, harmonious and healthy global environment that supports the well-being of both human and non-human Nature ...
- DECLARES that all wetlands are entities entitled to inherent and enduring rights, which derive from their existence as members of the Earth community and should possess legal standing in courts of law. These inherent rights include the following:

1. The right to exist.
2. The right to their ecologically determined location in the landscape.
3. The right to natural, connected and sustainable hydrological regimes.
4. The right to ecologically sustainable climatic conditions.
5. The right to have naturally occurring biodiversity, free of introduced or invasive species that disrupt their ecological integrity.
6. The right to integrity of structure, function, evolutionary processes and the ability to fulfil natural ecological roles in the Earth's processes.
7. The right to be free from pollution and degradation.

8. The right to regeneration and restoration.

Sources: 'Universal Declaration of the Rights of Wetlands', note 121; Davies et al, note 7, 598.

Since the proposed Declaration was first published online, more than 175 individuals and over 25 organisations have endorsed it, including the Wildfowl and Wetlands Trust and Wetlands International – two Ramsar International Organization Partners – as well as SWS, CELDF, the Gaia Foundation, the National Community Rights Network, and the Society for Ecological Restoration.[130] The authors of the proposed Declaration and others are working to garner additional acceptance of and support for the rights of wetlands and the proposed Declaration, which already 'has been used to support efforts for improved governance of Lake Tota in Colombia'.[131] The authors of the proposed Declaration have been actively collaborating with organisations and individuals to promote it and the rights of wetlands through various means, including targeted outreach, social media efforts, follow-up publications,[132] and in-person and online presentations at conferences, such as the 2021 International Association for Ecology (INTECOL) International Wetlands Conference and the 2021 World Lake Conference. Notably, the Ōtautahi Declaration on Wetlands – adopted at the 11th INTECOL International Wetlands Conference in 2021 – included a specific Resolution to Promote the Rights of Wetlands, and delegates endorsed and supported the Universal Declaration of the Rights of Wetlands.[133]

---

[130] 'Organisations that endorse the Universal Declaration of the Rights of Wetlands', at: https://www.rightsofwetlands.org/support.

[131] Finlayson et al, note 128, 1281.

[132] See, eg, ibid. Bridgewater published a follow-up article that raises some concerns about the Declaration. See P Bridgewater, 'Comment on Davies et al., "Towards a Universal Declaration of the Rights of Wetlands"' (2021) 72 *Marine and Freshwater Research* 1397. Davies et al responded to some of those concerns in GT Davies et al, 'Reply to Bridgewater (2021) "Response to Davies et al., 'Towards a Universal Declaration of the Rights of Wetlands'"' (2021) 72 *Marine and Freshwater Research* 1401.

[133] INTECOL, *The Ōtautahi Declaration on Wetlands*, Final Resolutions Adopted at the 11th INTECOL International Wetlands Conference, Ōtautahi, Christchurch, Aotearoa New Zealand, 10–15 October 2021. The conference included over 470 participants from 28 countries. Ibid.

4.2 The Potential Role of the Ramsar Convention

The rights of wetlands and the proposed Declaration are consistent with, and could be supported by and used to further, the mission of the Ramsar Convention.[134] The Ramsar Convention seeks to promote the conservation and wise use of wetlands worldwide.[135] The 'three pillars' of the Convention include 'the conservation of Wetlands of International Importance, the wise use of all wetlands, and international cooperation'.[136] Recognising and supporting the rights of wetlands seems to align with, and could benefit from and provide additional support for, the Convention's mission and the three pillars.[137]

There are a variety of ways through which the Ramsar Convention might acknowledge, support, and/or promote the rights of wetlands and the proposed Declaration. One of the most impactful routes would be adopting or recognising the proposed Declaration through a resolution at a future Conference of the Parties (COP). This could take the form of a standalone resolution focused on rights of wetlands or simply a reference in a broader resolution. Another option could be to incorporate (explicitly or implicitly) the rights of wetlands in the Convention's strategic plan, which is adopted by resolution;[138] the next strategic plan should be up for adoption at COP15 (expected to be held in 2025).[139] Other possibilities in the context of a Ramsar COP include promotion through a COP information paper (which can be used to provide

---

[134] See Finlayson et al, note 128, 1280–81.

[135] See Convention on Wetlands, note 6.

[136] Ramsar Convention Secretariat, *An Introduction to the Ramsar Convention on Wetlands* (Ramsar Handbooks, 5th edition, Ramsar Convention Secretariat, Gland, 2016) 37.

[137] As explained by Finlayson et al, the rights of wetlands in the proposed declaration 'are similar to the premises that underpin the Ramsar Convention's concept of ecological character'. Finlayson et al, note 128, 1280–81. See also Gillian Davies et al, 'Rights of Nature in Wetlands – Transformative Change for Securing the Future of Wetlands Through Effective Restoration' (April 2024) 42 *Wetland Science & Practice* 185, 190 (suggesting that the rights in the proposed Universal Declaration of the Rights of Wetlands 'are implicit in the [Ramsar] Convention').

[138] The contracting parties adopted the current strategic plan for 2016–24 through Resolution XII.2. See Ramsar Convention, *Resolution XII.2: The Ramsar Strategic Plan 2016–2024* (2015). Because the parties adopt the strategic plan by resolution, it would be subject to the tradition of consensus. See Chapter 2 in this volume.

[139] See Ramsar Convention, *Resolution XIV.4: Review of the fourth Strategic Plan of the Convention on Wetlands, additions for the period COP14–COP15 and framework for the fifth Strategic Plan* (2022) [16].

background information, data, guidance, or proposals on a topic of relevance to the COP)[140] or a side event hosted by an organisation on a topic that may be of interest to participants.[141] Indeed, SWS hosted a side event on the rights of wetlands at COP14 in 2022,[142] and future COPs may include additional side events on the subject.

The rights of wetlands also could potentially fit within the work of the Scientific and Technical Review Panel (STRP),[143] depending on its work plan. The STRP also prepares briefing notes and policy briefs to share information and advice with the contracting parties, policymakers, wetland managers, and others on issues that may be important, urgent, or of general interest.[144] Past briefing notes and policy briefs have covered topics such as peatland restoration, blue carbon ecosystems, wetlands and agriculture, and environmental flows.[145] In addition to briefing notes and policy briefs, the Ramsar Convention produces other publications, including fact sheets and handbooks, which could conceivably include information about the rights of wetlands and the proposed Declaration to help build awareness.

Additionally, the Convention could collect and disseminate useful information about how the contracting parties are incorporating rights of wetlands at the national level by including a question about rights of wetlands in national reports, which the parties prepare and submit for each COP. Another option could be to include rights of wetlands case studies in the Ramsar Sites Management Toolkit to help inform those who manage Ramsar Sites and other wetlands.[146] The rights of wetlands and the proposed Declaration could also be

---

[140] For example, COP12 in 2015 included an information paper about assessing the effectiveness of a Ramsar Site's management. Ramsar Convention, *Ramsar COP12 DOC.20: Information Paper: Management effectiveness assessments for Ramsar Sites* (2015), at: https://www.ramsar.org/sites/default/files/documents/library/cop12_doc20_pame_e.pdf.

[141] The side events hosted at COP14 are listed in Ramsar Convention, *Side Events Schedule*, 5–13 November 2022, at: https://www.ramsar.org/sites/default/files/documents/library/side_events_schedule_-cop14_final7_1.pdf.

[142] See ibid.

[143] The STRP gives scientific and technical guidance to the Convention bodies. Ramsar Convention, *The Scientific and Technical Review Panel* at: https://www.ramsar.org/about/bodies/scientific-and-technical-review-panel.

[144] H MacKay, 'Introduction to the Briefing Notes series' *Ramsar Briefing Note 1* (Ramsar Convention Secretariat, Gland, Switzerland, 2012) 2.

[145] Ramsar Convention, *STRP Outputs* at: https://www.ramsar.org/about/bodies/scientific-and-technical-review-panel/strp-outputs.

[146] See Ramsar Convention, *Ramsar Sites Management Toolkit* at: https://www.ramsar.org/resources/capacity-building-tools/ramsar-sites-management-toolkit.

incorporated through the Convention's programme on communication, capacity building, education, participation, and awareness,[147] and the Convention could promote the rights of wetlands through its World Wetlands Day activities.[148] Furthermore, the contracting parties to the Ramsar Convention could help to implement and support the rights of wetlands at the national and sub-national levels, regardless of whether and how the Ramsar Convention may eventually support and promote the rights of wetlands and/or the proposed Declaration in any official capacity.

Encouragingly, the rights of Nature and wetlands are already starting to take root within the Convention. The *Global Wetland Outlook 2021* explicitly references the rights of Nature and wetlands, stating that '[a]n important dimension of justice, for humans and the planet, is the recognition of "rights of nature" within legal frameworks, including proposals for a universal "Rights of Wetlands" statement'.[149] The *Outlook* further explains that '[t]he changes needed to stabilize the environment over the next few years are profound and reach far beyond conventional ideas of conservation'.[150] COP14, as noted above, also featured a rights of wetlands side event to promote the Universal Declaration of the Rights of Wetlands.[151]

Hopefully, the rights of wetlands will find space to continue to grow within the Ramsar Convention. While the treaty has undoubtedly contributed in myriad positive ways to wetland conservation and wise use, current conservation and management methods and efforts – including those provided through the Convention – have been unable to halt wetland loss and degradation worldwide. Recognition of the rights of wetlands and the proposed Declaration may offer another tool to help complement existing efforts. And although recognising the rights of wetlands may not be a panacea for the world's wetland ills, perhaps promoting the rights of wetlands and the proposed Declaration is the just type of change that is needed, and the Ramsar Convention could provide a meaningful path to champion this change.

---

[147] Ramsar Convention, *The CEPA Programme* at: https://www.ramsar.org/our-work/activities/cepa-programme.

[148] See Ramsar Convention, *World Wetlands Day* at: https://www.ramsar.org/our-work/activities/world-wetlands-day.

[149] Courouble et al, note 117, 33.

[150] Ibid.

[151] Ramsar Convention, note 141.

Index

Adams, William 107
Adler, Mortimer 28
Advisory Missions 13
agreements *see* multi-area agreements
Aichi Targets (Strategic Plan for Biodiversity 2011–20) *see* biodiversity
Algeria 4, 84
Ali, M. Kenza 166, 169, 178
Antarctic 73–4, 80
aquafers *see* water law
aquatic environments *see* wetlands
Arctic 157, 161
Argentina 56, 60, 84
Australia 55–7, 93, 144, 275, 366, 370, 371
awareness *see* capacity building; education; participation; awareness

Bangladesh 364, 370, 380
Baron Le Roy *see* Le Roy de Boiseaumarié; Pierre Gabriel Vincent Ernest (Baron Le Roy)
Belgium 1
   see also Democratic Republic of Congo (DRC)
Bentham, Jeremy 108
biodiversity
   see also Convention on Biological Diversity; *Global Biodiversity Outlook*
   acceleration of decline of 49
   assessment mechanisms enhancement 226–8
   climate change threat to 16, 298–302
   co-operation with non-state actors, enhancement of 210–212
   co-operation with other environmental regimes 204
   cross-regime alignment of targets and plans 204, 207–17, 220–223
   current crises as to 48–9
   EU law, and 305–6
   future engagement with, by the Convention 17
   Global Biodiversity Framework 204
   global review of economics of 48
   governance enhancement 217–26
   inclusive multilateralism, concept of 210–212
   institutional enhancements 223–6
   international co-operation as to 6
   IPBES 53–4, 77–8
   Kunming-Montreal Global Biodiversity Framework 149, 204, 206, 207, 208, 211, 212–20, 223, 227, 228–30, 301, 374
   Liaison Group (BLG) 224–6
   major conventions 203–4
   monitoring-mechanisms enhancement 226–8
   national biodiversity strategies and action plans (NBSAPs) 204
   Ramsar and *see* Ramsar Convention
   reporting-mechanisms enhancement 226–8
   'Rights of Nature' movement, and 361
   Strategic Plan for Biodiversity 2011–20 (Aichi Targets) 149, 150, 204, 206, 218, 221–2, 301, 313
   treaties, influence of the Convention upon 8, 14
   treaty synergising agendas 14–15

UN Common Approach to
    Integrating Biodiversity and
    Nature-based Solutions for
    Sustainable Development
    into United Nations Policy
    and Program Planning and
    Delivery 215
wetland ecosystems, within 2
birds *see* European Union; waterfowl
Boer, B. 268
Bolivia 364, 365, 372, 375
Bonells, Marcela 13
Bowman, Michael 6, 11, 79, 123, 129, 183
Brazil 5, 68, 71–2, 276, 368
Bridgewater, P. 184, 271
budgets
    consensus as to approval of 58–9
    parties' contributions to 65
Burma *see* Myanmar

Caddell, Richard 16, 213
Canada 5, 21, 56, 100, 144, 370
    see also Kunming-Montreal Global
        Biodiversity Framework
capacity building, education,
    participation, and awareness
    (CEPA) 33, 120, 121, 165, 166
carbon dioxide (CO_2)
    EU removal target achievement 308
    forests capture of 294
    wetlands capture of 294
CBD *see* Convention on Biological
    Diversity
Chan, K.M.A. 289
China 58, 66, 204, 211, 257–8
    see also Kunming-Montreal Global
        Biodiversity Framework
Cittadino, F. 163
climate change
    author's approach to analysis of, in
        current book 292–4
    biodiversity and 298–302
    current crises as to 48
    effects on wetlands 310–311
    fight against, concerns as to role of
        wetlands in 302–9
    fight against, contribution by
        wetlands to 292, 294–5, 317

international climate change law,
    wetlands and 296–8
protected wetlands, management of
    designated areas 311–16
Ramsar references to 16, 293–4
threats to wetlands from 16
wetlands loss because of 317–18
Cliquet, An 16
CO_2. *see* carbon dioxide
Colombia 364, 371, 380, 381
communities *see* Indigenous peoples
compensation
    author's approach to analysis of, in
        current book 132–3
    author's assessment of, in current
        book 151–3
    compensatory mechanisms of the
        Convention 13
    controversies as to 138–40
    Convention compensatory
        mechanisms 13, 131–2
    COP and 73
    guidelines for 73
    ICJ judgments as to 99
    obligations as to 95, 133–8
    practice as to 141–6
    sustainability enhancement
        mechanism, as 146–51
    Wise Use, and 127–8
Comte, Auguste 26
Conference of the Parties (COP)
    compensation and 73
    'Conferences on the Conservation of
        Wetlands and Waterfowl' 44
    consensus approach by 52, 53
    consensus decision-making by 54–9
    decision-making function of 7
    extraordinary COPs 58–9
    meetings of 7
    'organic' institutional framework
        of 41
    purpose of 7
    Ramsar Sites, and 80, 89–90
    resolutions 41
    UN legal opinion as to 53
consensus
    author's approach to analysis of, in
        current book 53
    budget approvals, as to 58

consensus approach by the
    Convention 10
consensus decision-making 54–9
    definition of 55
    extraordinary COPs, at 58–9
    leveraging of 53
    objectives of the Convention, and 53
    other treaty regimes, within 53
    practice, and tradition of, the
        Convention 52
    reservations, as to 68–70
    resolutions, as to 10, 52, 53, 55,
        71–2, 76
    statements (for the record) as to 53,
        56–7, 70–73
    strengths, and weaknesses, of 75–8
    UN legal opinion as to 53
    voting, as to 59–66
    withdrawal of draft resolutions, as
        to 73–5
conservation
    see also Wise Use
    'Conferences on the Conservation of
        Wetlands and Waterfowl' 44
    conservation emphasis of the
        Convention 12
    conservation tools of the Convention
        13
    EU and *see* European Union
    legal definition of 123–4
    obligation of result, as 125–9
    'Rights of Nature' movement, and
        361
    treaties *see* multi-area agreements
Convention on Biological Diversity
    (CBD)
    see also Global Biodiversity
        Framework
    assessment mechanisms
        enhancement 226–8
    biodiversity governance
        enhancement 210–212
    Biodiversity Liaison Group (BLG),
        and 224–6
    co-operation with non-state actors,
        enhancement of 210–212
    COP15 204, 205, 228–30
    cross-regime alignment of targets
        and plans 207–17, 220–223

inclusive multilateralism, concept of
    210–212
Informal Advisory Group on
    Synergies 226
institutional enhancements 223–26
Model Memorandum of
    Co-operation 219–220
monitoring-mechanisms
    enhancement 226–8
pre-eminent status of 204
Ramsar and *see* Ramsar Convention
reporting-mechanisms enhancement
    226–8
Subsidiary Body for Implementation
    228
Convention on Wetlands of International
    Importance especially as
    Waterfowl Habitat 1971 (Ramsar
    Convention or Ramsar) *see*
    Ramsar Convention
co-operation *see* international
    co-operation
Costa Rica 66, 98–9, 259, 368
Croatia 71
cultural services
    author's approach to analysis of, in
        current book 267–9
    challenges for 285–9
    classification of 288–9
    complexities of scale as to 286–8
    concept of culture, acceptance of
        269–71
    cultural significance of wetlands
        265–7
    cultural values, distinction from
        276–7
    cultural values, recognition of
        277–80
    historical progression of Ramsar's
        approach to culture 271–5
    obligations as to values, and as to
        services 282–5
    obstacles to Ramsar's recognition of
        culture 275–6
    protection of 276–85
    Ramsar and 269–76, 290–291
    recognition of 280–282
culture *see* cultural services; Indigenous
    peoples; traditional knowledge
Czechia (Czech Republic) 4

Darwin, Charles 26
Dasgupta, Partha 48
Davies, P. 378
decision-making
    consensus decision-making 54–9
    COP's decision-making role 7
    treaty enhancement of 43
de Klemm, C. 187
De Lucia, V. 212
Democratic Republic of Congo (DRC) 5, 109
Denmark 71, 325, 336
Domica-Baradla Cave System (Hungary/Slovakia) 101
Doñana National Park (Spain) 351, 359
DRC *see* Democratic Republic of Congo

ecological character *see* wetlands; Wetlands of International Importance
ecosystems *see* biodiversity; wetlands; Wetlands of International Importance
Ecuador 363, 364, 365, 367, 371, 375, 376, 379
education *see* capacity building; education; participation; awareness
Edwards, Hugh 115
Eilean na Muice Duibhe/Duich Moss wetland (Scotland) 337, 350
Elmenteita, Lake (Kenya) 200
emissions *see* carbon dioxide
England 3, 21, 371
environment *see* biodiversity; carbon dioxide; conservation; multi-area agreements; powers, rights, and duties; water law
environmental principles *see* Wise Use principle
European Union (EU)
    author's approach to analysis of, in current book 321–2
    author's assessment of, in current book 359–60
    Biodiversity Strategy 305–6
    Birds Directive 321, 322–47, 359–60
    carbon dioxide removal target, achievement of 308

EU/EU Member States membership of Ramsar Convention 320
    forests protection 305
    Habitats Directive 321, 348–57, 360
    historical background to Birds Directive 322–30
    history of wetland management, and of regulation of 320
    implementation of Birds Directive 331–42
    importance of wetlands 319
    judicial interpretations of Ramsar by 17
    Land-Use, Land-Use Change and Forestry Regulation, amendment of 308
    lead shot ban 17
    lead shot restrictions 342–7
    'Natura 2000' protected areas 321
    nature conservation Directives 321
    Nature Restoration Law, proposal for 306–8
    nature restoration plan 306
    primacy of EU nature conservation law 350
    Ramsar and 17, 357–60
    'Taxonomy Regulation' 308–9
    waterbirds, and waterbird habitats protection 320–321
    'wetland', definition of 192
    wetland coverage within 319
    wetland loss within 319–20
extraordinary COPs *see* Conference of the Parties

Fajardo, Teresa 15–16
Falkland Islands 84
Farrier, D. 124
finance *see* budgets
'first pillar' of Ramsar *see* Wetlands of International Importance
forests
    climate change, and 294
    CO_2 capture by 294
    cultural value of 280
    EU protection of 305
    US conservation of 108
France 1, 3, 26, 144, 335, 372

Gardner, Royal C. 10, 13, 214
Germany 144, 150, 270, 320, 337, 359
Global Biodiversity Framework
    adoption of 206, 208, 219, 223, 227, 301, 374
    alignment with Ramsar 204
    challenges from 228
    co-operation as to 207
    final outcome of 220
    Framework Fund 228
    future effect of 230
    implementation of 224
    Monitoring Framework 227
    negotiation of 218
    shared principles for 212–17
    targets of 149, 223
Global Biodiversity Outlook 208, 221, 301, 314, 317
Global Wetland Outlook 8, 11, 53–4, 77, 152, 198, 205, 208, 294–5, 310, 314, 317, 384
Goodwin, Edward J. 6, 12
greenhouse gas emissions *see* carbon dioxide
Gruber, S. 268
Guterres, Antonio 48

habitats, protection of *see* European Union; Ramsar Convention; waterfowl
Haeckel, Ernst 27
Hamman, Evan 11–12
Hegel, G.W.F. 27
Hoffmann, Luc 2–3, 127
Humboldt, Alexander von 27
Humedal Caribe Noreste (Costa Rica) 98–9
Hungary 101, 231, 320

ICJ *see* International Court of Justice
ILC *see* International Law Commission
India 57, 71, 97, 125, 370
Indigenous peoples
    author's approach to analysis of, in current book 154, 156
    author's assessment of, in current book 177–8
    'community', definition of 159
    contextual international law as to 160–164
    Convention acknowledgement of wetlands management role of 154
    Convention shortcomings as to 155
    cultural values, and 167–9
    definitional controversies as to 156–9
    ILO definition of 156–7
    knowledge/know-how *see* traditional knowledge
    local and Indigenous communities, valued role of 12
    other treaty regimes, within 172–7
    Outstanding Universal Value assessment as to land use 176
    overlap with local communities 159
    Permanent Forum on Indigenous Issues (UNPFII) 158
    Ramsar definition of 158–9
    rights, participation, and roles 160–167
    role as to Ramsar processes 164–7
    UN Declaration on the Rights of Indigenous Peoples (UNDRIP) 156
    Wise Use, and 169–72
intellectual property *see* traditional knowledge
Intergovernmental Science-Policy Platform on Biodiversity and Ecosystem Services (IPBES) 53–4, 77–8
international co-operation
    Ramsar Convention and 15–16
    'third pillar' of Ramsar 6–7
International Court of Justice (ICJ) 47, 95, 98–9, 246–7
international law *see* international co-operation; multi-area agreements
International Law Commission (ILC), Draft Articles on Transboundary Aquifers 233
International Organization Partners (IOPs) 7–8
International Waterfowl and Wetlands Research Bureau (IWRB) 23, 44, 110

International Waterfowl Research Bureau (IWRB) (now Wetlands International) 4
international water law *see* water law
IOPs *see* International Organization Partners
IPBES *see* Intergovernmental Science-Policy Platform on Biodiversity and Ecosystem Services
Iran 4–5, 60, 67, 260
Iraq 60, 260
Ireland 320, 325, 371
Israel 53, 59–62
Italy 7, 144
IWRB *see* International Waterfowl and Wetlands Research Bureau; International Waterfowl Research Bureau

Japan 57, 258
Jordan 60, 97

Kant, Immanuel 28
Kariba Dam (Zambia and Zimbabwe) 106–7
Kazakhstan 84
Kenya 125–6, 200
Kim, R.E. 184, 271
Kingsford, R.T. 91
knowledge *see* traditional knowledge
Korea 102, 258
Krämer, L. 342
Krug, Julius 108
Krutilla, John 108
Kunming-Montreal Global Biodiversity Framework *see* Global Biodiversity Framework

land use
 CO_2 emissions from 308
 decision-making as to 265
 ecosystem approach to 246
 EU law, and 308
 international restrictions on national land use policies 111
 land use plans 100
 Outstanding Universal Value assessment as to 176
 planning 246
 sustainable 174
 wetlands and 246
Lang, John Temple 334–6
lead shot
 EU restrictions 342–7
 Ramsar ban 17
Lee, J. 261–2
Leopold, Aldo 108
Le Roy de Boiseaumarié, Pierre Gabriel Vincent Ernest (Baron Le Roy) 3
Lewis, Melissa 15–16
Liaison Group of the Biodiversity-related Conventions (BLG) 224–6
Lippens, Léon Count 1
Lithuania 320
local and Indigenous communities *see* Indigenous peoples
local knowledge *see* traditional knowledge
Locke, John 27
LOSC *see* Convention on the Law of the Sea 1982

MA *see* Millennium Ecosystem Assessment
Maljean-Dubois, S. 226
management plans 91
MAR Conference
 convening of 3
 follow-up meetings 4
 naming of 'Project MAR' 3
 recommendation for establishment of wetlands convention 3–4
Marsden, Simon 12
Matthews, G.V.T. 3, 59, 111, 113–16
Mayer, Benoit 118–19, 125, 129
McIntyre, Owen 15–16
memorandums
 CBD-Ramsar Memorandum of Co-operation (MOC) 220
 Model Memorandum of Co-operation 219–20
Mendel, Gregor 26
Mexico 73, 367
migratory waterbirds *see* waterfowl
Millennium Ecosystem Assessment (MA) 89–90
MOCs (memorandum of co-operation) *see* memorandums

Mongolia 258
Montreux Record
  role of 13
  Wetlands of International Importance ('Ramsar Sites') and 96
Morocco 4
Mozambique 199–200
Muir, John 107, 124, 126
multi-area agreements
  consensus within 53
  cross-regime alignment of targets and plans 204, 207–17
  decision-making structures of 15, 43
  Indigenous peoples, within 172–7
  institutional frameworks of 40–42
  legal instruments, as 40
  MEA secretariat-services provision by UNEP 224
  'organic' style of 40
  Ramsar Convention relationship with other international environmental law regimes 14–15, 17
  Ramsar Convention's influence on later agreements 16–17
  traditional knowledge, within 12, 173
  treaty regime communities 42–7
  whaling treaties as forerunners of the Convention 9–10
Myanmar 287

Natron, Lake (Tanzania) 199
nature conservation *see* conservation
Netherlands 4, 21, 60, 111–14
New Zealand 275, 287, 366, 368
Ngiri-Tumba-Maindombe (Democratic Republic of Congo) 5
Nicaragua 98–9, 259
Noordwijk Conference 1966 (Netherlands) 4, 111
Northern Rhodesia *see* Zambia

obligations *see* powers, rights, and duties
Okuno, Erin 13
other treaty regimes *see* multi-area agreements
Oviedo, G 166, 169, 178

Panama 365, 367
Papayannis, T. 266, 275
participation *see* capacity building, education, participation, and awareness
Penca, Jerneja 13
Pinchot, Gifford 108–9, 114, 126
powers, rights, and duties
  changes in ecological character, as to 91–2
  compensation, as to 95, 133–8
  conservation, as obligation of result 125–9
  Indigenous peoples, as to 160–167
  obligations as to wetlands 88–95
  Ramsar Convention as a legal obligation 34
  reporting requirements as to Ramsar Sites 91–9
  'three pillars' of Ramsar 5–6
  traditional knowledge, as to 166
Pratt, Katherine 13
principles *see* Wise Use principle
Pritchard, David 117, 266, 275, 287
Project MAR *see* MAR Conference
public awareness *see* capacity building, education, participation, and awareness

Queen Maud Gulf (Canada) 5

RAMs 97–8
Ramsar Advisory Missions *see* Advisory Missions
Ramsar Conference 1971 (Iran) 4–5
Ramsar Convention (the Convention) 276–85
  see also reservations
  adoption of 7
  Advisory Missions 13
  alignment with Global Biodiversity Framework 204
  applicability of 234–9
  awareness of existence of 19
  biodiversity and 204–7
  Biodiversity Liaison Group (BLG), and 224–6
  budget and finance *see* budgets

capacity building, education, participation, and awareness (CEPA) 33, 120, 121, 165, 166
CBD and 203–7, 228–9
CBD-Ramsar Memorandum of Co-operation (MOC) 220
climate change, and *see* climate change
compensation provisions *see* compensation
compilation of wetland sites 3–4
Conference of the Parties (COP) 7
consensus approach by 10
conservation emphasis of 12
conservation tools of 13
content, and, structure, of current book 9
contracting parties to 5
contributors to conceiving of 2–3
core values of 10–11
cross-regime alignment of biodiversity targets and plans 204, 220–223
cultural context of adoption of 22–31
definition of 'wetlands' 5
distinctive features of 10–14
Dutch government's draft text 4
early history of 20–22
'ecosystem approach' by 254–6
education, as core value of 11
educational role of 31–47
EU and *see* European Union
expansion of ecological normativity 34–9
first fifty years of 47–51
focus on core mission of 17
future adaptiveness of 17
global significance of advent of 19–20
headquarters 7
historical background to 2–8
implementation of 8–17, 13–14
influence on later environmental regimes 16–17
international co-operation, and 15–16
international co-operation under ('third pillar' of Ramsar) 6–7

International Organization Partners (IOPs) 7–8
IWRB compromise, and final, draft texts 4
juridical impacts on nature regulation 34
lack of academic legal studies on 8–9
legal instrument, as 34, 40
legal obligation, as 34
lines of enquiry, by current book 9, 17
local and Indigenous communities, valued role of 12
main obligations under ('three pillars' of Ramsar) 5–6
methodological impacts on nature regulation 33–4
Montreux Record 13
new vision for management of the natural world 33–4
non-environmental values of 11–12
Noordwijk Conference 1966 (Netherlands) 4
normative requirements of 239–45
'organic' institutional framework of 40–41
organisational structure of 7
other international environmental law regimes, and 14–15, 17, 40–42
outreach, as core value of 11
pioneer role of 7–8, 9
potential tension as to focus of 17
public awareness role of 31–3
Ramsar Conference 1971 (Iran) 4–5
recommendation for establishment of 3–4
'Rights of Nature' movement, and *see* 'Rights of Nature' movement
science, as core value of 11
Scientific and Technical Review Panel (STRP) 7
Secretariat (was Bureau) 7, 65–6, 223, 224
Secretary General 7, 223
Soviet draft text 4
Standing Committee 7

St Andrews Conference 1963
(Scotland) 4
Stockholm Conference 1972, and 8
Strategic Plan 33, 121, 166, 169,
171, 184, 195–6, 204
success of 15, 17
UNEP, and 223–4
water law and *see* water law
Wetlands of International
Importance ('Ramsar Sites')
('first pillar' of Ramsar) 5–6
whaling treaties as forerunners of
9–10
Wise Use emphasis of 12
Wise Use principle ('second pillar'
of Ramsar) 6
'Ramsar Sites' *see* Wetlands of
International Importance
Refugio de Vida Silvestre Río San Juan
(Nicaragua) 98–9
reservations, consensus as to 53, 56,
68–70
resolutions
climate change, as to 71–2
consensus as to 10, 52, 53, 55, 71–2,
76
consensus as to withdrawal of draft
resolutions 73–5
institutional frameworks as to 41
Rhodesia (Northern) *see* Zambia
Rhodesia (Southern) *see* Zimbabwe
Rieu-Clarke, A. 251
rights *see* powers, rights, and duties
'Rights of Nature' movement
author's approach to analysis of, in
current book 362
challenges for 375–7
descriptive overview of 362–75
emergence of 361
enforcement approaches 374–5
forms of recognition 367–74
historical background to 362–4
philosophy of 361–2
potential role for Ramsar within
381–4
Ramsar and 377–84
recognition of 361
rights holders, and rights 364–7
Universal Declaration of the Rights
of Wetlands 378–81

Rosenblum, Z.H. 102
Rosenne, S. 40
Russia (inc. Soviet Union) 4, 52, 60,
66, 80, 97–8, 103, 113–14, 183,
257–8

Sands, Philippe 117
Schmeier, S. 102
science
see also capacity building,
education, participation,
and awareness; traditional
knowledge
core value of the Convention, as 11
Scientific and Technical Review Panel
(STRP)
see also Global Wetland Outlook
'organic' institutional framework
of 41
purpose of 7
Ramsar Sites, and 89–90
Scotland 4, 337
Scott, Karen 225
Sebkhet Sejoumi Ramsar Site (Tunisia)
200
'second pillar' of Ramsar *see* Wise Use
principle
Secretariat *see* Ramsar Convention
Serengeti National Park 126
Slovakia 101
Smith, Adam 29, 49
Songor, Lake (Ghana) 100
South Africa 66, 107, 125, 142–3, 194
Southern Rhodesia *see* Zimbabwe
Spain 3, 56, 145, 339, 351, 359
Spray, C. 251
Standing Committee
'organic' institutional framework
of 41
role of 7
St Andrews Conference 1963 (Scotland)
4
statements (for the record), consensus as
to 53, 56–7, 70–73
Strategic Plan for Biodiversity 2011–20
(Aichi Targets) *see* biodiversity
Suietnov, Y. 214
Sweden 8, 74
Switzerland 2, 7, 59, 64, 73–4
Syria 61–2

Szigetkötz (Hungary) 231

Tanzania (Tanganyika) 125, 126, 199
'third pillar' of Ramsar *see* international co-operation
traditional knowledge
    author's approach to analysis of, in current book 156
    Convention acknowledgement of role of 154
    cultural values, and 167
    multi-area agreements, within 12
    obligations as to 166
    other treaty regimes, within 173
    recognition and respect for 161
    'Traditional' or 'Indigenous knowledge', definition of 157–8
    wetlands management, and 166
    Wise Use, and 168–72
Transboundary Ramsar Sites 101–2
treaties *see* multi-area agreements
Tucker, Joshua A. 124
Tunisia 4, 102, 200
Turkey 4, 68–70, 260

UAE *see* United Arab Emirates
Uganda 365, 367
Ukraine 52, 60, 80, 97–8, 103, 260
UN Convention on the Law of the Sea 1982 (UNCLOS) 118, 332
United Arab Emirates (UAE) 62, 66
United Kingdom (UK) 5, 33, 48, 50, 60, 84, 97, 107, 125, 144, 325, 337, 339, 350
    see also England; Scotland; Wales
United Nations (UN)
    Chief Executives Board for Coordination of the UN 215–16
    Common Approach to Integrating Biodiversity and Nature-based Solutions for Sustainable Development into United Nations Policy and Program Planning and Delivery 215
    Convention on the Law of the Sea 1982 (UNCLOS) 118, 332

Declaration on the Rights of Indigenous Peoples (UNDRIP) 156
Economic and Social Council 108
Economic Commission for Europe (UNECE) 232
Educational, Scientific and Cultural Organization (UNESCO) 23, 68, 83, 269
environmental issues at 23
Environment Programme (UNEP) 65–6, 217, 219, 223–4, 224, 227
Framework Convention on Climate Change (UNFCCC) 7, 55, 57, 72, 209, 296–7
General Assembly, Harmony with Nature Resolution 372
Geospatial Network on the List of Wetlands of International Importance 84
institutional structures 8
legal opinion as to COP 53
LOSC *see* Convention on the Law of the Sea 1982
Office of Legal Affairs 53
Permanent Forum on Indigenous Issues (UNPFII) 158
policy-making processes 204
REDD programme 297
Scientific Conference on the Conservation and Utilization of Resources 108
UNCLOS *see* UN Convention on the Law of the Sea 1982
UNECE Water Convention 232, 233, 235, 237, 240–241, 249, 253–4
see also water law
Watercourses Convention 232, 233, 235, 236, 237, 240–241, 246, 247, 248, 249, 253–4, 257, 260, 261
United States (US) 23, 28–9, 57–8, 70–71, 100, 107–8, 125–6, 135, 144, 363, 364, 366, 367–8, 369–70, 371–2, 376
UN Watercourses Convention *see* water law

values *see* cultural services
Velázquez Gomar, J.O. 214
voting
    consensus as to 59–66
    COP agreement as to voting procedure 21–2

Wales 304, 371
watercourses *see* water law; wetlands
waterfowl
    Advisory Missions 199–200
    alignment between migratory waterbird regimes 191–3
    author's approach to analysis of, in current book 181
    author's assessment of, in current book 201–2
    collaborative projects 200–201
    compensation, and 133
    'Conferences on the Conservation of Wetlands and Waterfowl' 44
    Convention contribution to international water law 16
    Convention emphasis on 35, 180–181, 182–6
    co-operation as to population estimates 193–4
    definition of 154, 191
    EU and *see* European Union
    expert representation at COP 43, 154
    flyway initiatives 189–91, 197
    flyway instruments' promotion of Ramsar designation 194–5
    habitat protection, Convention emphasis on 4–5, 184–5
    habitat protection, data collection as to 184–5
    historical progression of Ramsar's emphasis on 182–4
    interflyway collaboration 197
    international migratory waterbird regimes, emergence and role of 186–9
    IWRB 23, 44, 110
    joint work plans as to 195–6
    MOUs as to 195–6
    priority measures for conservation of 181–2
    Ramsar collaboration with CMS, and with AEWA 195–201
    Ramsar Site listing, and 81–2
    relevance of species-based conservation approaches 185–6
    removal from Convention core agenda 14
    scientific collaboration, technical collaboration, task force collaboration 197–9
waterfowl hunters
    contribution to conceiving of the Convention 2–3
    lead shot banning/restricting, and 17, 342–7
water law
    applicability of 234–9, 239–45
    aquatic ecosystems, focus on 232
    author's approach to analysis of, in current book 232–4
    'ecosystem approach' by 245–9
    ecosystem services 249–51
    environmental flows 251–3
    ILC Draft Articles on Transboundary Aquifers 233, 235, 248
    improvements in convention coverage of 232
    interaction with Ramsar, complementarity with Ramsar 256–63
    other environmental treaty regimes, and 233
    Ramsar Convention and 15–16, 231–4, 263–4
    technical guidance 253–54
    UNECE Water Convention 232, 233, 235, 237, 240–241, 249, 253–4
    UN Watercourses Convention 232, 233, 235, 236, 237, 240–241, 246, 247, 248, 249, 253–4, 257, 260, 261
    wetlands protection, and 231–4
Wetland City Accreditation 102–3
wetlands
    see also Global Wetland Outlook
    climate change, and *see* climate change

CO_2 capture by 294
'Conferences on the Conservation of Wetlands and Waterfowl' 44
contributions to human survival 2
Convention definition of 5, 154
critical ecosystems, as 1–2
dangers from eradication, and degradation of 1
definition of 192
EU Biodiversity Strategy, and 305–6
EU Birds Directive 321, 322–47, 359–60
EU Habitats Directive, and 321, 348–57, 360
EU law, and 192
importance of 1–2
Indigenous peoples, and *see* Indigenous peoples
lead shot ban 17
obligations as to 154
protected wetlands, management of designated areas 311–16
rights *see* powers, rights, and duties
'Rights of Nature' movement, and 361
see also 'Rights of Nature' movement
traditional knowledge, and 166
UN Geospatial Network on the List of Wetlands of International Importance 84
Universal Declaration of the Rights of Wetlands 378–81
UN REDD programme 297
'waterfowl', definition of 154
Wetlands International *see* International Waterfowl Research Bureau
Wetlands of International Importance ('Ramsar Sites')
author's approach to analysis of, in current book 80–81
author's assessment of, in current book 103–4
benefits of designation as 100–101
concept of 79
conservation of *see* conservation; Wise Use
COP and 80, 89–90
designation and listing process 80–81

designation process 81–8
'ecological character', concept/ definition of 89–90
ecological character, ICJ judgment as to 98–9
ecological character, responses to adverse change in 95–9
ecological character, tracking of reported changes in 94–5
'ecosystem services', notion of 89–90
'first pillar' of Ramsar 5–6
largest protected area network, whether 84–8
listing of, criteria for 81–3
listing process 81–8
management plans 91
Millennium Ecosystem Assessment (MA) of 89–90
Montreux Record and 96
obligations as to 88–95
obligation to report as to changes in ecological character 91–2
other types of site-based recognitions 101–3
protected areas, provisions in prior conventions 79–80
RAMs and 97–8
reports procedure 93–4
sovereignty as to 83–4
STRP and 89–90
timing as to submission of reports 92–3
Transboundary Ramsar Sites 101–2
Wetland City Accreditation 102–3
whaling treaties, as forerunners of the Convention 9–10
Wise Use
author's approach to analysis of, in current book 105–6
author's assessment of, in current book 129–30
CEPA and 33
clarification of concept of 115–17
commitment as to advancement of 12, 32
compensation and 127–8
consensus and 76
conservation, as obligation of result 125–9

'conservation', legal definition of 123–4
conservation of listed wetlands, and 122–9
COP and 105
guidance/handbooks as to 31, 37
historical progression of 106–17
implementation of 17, 32
Indigenous peoples, and 169–72
legal conception of 117–22
linking of 'conservation' with 124–5
mainstreaming of 33
negotiations as to adoption of principle of 106–15
redefinition of 35–6, 47
'second pillar' of Ramsar 6
status of, within Convention regime 105
traditional knowledge, and 168–72
Wetland City Accreditation and 101, 103

Zambia (Northern Rhodesia) 59, 107
Zimbabwe (Southern Rhodesia) 7, 107